T0212986

Lecture Notes in Artificial Intelligence 9880

Subseries of Lecture Notes in Computer Science

LNAI Series Editors

Randy Goebel
University of Alberta, Edmonton, Canada
Yuzuru Tanaka
Hokkaido University, Sapporo, Japan
Wolfgang Wahlster
DFKI and Saarland University, Saarbrücken, Germany

LNAI Founding Series Editor

Joerg Siekmann
DFKI and Saarland University, Saarbrücken, Germany

More information about this series at http://www.springer.com/series/1244

Vicenç Torra · Yasuo Narukawa
Guillermo Navarro-Arribas · Cristina Yañez (Eds.)

Modeling Decisions
for Artificial Intelligence

13th International Conference, MDAI 2016
Sant Julià de Lòria, Andorra, September 19–21, 2016
Proceedings

Springer

Editors
Vicenç Torra
University of Skövde
Skövde
Sweden

Yasuo Narukawa
Toho Gakuen
Kunitachi, Tokyo
Japan

Guillermo Navarro-Arribas
Universitat Autònoma de Barcelona
Bellaterra
Spain

Cristina Yañez
Universitat d'Andorra
Sant Julià de Lòria
Andorra

ISSN 0302-9743   ISSN 1611-3349   (electronic)
Lecture Notes in Artificial Intelligence
ISBN 978-3-319-45655-3   ISBN 978-3-319-45656-0   (eBook)
DOI 10.1007/978-3-319-45656-0

Library of Congress Control Number: 2016949568

LNCS Sublibrary: SL7 – Artificial Intelligence

Printed on acid-free paper

This Springer imprint is published by Springer Nature
The registered company is Springer International Publishing AG Switzerland

Preface

This volume contains papers presented at the 13th International Conference on Modeling Decisions for Artificial Intelligence (MDAI 2016), held in Sant Julià de Lòria, Andorra, September 19–21, 2016. This conference followed MDAI 2004 (Barcelona, Spain), MDAI 2005 (Tsukuba, Japan), MDAI 2006 (Tarragona, Spain), MDAI 2007 (Kitakyushu, Japan), MDAI 2008 (Sabadell, Spain), MDAI 2009 (Awaji Island, Japan), MDAI 2010 (Perpignan, France), MDAI 2011 (Changsha, China), MDAI 2012 (Girona, Spain), MDAI 2013 (Barcelona, Spain), MDAI 2014 (Tokyo, Japan), and MDAI 2015 (Skövde, Sweden) with proceedings also published in the LNAI series (Vols. 3131, 3558, 3885, 4617, 5285, 5861, 6408, 6820, 7647, 8234, 8825, and 9321).

The aim of this conference was to provide a forum for researchers to discuss theory and tools for modeling decisions, as well as applications that encompass decision-making processes and information-fusion techniques.

The organizers received 36 papers from 13 different countries, 22 of which are published in this volume. Each submission received at least two reviews from the Program Committee and a few external reviewers. We would like to express our gratitude to them for their work. The plenary talks presented at the conference are also included in this volume.

The conference was supported by the University of Andorra, the city of Sant Julià de Lòria (Andorra), Andorra Turisme, INNTEC – Jornades de la innovació i les noves tecnologies, the Catalan Association for Artificial Intelligence (ACIA), the Japan Society for Fuzzy Theory and Intelligent Informatics (SOFT), the European Society for Fuzzy Logic and Technology (EUSFLAT), the UNESCO Chair in Data Privacy, and the project TIN2014-55243-P from the Spanish MINECO.

July 2016

Vicenç Torra
Yasuo Narukawa
Guillermo Navarro-Arribas
Cristina Yáñez

Organization

Modeling Decisions for Artificial Intelligence – MDAI 2016

General Chairs

Guillermo Navarro-Arribas	Universitat Autònoma de Barcelona, Spain
Cristina Yáñez	Universitat d'Andorra, Andorra

Program Chairs

Vicenç Torra	University of Skövde, Skövde, Sweden
Yasuo Narukawa	Toho Gakuen, Tokyo, Japan

Advisory Board

Didier Dubois	Institut de Recherche en Informatique de Toulouse, CNRS, France
Lluis Godo	IIIA-CSIC, Spain
Kaoru Hirota	Tokyo Institute of Technology, Japan
Janusz Kacprzyk	Systems Research Institute, Polish Academy of Sciences, Poland
Sadaaki Miyamoto	University of Tsukuba, Japan
Michio Sugeno	European Centre for Soft Computing, Spain
Ronald R. Yager	Machine Intelligence Institute, Iona Collegue, NY, USA

Program Committee

Eva Armengol	IIIA-CSIC, Spain
Edurne Barrenechea	Universidad Pública de Navarra, Spain
Gleb Beliakov	Deakin University, Australia
Gloria Bordogna	Consiglio Nazionale delle Ricerche, Italy
Humberto Bustince	Universidad Pública de Navarra, Spain
Francisco Chiclana	De Montfort University, UK
Anders Dahlbom	University of Skövde, Sweden
Susana Díaz	Universidad de Oviedo, Spain
Josep Domingo-Ferrer	Universitat Rovira i Virgili, Spain
Jozo Dujmovic	San Francisco State University, USA
Yasunori Endo	University of Tsukuba, Japan

Zoe Falomir	Universität Bremen, Germany
Katsushige Fujimoto	Fukushima University, Japan
Michel Grabisch	Université Paris I Panthéon-Sorbonne, France
Enrique Herrera-Viedma	Universidad de Granada, Spain
Aoi Honda	Kyushu Institute of Technology, Japan
Masahiro Inuiguchi	Osaka University, Japan
Yuchi Kanzawa	Shibaura Institute of Technology, Japan
Petr Krajča	Palacky University Olomouc, Czech Republic
Marie-Jeanne Lesot	Université Pierre et Marie Curie (Paris VI), France
Xinwang Liu	Southeast University, China
Jun Long	National University of Defense Technology, China
Jean-Luc Marichal	University of Luxembourg, Luxembourg
Radko Mesiar	Slovak University of Technology, Slovakia
Andrea Mesiarová-Zemánková	Slovak Academy of Sciences, Slovakia
Tetsuya Murai	Hokkaido University, Japan
Toshiaki Murofushi	Tokyo Institute of Technology, Japan
Guillermo Navarro-Arribas	Universitat Autònoma de Barcelona, Spain
Gabriella Pasi	Università di Milano Bicocca, Italy
Susanne Saminger-Platz	Johannes Kepler University Linz, Austria
Sandra Sandri	Instituto Nacional de Pesquisas Espaciais, Brazil
László Szilágyi	Sapientia-Hungarian Science University of Transylvania, Hungary
Aida Valls	Universitat Rovira i Virgili, Spain
Vilem Vychodil	Palacky University, Czech Republic
Zeshui Xu	Southeast University, China
Yuji Yoshida	University of Kitakyushu, Japan

Local Organizing Committee Chair

| Cristina Yáñez | Universitat d'Andorra, Andorra |

Local Organizing Committee

Meri Sinfreu	Universitat d'Andorra
Aleix Dorca	Universitat d'Andorra
Univers Bertrana	Universitat d'Andorra
Sergi Delgado-Segura	Universitat Autònoma de Barcelona

Additional Referees

Marco Viviani	Fco Javier Fernandez
Julián Salas	Laura De Miguel
Javier Parra	

Supporting Institutions

University of Andorra
The city of Sant Julià de Lòria (Andorra)
Andorra Turisme
INNTEC – Jornades de la innovació i les noves tecnologies
The Catalan Association for Artificial Intelligence (ACIA)
The Japan Society for Fuzzy Theory and Intelligent Informatics (SOFT)
The European Society for Fuzzy Logic and Technology (EUSFLAT)
The UNESCO Chair in Data Privacy

Contents

Clustering and Classification

Data Privacy and Security

Data Mining and Applications

Invited Papers

Aggregation Operators to Support Collective Reasoning

Juan A. Rodriguez-Aguilar[1]([✉]), Marc Serramia[2], and Maite Lopez-Sanchez[2]

[1] Artificial Intelligence Research Institute (IIIA-CSIC),
Campus UAB, Bellaterra, Spain
`jar@iiia.csic.es`
[2] Mathematics and Computer Science Department, University of Barcelona (UB),
Gran Via de les Corts Catalanes 585, Barcelona, Spain

Abstract. Moderation poses one of the main Internet challenges. Currently, many Internet platforms and virtual communities deal with it by intensive human labour, some big companies –such as YouTube or Facebook– hire people to do it, others –such as 4chan or fanscup– just ask volunteer users to get in charge of it. But in most cases the policies that they use to decide if some contents should be removed or if a user should be banned are not clear enough to users. And, in any case, typically users are not involved in their definition.

Nobel laureate Elinor Ostrom concluded that societies –such as institutions that had to share scarce resources– that involve individuals in the definition of their rules performed better –resources lasted more or did not deplete– than those organisations whose norms where imposed externally. Democracy also relies on this same idea of considering peoples' opinions.

In this vein, we argue that participants in a virtual community will be more prone to behave correctly –and thus the community itself will be "healthier"– if they take part in the decisions about the norms of coexistence that rule the community. With this aim, we investigate a collective decision framework that: (1) structures (relate) arguments issued by different participants; (2) allows agents to express their opinions about arguments; and (3) aggregates opinions to synthesise a collective decision. More precisely, we investigate two aggregation operators that merge discrete and continuous opinions. Finally, we analyse the social choice properties that our discrete aggregator operator satisfies.

1 Introduction

With the advent of the Internet, a plethora of on-line communities, such as social networks, have emerged to articulate human interaction. Nonetheless, interactions are not frictionless. Thus, for instance, users may post inappropriate or offensive contents, or spam ads. Thus, typically the owners of on-line communities establish their own norms (terms and policies) to regulate interactions without the involvement of its participants. Moderators become then in charge

V. Torra et al. (Eds.): MDAI 2016, LNAI 9880, pp. 3–14, 2016.
DOI: 10.1007/978-3-319-45656-0_1

of guaranteeing the enforcement of such norms disregarding what users may deem as fair or discomforting.

Here we take the stance that the participants in a social network must decide the norms that govern their interactions. Thus, we are in line with Nobel-prize winner Ostrom [4], who observed that involving a community's participants in their decisions improves its long-term operation. Then, there is the matter of helping users agree on their norms. As argued in [2,3], argumentative debates are a powerful tool for reaching agreements in open environments such as on-line communities. On-line debates are usually organised as threads of arguments and counter-arguments that users issue to convince others. There are two main issues in the management of large-scale on-line debates. On the one hand, as highlighted by [2,3], there is simply too much noise when many individuals participate in a discussion, and hence there is the need for *structuring* it to keep the focus. On the other hand, the preferences on arguments issued by users must be aggregated to achieve a collective decision about the topic under discussion [1].

Against this background, here we consider that structured argumentative debates can also be employed to help users of a virtual community jointly agree on the norms that rule their interactions. With this aim, we present the following contributions:

- Based on the work in [3], we introduce an argumentative structure, the so-called *norm argument map*, to structure a debate focusing on the acceptance or rejection of a target norm. Figure 1 shows one example in an online sports community.
- A novel aggregation method to assess the *collective support for a single argument* by aggregating the preferences (expressed as ratings) issued by the participants in a discussion. Such method will consider that the impact of a single rating on the overall aggregated value will depend on the distance of that rating from neutrality. More precisely, our aggregation method abides by the following design principle: the farther a rating is from neutrality, the stronger its importance when computing the collective support for an argument.
- A novel aggregation method to compute the *collective support for a norm* based on the arguments issued by the participants in a discussion. This method is based on the following design principles: (1) the larger the support for an argument, the larger its importance on the computation of the collective support for a norm; and (2) only those arguments that are *relevant enough* (count on *sufficient support*) are worth aggregating. Technically, this method is conceived as a WOWA operator [7] because it allows to consider both the values and the information sources when performing the aggregation of argument supports.
- We compared our aggregation method with a more naive approach that simply averages participants' preferences on a collection of prototypical argumentation scenarios. We observe that our method obtains support values for norms that better capture the collective preference of the participants.

Fig. 1. Example of a norm argument map. Rated positive/negative arguments in favor/against a norm prohibiting to upload spam content at a social network forum.

The paper is organised as follows. Section 2 introduces some background on the aggregation operators employed, Sects. 3, 4, 5, and 6 introduce our formal notion of norm argument map and our functions to compute the support for an argument, a set of arguments and a norm. Section 7 details the analysis of our support functions on argumentation scenarios. Finally, Sect. 8 draws conclusions and sets paths to future research.

2   Background

As previously stated, the main goal of this work is to compute an aggregated numerical score for a norm from its arguments and opinions[1]. Hence, aggregation operators become necessary to fuse all the numerical information participants provide. Next, we introduce the aggregation operators employed in this work, namely the *standard weighted mean (WM)* and the *weighted ordered weighted average (WOWA, an OWA's* [8] *variation)* from Torra [7], to compute the collective support for a norm.

Definition 1. *A **weighting vector** w is a vector such that if $w = (w_1, \ldots, w_n) \in \mathbb{R}^n$ then $w_i \in [0,1]$ and $\sum_{i=1}^{n} w_i = 1$.*

Definition 2. *Let $w = (w_1, \ldots, w_n) \in \mathbb{R}^n$ be a weighting vector and let $e = (e_1, \ldots, e_n) \in \mathbb{R}^n$ be the vector of elements we want to aggregate. A **weighted mean** is a function $WM_w(e) : \mathbb{R}^n \to \mathbb{R}$, defined as $WM_w(e) = \sum_{i=1}^{n} w_i e_i$.*

Notice that WM weighs the position of the elements, which amounts to concede different importance degrees to each particular (information) source. In order to weigh the values of aggregated elements in e we need an alternative operator.

---

[1] An argument's opinions are numerical values that, in the case of Fig. 1, take the form of number of stars awarded to each argument.

Definition 3. *Let* $w = (w_1, \ldots, w_n) \in \mathbb{R}^n$ *and* $q = (q_1, \ldots, q_n) \in \mathbb{R}^n$ *be two weighing vectors and let* $e = (e_1, \ldots, e_n) \in \mathbb{R}^n$ *be the vector of elements we want to aggregate. A **weighted ordered weighted average, weighted OWA** or **WOWA** is a function* $WOWA_{w,q}(e) : \mathbb{R}^n \to \mathbb{R}$ *defined as:*

$$WOWA_{w,q}(e) = \sum_{i=1}^{n} p_i e_{\sigma(i)}, \qquad p_i = f^*\left(\sum_{j \leq i} w_{\sigma(j)}\right) - f^*\left(\sum_{j < i} w_{\sigma(j)}\right),$$

where σ *is a permutation of the elements in* e *so that* $e_{\sigma(i)}$ *is the* i^{th} *largest element in* e *and* f^* *is a non-decreasing interpolation function of the points:* $\{(i/n, \sum_{j \leq i} q_j)\}_{i=1,\ldots,n} \cup \{(0,0)\}$ *that has to be a straight line when the points can be interpolated that way.*

Note that w acts as the vector in the weighted mean, weighing the information source, while q weighs the value of the aggregated elements. For instance, $q_1 = q_n > q_2, \ldots, q_{n-1}$, gives more importance to extreme (i.e., the highest and lowest) values.

3 Norm Argument Map

Next we formalise the notion of norm argument map as the argumentative structure that contains all arguments and opinions about a norm.

Definition 4. *A **norm** is pair* $n = (\phi, \theta(\alpha))$, *where* ϕ *is the norm's precondition,* θ *is a deontic operator[2] and* α *is an action that participants can perform.*

Definition 5. *An **argument** is a pair* $a_i = (s, \boldsymbol{O}_{a_i})$ *composed of a statement* s, *the argument itself, and a vector of opinions* $\boldsymbol{O}_{a_i}$ *that contains all the opinion values participants issued.*

Henceforth we will note the vector of opinions as $\boldsymbol{O}_{a_i} = (o_1^i, \ldots, o_{n_i}^i)$, where o_j^i is the jth opinion about argument a_i.

Definition 6. *Given a norm* n, *the **argument set** for* n *is a non-empty collection of arguments* A_n *containing both arguments supporting and attacking the norm.*

We will note the vector of all the opinions of the arguments in A_n as $\boldsymbol{O}_{A_n}$. For the sake of simplicity, we assume that all arguments in the argument set A_n of a norm are different. We divide the argument set of a norm into two subsets: the arguments in favor of the norm and the arguments against it.

We are now ready to define our argumentative structure as follows:

Definition 7. *A **norm argument map** $M = (n, A_n, \kappa)$ is a triple composed of a norm* n, *a norm argument set* A_n, *and a function* κ *that classifies the arguments of* A_n *between the ones that are in favor of the norm and the ones that are against it.*

---

[2] A deontic operator stands for either prohibition, permission, or obligation.

Hereafter we will refer to the positive arguments of norm n as the set of arguments in favor of the norm and to its negative arguments as the set of arguments against the norm. These argument sets will be noted as A_n^+ and A_n^- respectively. A negative argument is distinguished from positive arguments by adding a bar over the argument (e.g. $\bar{a}_i \in A_n^-$).

Finally, we also define a framework wherein participants can simultaneously discuss over multiple norms.

Definition 8. *A **norm argument map framework** $F = (P, N)$ is a pair of a set of participants P and a set of norm argument maps N, so that participants in P can deliberate about different norms by means of the norm argument maps in N.*

4 Argument Support

Having defined the norm argument map we aim now at aggregating arguments' opinions to calculate the support for each argument. In our case opinions will be numerical values defined in an opinion spectrum.

Definition 9. *An **opinion spectrum** is a set of possible numerical values individual participants can assign to each argument meaning her opinion about the argument.*

The spectrum will be considered a closed real number interval, and thus there exist a maximum, a minimum and a middle opinion values. Figure 2 shows an example of the opinion spectrum semantics considering $\lambda = [1, 5]$. Since opinions will have different values, we consider different semantics for them. The opinion spectrum will be divided into three subsets of opinions. Given an opinion spectrum $\lambda = [lb, ub]$ such that $lb, ub \in \mathbb{R}$ and $lb < ub$: $[lb, \frac{lb+ub}{2})$ contains the values for negative opinions, $(\frac{lb+ub}{2}, ub]$ contains the values for positive opinions, and $\{\frac{lb+ub}{2}\}$ contains the value for the neutral opinion. Note that an opinion $o_j^i = lb$ is the most extreme opinion against argument a_i, while another opinion $o_k^i = ub$ would represent the most extreme opinion in favor of the argument. Additionally, we consider the opinion laying in the middle of the spectrum $\frac{lb+ub}{2}$ as a neutral opinion.

Since different opinions in an opinion spectrum have different meanings and we aim at aggregating them in order to calculate the support for an argument, we need a function that weighs the importance of each opinion. Such *importance*

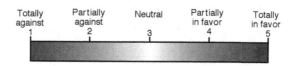

Fig. 2. Semantics of the opinion spectrum $\lambda = [1, 5]$.

function will consider neutral opinions less important than the extreme (strongly stated) ones. Formally,

Definition 10. *Given an opinion spectrum $\lambda = [lb, ub]$, we say that a function $I : \lambda \to [0, 1]$ is an importance function iff it satisfies the following conditions:*

(C1) *I continuous and piecewise differentiable*
(C2) *$I(ub) = I(lb) = 1$*
(C3) *$I(\frac{lb+ub}{2}) = 0$*

(C4) $\begin{cases} I'(x) < 0 & \text{if } x \in [lb, \frac{lb+ub}{2}) \text{ and } I \text{ is differentiable in } x \\ I'(x) = 0 & \text{if } x = \frac{lb+ub}{2} \text{ and } I \text{ is differentiable in } x \\ I'(x) > 0 & \text{if } x \in (\frac{lb+ub}{2}, ub] \text{ and } I \text{ is differentiable in } x \end{cases}$

Given a opinion spectrum, we can construct an importance function either by interpolation or geometrically (parabola case). Here we follow the first approach. Below we formally define the importance function that we propose in this paper, which is graphically depicted in Fig. 3.

$$I(x) = \begin{cases} \frac{ub^2 - 1.8\ ub\ lb - 0.2\ ub\ x - 4lb^2 + 9.8\ lb\ x - 4.8x^2}{(lb-ub)^2} & \text{if } x \in \left[lb, \frac{ub+3lb}{4}\right] \\ \frac{1.45ub + 1.75lb - 3.2x}{ub - lb} & \text{if } x \in \left(\frac{ub+3lb}{6}, \frac{3ub+5lb}{8}\right) \\ \frac{4ub^2 + 8\ ub\ lb - 16\ ub\ x + 4lb^2 - 16\ lb\ x + 16x^2}{(lb-ub)^2} & \text{if } x \in \left[\frac{3ub+5lb}{8}, \frac{5ub+3lb}{8}\right] \\ \frac{1.75ub + 1.45lb - 3.2x}{lb - ub} & \text{if } x \in \left(\frac{5ub+3lb}{8}, \frac{3ub+lb}{4}\right) \\ \frac{-4ub^2 - 1.8\ ub\ lb + 9.8\ ub\ x + lb^2 - 0.2\ lb\ x - 4.8x^2}{(lb-ub)^2} & \text{if } x \in \left[\frac{3ub+lb}{4}, ub\right] \end{cases}$$

We can now weigh the importance of each opinion with our importance function to calculate the support of an argument as the weighted mean of its opinions.

Definition 11. *Given an opinion spectrum $\lambda = [lb, ub]$, an **argument support function** $S_{arg} : A \to \lambda$ is a function that yields the collective support for each argument $a_i \in A$ as: $S_{arg}(a_i) = WM_w(O_{a_i})$, where $w = \left(\frac{I(o_1^i)}{l_i}, \ldots, \frac{I(o_{n_i}^i)}{l_i}\right)$*

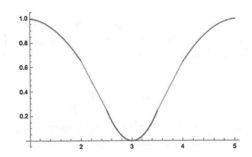

Fig. 3. Importance function (I) plot when $\lambda = [1, 5]$.

stands for a weighting vector for the opinions in O_{a_i}, I is an importance function, and $l_i = \sum_{j=1}^{n_i} I(o_j^i)$.

Notice that o_j^i is the jth opinion of argument a_i and l_i stands for the overall addition of all importance values associated to all opinions about argument a_i. Since the elements in w add up to one, w is a weighting vector.

5   Argument Set Support

So far we have learned how to aggregate an argument's opinions to compute its support. Next we face the problem of calculating the support for an argument set. To motive the choice of our aggregation function, we start with an example.

Example. *Consider a norm n with positive and negative arguments with opinions in the spectrum $\lambda = [1, 5]$. Say that there are three positive arguments a_1, a_2, a_3, and a single negative argument $\bar{a}_4$. On the one hand, in the set of positive arguments a_1 has a support of 5, which comes from a single opinion while both a_2 and a_3 have a support of 1, which comes from aggregating 100 opinions. On the other hand, on the set of negative arguments, $\bar{a}_4$'s support is 5, which comes from aggregating 30 opinions:*

A_n^+	$S_{arg}(a_i)$	$dim(O_{a_i})$
a_1	5	1
a_2	1	100
a_3	1	100

A_n^-	$S_{arg}(\bar{a})$	$dim(O_{\bar{a}})$
$\bar{a}_4$	5	30

What should we consider to give the support for A_n^+ on this extreme case? We should discard a_2 or a_3 because they have bad (the minimum) support. People have decided these arguments are not appropriate or do not provide a valid reason to defend the norm under discussion. Since opinions' semantics can be applied to argument support: arguments with supports outside $(\frac{lb+ub}{2}, ub]$ are not accepted by participants and, therefore, should not be considered as valid arguments. We cannot consider a_1 either because, although it has the maximum possible support, it has only been validated by one person, hence it is negligible in front of the other arguments. Therefore, we propose to filter out arguments by just considering those having at least a number of opinions that corresponds to a significant fraction of the number of opinions of the argument with the largest number of opinions.

Thus, we tackle this argument relevance problem by creating a new subset of arguments containing only the arguments considered to be α-relevant (namely, relevant enough) and by defining the criteria needed to be considered as such:

Definition 12. *Let A be a set of arguments and $\lambda = [lb, ub]$ an opinion spectrum, we say that an argument $a_i \in A$ is **relevant** iff $S_{arg}(a_i) > \frac{lb+ub}{2}$.*

Definition 13. *Let A be a set of arguments, $\lambda = [lb, ub]$ an opinion spectrum, $\alpha \in [0, 1]$ a relevance level, and $a_k \in A$ the argument with the largest number of opinions. We say that a relevant argument $a_i \in A$ is **α-relevant** iff $dim(O_{a_i}) \geq \alpha \, dim(O_{a_k})$.*

Henceforth, $R_\alpha(A) = \{a_1^\alpha, \ldots, a_r^\alpha\}$ will denote the set of α-relevant arguments in A. Notice that $r \leq |A|$ and that in general a_i^α is not equal to a_i.

We propose to aggregate the set of α-relevant arguments by weighting their supports with the importance function previously introduced in order to weight more those arguments that have received greater support than others. Moreover, since arguments count on different numbers of opinions, we consider the sum of importances of their opinions so that important opinions account for more weight that neutral opinions.

To aggregate the supports of the arguments weighting these two values we will use a WOWA operator. Hence, we define the argument set support function as follows:

Definition 14. *Let λ be an opinion spectrum, an **argument set support function** S_{set} is a function that takes a non-empty argument set A, with $R_\alpha(A) \neq \emptyset$, and yields its support in λ as:*

$$S_{set}(A) = S_{set}(R_\alpha(A)) = WOWA_{w,q}(S_{arg}(a_1^\alpha), \ldots, S_{arg}(a_r^\alpha)),$$

where $R_\alpha(A) = \{a_1^\alpha, \ldots, a_r^\alpha\}$, $w = \left(\dfrac{\sum_{j=1}^{dim(\boldsymbol{O}_{a_1^\alpha})} I(o_j^1)}{\mathcal{I}_A^o}, \ldots, \dfrac{\sum_{j=1}^{dim(\boldsymbol{O}_{a_r^\alpha})} I(o_j^r)}{\mathcal{I}_A^o} \right),$

$$\mathcal{I}_A^o = \sum_{i=1}^{r} \left(\sum_{j=1}^{dim(\boldsymbol{O}_{a_i^\alpha})} I(o_j^i) \right) \text{ with } o_j^i \in \boldsymbol{O}_{a_i^\alpha} = \{o_1^i, \ldots o_{n_i}^i\}$$

stands for the overall importance of all the opinions over arguments in A,

$$q = \left(\frac{I(S_{arg}(a_{\sigma(1)}^\alpha))}{\mathcal{I}_A^{arg}}, \ldots, \frac{I(S_{arg}(a_{\sigma(r)}^\alpha))}{\mathcal{I}_A^{arg}} \right)$$

$\mathcal{I}_A^{arg} = \sum_{i=1}^{r} I(S_{arg}(a_{\sigma(i)}^\alpha))$ stands for the overall importance of the collective supports received by the arguments in A, $a_{\sigma(i)}^\alpha \in R_\alpha(A) = \{a_1^\alpha, \ldots, a_r^\alpha\}$, and $a_{\sigma(i)}^\alpha$ is the α-relevant argument with the ith largest support.

Notice that, if there are no α-relevant arguments then we cannot asses the support for the set, hence we consider $S_{set}(\emptyset)$ to be not defined.

Also note that the w vector is used to weigh the importance of the arguments as the sum of the importances of its opinions. After that we have to divide by $\mathcal{I}_A^o$ so we get a weighting vector. The q vector uses the importance of the supports for the arguments. We have to order the arguments with the σ permutation because the WOWA orders the values being aggregated. This way each weight in the q vector weighs its corresponding element. With this modification, we get the WOWA to aggregate the elements using two weighting vectors. Note that the weighting vector w does not have to be ordered because the WOWA itself orders it.

6 Computing the Collective Support for a Norm

To compute the collective support for a norm, we will use the support for its positive and negative argument sets, namely $S(A_n^+)$ and $S(A_n^-)$. In general, a large support for the negative arguments of a norm is expected to negatively impact the norm's support. Thus, instead of directly aggregating $S(A_n^-)$, we will aggregate the symmetric value of the support in the spectrum with respect to the center of the spectrum, namely $ub + lb - S(A_n^-)$.

Analogously to the computation of the support for an argument set, here we have to weigh the importance of the values aggregated as well as the importance of each argument set as information source. Thus, we will also employ a WOWA operator to compute the collective support for a norm, which we define as follows:

Definition 15. *A **norm support function** is a function S_{norm} that takes a norm n, and uses the supports of its positive and negative arguments to obtain the support for the norm in $\lambda = [lb, ub]$. If $R_\alpha(A_n^+) \neq \emptyset$ and $R_\alpha(A_n^-) \neq \emptyset$, the function is defined as follows:*

$$S_{norm}(n) = WOWA_{w,q}(S_{set}(A_n^+), ub + lb - S_{set}(A_n^-))$$

such that the information source is weighed by

$$w = \left(\frac{\sum_{i=1}^{|R_\alpha(A_n^+)|}(\sum_{j=1}^{n_i} I(o_j^i))}{\mathcal{I}_n^o}, \frac{\sum_{i=1}^{|R_\alpha(A_n^-)|}(\sum_{j=1}^{\overline{n_i}} I(\overline{o}_j^i))}{\mathcal{I}_n^o} \right)$$

and the aggregated values are weighed by

$$q = \left(\frac{I(S_{set}(A_n^+))}{\mathcal{I}_n^{set}}, \frac{I(ub + lb - S_{set}(A_n^-))}{\mathcal{I}_n^{set}} \right)$$

where $\mathcal{I}_n^o = \sum_{i=1}^{|R_\alpha(A_n^+)|}(\sum_{j=1}^{n_i} I(o_j^i)) + \sum_{i=1}^{|R_\alpha(A_n^-)|}(\sum_{j=1}^{\overline{n_i}} I(\overline{o}_j^i))$, o_j^i is the jth opinion in $O_{a_i^\alpha} = \{o_1^i, \ldots, o_{n_i}^i\}$, $a_i^\alpha \in R_\alpha(A_n^+) = \{a_1^\alpha, \ldots, a_{k_1}^\alpha\}$, $\overline{o}_j^i$ is the jth opinion in $O_{\overline{a}_i^\alpha} = \{\overline{o}_1^i, \ldots, \overline{o}_{\overline{n_i}}^i\}$, $\overline{a}_i^\alpha \in R_\alpha(A_n^-) = \{\overline{a}_1^\alpha, \ldots, \overline{a}_{k_2}^\alpha\}$, and $\mathcal{I}_n^{set} = I(S_{set}(A_n^+)) + I(ub + lb - S_{set}(A_n^-))$.

If one or both relevant argument sets are empty the function is defined as follows:

$$S_{norm}(n) = \begin{cases} ub + lb - S_{set}(A_n^-) & \text{if } R_\alpha(A_n^+) = \emptyset \text{ and } R_\alpha(A_n^-) \neq \emptyset \\ S_{set}(A_n^+) & \text{if } R_\alpha(A_n^+) \neq \emptyset \text{ and } R_\alpha(A_n^-) = \emptyset \\ \text{not defined} & \text{if } R_\alpha(A_n^+) = \emptyset \text{ and } R_\alpha(A_n^-) = \emptyset \end{cases}$$

At this point, once we compute the collective support for a norm, we can decide whether the norm should be enacted or not. Given a predefined norm acceptance level μ, a norm will be enacted if $S_{norm}(n) > \mu$. For the norm to be enacted, its support should be laying on the positive side of the spectrum, hence μ should be picked so that $\mu \in (\frac{lb+ub}{2}, ub]$.

7   Case Study: A Virtual Community

In this section we qualitatively compare the outcome of our norm support function with that of a naive average support function. This naive average support function obtains the support for a norm n as $S_{avg}(n) = \frac{1}{dim(\mathbf{O}_{A_n})} \left(\sum_{i=1}^{|A_n^+|} \sum_{j=1}^{n_i} o_j^i + \sum_{i=1}^{|A_n^-|} \sum_{j=1}^{\overline{n_i}} ub + lb - \overline{o}_j^i \right)$. Our comparison encompasses a collection of Norm Argument Maps (NAM) that we characterise based on the opinions about their positive and negative arguments. Table 1 summarises the results of our comparison, which we detail next through some examples which invoke our norm support function[3] with an opinion spectrum $\lambda = [1, 5]$, the importance function I, and a relevance level $\alpha = 0.3$.

NAM 1. Consider a norm n with one positive argument which is highly supported by opinions (e.g. with values 5, 5, 5), and three negative arguments that count on neutral supports (e.g. one with opinions 3.15, 3.2, 2.8; another one with opinions 3, 3.5, 2.6; and a third one with opinions 2.5, 3.5, 3.2). Thus, while the average support function would yield a rather neutral norm support ($S_{avg}(n) = 3.5375$), our norm support function would compute a strong support ($S_{norm}(n) = 4.9842$). Note that, since participants have not issued negative arguments that are strong enough to attack the norm, whereas they have found a strong argument to support it, the norm support should be favorable to the enacting of the norm. This is captured by our norm support function, while the average support function remains neutral. This happens because it is fundamental to weigh the importance of the arguments as well as the importance of the argument sets. In this way neutral arguments do not weigh much in the overall norm support. **NAM 2.** Consider the case of a norm with one positive argument with neutral opinions (e.g. 3.5, 3.25, 3.5, 3, 2.5) and one negative argument with a similar number of opinions but with weak support (e.g. opinions with values 1, 1, 1.2, 1.3, 1.25). The average support function would yield a strong support for the norm ($S_{avg}(n) = 4$) because of the weak support received by the negative argument. Unlike the average, our support function would obtain a neutral support for the norm ($S_{norm}(n) = 3.1731$) because the negative arguments are weakly supported and the positive one counts on neutral support.

Table 1. Norm supports computed by the average approach S_{avg} and our approach S_{norm}.

Norm argument map	Argument sets		Norm support	
	Positive arguments	Negative arguments	S_{avg}	S_{norm}
NAM 1	One strong argument	Several neutral arguments	Neutral	**Strong**
NAM 2	One neutral argument	One weak argument	**Strong**	Neutral
NAM 3	Weak arguments	None	Weak	Undefined
NAM 4	None	Weak arguments	**Strong**	Undefined
NAM 5	Strong with few opinions	Weak with lots of opinions	**Strong**	undefined

---

[3] This implementation is based on [6] and we have made it publicly available in [5].

The two cases above show that the norm support of our method is in line with positive arguments because negative arguments are not strong enough. The next three cases show the importance of counting on relevant arguments. **NAM 3.** Consider now a norm with weak positive arguments and no negative arguments. The average approach would yield weak support for the norm. However, notice that the lack of strongly-supported positive arguments does not imply that the norm is not good. If the norm was not good, we should expect that participants eventually issue strongly-supported negative arguments. Since there is not enough relevant information to decide whether the norm is good or not, the norm support would be undefined for our norm support function. This seems more reasonable than the weak support computed by the naive approach. **NAM 4.** Consider now the dual of our last NAM: a norm counts on weak negative arguments and no positive arguments. Here the average support function would obtain a strong norm support. Again, like in the previous case, our norm support function would be undefined, which seems more adequate due to the lack of relevance of the arguments issued so far. **NAM 5.** Consider the case of a norm with positive arguments, each one counting with a few high-valued opinions, and negative arguments, each one counting on a much larger number of low-valued opinions. The average support function $S_{avg}(n)$ would produce a strong norm support. However, notice that weak negative arguments should not favorably support a norm. Moreover, the positive arguments count on few opinions. If the norm was good enough, we should have expected to receive more supporting opinions, which is not the case. This is why our norm support function $S_{norm}(n)$ yiedls an undefined support. Overall, the three last examples show that the lack of enough relevant information leads our norm support function to an undefined norm support, which seems more reasonable (and cautious) than that of a naive average support function.

7.1 A Test with Human Users

We conducted a test to evaluate the functionality of the norm argument map. Our test encompassed eleven people debating on norms similar to the one in Fig. 1 within a prototyped football social network. Users debated normally for several rounds and, afterwards, a satisfaction survey asked them if resulting aggregated ratings were reasonable. In a scale from 1 to 5, the answers' mean was 3.36, which we can consider as a positive preliminary result if we take into account the usability deficiencies of our prototype.

8 Conclusions and Future Work

To provide a more democratic way of moderating virtual communities, we propose a new argumentative structure, the so-called norm argument map. We also faced the problem of computing the collective support for a norm from the opinions of an argument's participants. We have identified two core concepts when computing a norm's support: the relevance of arguments and their importance.

Thus, we argue that we must only consider relevant enough arguments and weigh opinions based on their importance (strength).

As to future work, we are currently working on identifying similar arguments that should be colapsed, but some other issues, such as when to close the argumentation process or how to define the norm acceptance level μ, still need to be studied. Moreover, we also plan to apply it to other social participation situations such as direct democracy.

Acknowledgments. Work funded by Spanish National project CollectiveWare code TIN2015-66863-C2-1-R (MINECO/FEDER).

References

1. Awad, E., Booth, R., Tohmé, F., Rahwan, I.: Judgment aggregation in multi-agent argumentation. CoRR, abs/1405.6509 (2014)
2. Gabbriellini, S., Torroni, P.: Microdebates: structuring debates without a structuring tool1. AI Commun. **29**(1), 31–51 (2015)
3. Klein, M.: Enabling large-scale deliberation using attention-mediation metrics. Comput. Support. Coop. Work **21**(4–5), 449–473 (2012)
4. Ostrom, E.: Governing the Commons: The Evolution of Institutions for Collective Action. Cambridge University Press, Cambridge (1990)
5. Serramia, M.: Java implementation for the norm argument map (2016). https://bitbucket.org/msamsa/norm-argument-map.git
6. Torra, V.: Java implementation of the WOWA, OWA and WM aggregation operators (2000–2004). http://www.mdai.cat/ifao/wowa.php
7. Torra, V., Narukawa, Y.: Modeling Decisions - Information Fusion and Aggregation Operators. Springer, Heidelberg (2007)
8. Yager, R.R.: On ordered weighted averaging aggregation operators in multicriteria decisionmaking. IEEE Trans. Syst. Man Cybern. **18**(1), 183–190 (1988)

Fundamentals of Risk Measurement and Aggregation for Insurance Applications

Montserrat Guillen$^{(\boxtimes)}$, Catalina Bolancé, and Miguel Santolino

Department of Econometrics, Riskcenter-IREA, Universitat de Barcelona,
Av. Diagonal, 690, 08034 Barcelona, Spain
mguillen@ub.edu
http://www.ub.edu/riskcenter

Abstract. The fundamentals of insurance are introduced and alternatives to risk measurement are presented, illustrating how the size and likelihood of future losses may be quantified. Real data indicate that insurance companies handle many small losses, while large or extreme claims occur only very rarely. The skewness of the profit and loss probability distribution function is especially troublesome for risk quantification, but its strong asymmetry is successfully addressed with generalizations of kernel estimation. Closely connected to this approach, distortion risk measures study the expected losses of a transformation of the original data. GlueVaR risk measures are presented. The notions of subadditivity and tail-subadditivity are discussed and an overview of risk aggregation is given with some additional applications to insurance.

Keywords: Risk analysis · Extremes · Quantiles · Distortion measures

1 Introduction and Motivation

The insurance market is made up of customers that buy insurance policies and shareholders that own insurance companies. The latter are typically concerned about adverse situations and seek to maximize their profits, while the former search for the best market price, although they also need reassurance that they have opted for a solvent company.

Every insurance contract has an associated risk. Here, we analyse the caveats of measuring risk individually when we consider more than one contract and more than one customer, i.e., the aggregate risk in insurance.

Risk quantification serves as the basis for identifying the appropriate price for an insurance contract and, thus, guaranteeing the stability and financial strength of the insurance company. The aim of this article is to provide some fundamentals on how best to undertake this analysis. Once the individual risk associated with each contract has been calculated, the sum of the risk of all contracts provides an estimate of the overall risk. In this way, we also provide an overview of risk aggregation.

© Springer International Publishing Switzerland 2016
V. Torra et al. (Eds.): MDAI 2016, LNAI 9880, pp. 15–25, 2016.
DOI: 10.1007/978-3-319-45656-0_2

1.1 Basic Risk Quantification in Insurance

Let us consider a client who buys a car insurance policy that covers the risk of losses caused by accidents involving that vehicle for a period of one year. The insurance company needs to cover its expenses attributable to administration costs, regulatory mandates, advertising and IT systems. In other words, the company needs to fix a minimum price to cover the general expenses derived from its ordinary operations. The contract price is known as the premium.

The premiums collected can then be invested in the financial market, producing returns for the company before its financial resources are required for paying out compensation to its customers. A company that sells car insurance may sell thousands of one-year contracts but only those clients that suffer an accident, and who are covered, are compensated.

Each insurance contract has an associated profit or loss outcome, which can only be observed at the end of the contract. Two problems emerge when measuring this outcome. First, from an economic point of view, the production process of an insurance contract follows what is known as an inverted cycle, i.e., the price has to be fixed before the cost of the product is fully known. In a normal factory production process, the first step is to create and manufacture the product and, then, according to existing demand and the expenses incurred, a minimum price is fixed for the product. In the insurance sector, however, information on costs is only partial at the beginning of the contract, since accidents have yet to occur. Moreover, uncertainty exists. The eventual outcome of an insurance contract depends, first, on whether or not the policyholder suffers an accident and, second, on its severity. If an accident occurs, then the company has to compensate the insured party and this amount may be much greater than the premium initially received. Thus, the cost of any one given contract is difficult to predict and the eventual outcome may be negative for the insurer.

Despite the large financial component involved in the management of an insurance firm, insurance underwriting is based primarily on the analysis of historical statistical data and the law of large numbers. Here, recent advances in the field of data mining allow massive amounts of information to be scrutinized and, thus, they have changed the way insurance companies address the problem of fixing the correct price for an insurance contract. This price, moreover, has to be fair for each customer and, therefore, premium calculation requires a sophisticated analysis of risk. In addition, the sum of all prices has to be sufficient to cover the pool of insureds.

Insurance companies around the world are highly regulated institutions. An insurance company cannot sell its products unless they have been authorized by the corresponding supervisor. In Spain, supervision is carried out by the *Direccion General de Seguros y Fondos de Pensiones*, an official bureau that depends on the Ministry of Economics and which has adhered to European guidelines since January 2016. Under the European directive known as *Solvency II*, no company is allowed to operate in European territory unless it complies with strict legal requirements. This directive is motivated by the need to provide an overall assessment of the companys capacity to face its aggregate risk, even in the worst case scenario.

The choice of loss models and risk measures is crucial, as we shall illustrate in the sections that follow. We start by providing definitions and notations and include a simple example that illustrates the definition of losses and the risk measure. We present distortion risk measures and report key findings about their behaviour when aggregating losses. We then present a special family of distortion risk measures. The non-parametric approach to the estimation of distribution functions is discussed. An example using data from car insurance accidents is analysed and we conclude with a discussion of some possible lines of future research.

1.2 Notation

Consider a probability space and the set of all random variables defined on this space. A risk measure ρ is a mapping from the set of random variables to the real line [26].

Definition 1. Subadditivity. *A risk measure is subadditive when the aggregated risk, which is the risk of the sum of individual losses, is less than or equal to the sum of individual risks.*

Subadditvity is an appealing property when aggregating risks in order to preserve the benefits of diversification.

Value-at-Risk (VaR) has been adopted as a standard tool to assess risk and to calculate legal requirements in the insurance industry. Throughout this discussion, we assume without loss of generality that all data on costs are non-negative, so we will only consider non-negative random variables.

Definition 2. Value-at-Risk. *Value-at-Risk at level α is the α-quantile of a random variable X (which refers to a cost, a loss or the severity of an accident in our context), so*

$$\mathrm{VaR}_\alpha\left(X\right) = \inf\left\{x \mid F_X\left(x\right) \geq \alpha\right\} = F_X^{-1}\left(\alpha\right),$$

where F_X is the cumulative distribution function (cdf) of X and α is the confidence or the tolerance level $0 \leq \alpha \leq 1$.

VaR has many pitfalls in practice [23]. A major disadvantage when using VaR in the insurance context is that this risk measure does not always fulfill the subadditivity property [1,3]. So, the VaR of a sum of losses is not necessarily smaller than or equal to the sum of VaRs of individual losses. An example of such a case is presented in Sect. 5. VaR is subadditive for elliptically distributed losses [25].

Definition 3. Tail Value-at-Risk. *Tail Value-at-Risk at level α is defined as:*

$$\mathrm{TVaR}_\alpha\left(X\right) = \frac{1}{1-\alpha}\int_\alpha^1 VaR_\lambda\left(X\right)d\lambda.$$

Roughly speaking, the TVaR is understood as the mathematical expectation beyond the VaR. The TVaR risk measure is subadditive and it is a coherent risk measure [18].

Since we are mainly concerned with extreme values, we consider the definition of *tail-subadditivity*. This means that we only examine the domain of the variables that lies beyond the VaR of the aggregate risk.

Definition 4. *Tail-Subadditivity.* *A risk measure is tail-subadditive when the aggregated risk (risk of the sum of losses) is less than or equal to the sum of individual risks, only in the domain defined by the VaR of the sum of losses.*

Additional information on the algorithm to rescale the risk measure in the tail is given below.

1.3   Exposure to Risk: A Paradox

An additional problem of measuring risk in insurance is that of exposure. The following simple example shows the importance of defining losses with respect to a certain level of exposure. For this purpose, we compare flying vs. driving.

There is typically much discussion as to whether flying is riskier than driving. In a recent paper published in *Risk Analysis* [24], a comparison of the risks of suffering a fatal accident in the air and on the highway illustrates that the construction and interpretation of risk measures is crucial when assessing risk. However, this example does not discuss the paradox that is described in [20], which argues that risk quantification also depends on how exposure is measured.

MacKenzie [24] calculates the probability of a fatal incident by dividing the total number of fatal incidents by the total number of miles travelled in the United States. He also approximates the distributions of the number of victims given a fatal incident occurs. The probabilities of a fatal incident per one million miles travelled compared to those calculated by Guillen [20] for 10,000 hours of travel (in parentheses) are 0.017 % (0.096 %) for air carriers, 22.919 % (45.838 %) for air taxis and commuters, and 1.205 % (0.843 %) for highway driving. The two approaches produce different outcomes in the probability of an accident with fatalities, because speed is not homogeneous across all transportation modes. However, regardless of whether miles travelled or hours of travel are considered, we always conclude that the safest means of transport is flying with a commercial air carrier if we look solely at the probability of an incident occurring.

However, if the expected number of fatalities per one million miles or per 10,000 hours of travel is compared, a contradiction emerges. The average number of victims per one million miles is 0.003 if we consider distance in terms of commercial aviation trips, whereas the average number of victims is 0.013 if we consider distance driven on highways. However, if we consider the time spent on the commercial aviation trip, the average is 0.017 victims compared to 0.009 when driving on highways. The conclusion we draw here is that highway trips are safer than commercial airline flights. This contradiction with respect to the previous discussion is caused by the use of the mathematical expectation of two different loss functions. This simple example shows the importance of knowing how to define the losses and the implications of the choice of the risk measure.

2 Distortion Risk Measures

Distortion risk measures were introduced by Wang [29, 30] and are closely related to the distortion expectation theory [31]. A review of how risk measures can be interpreted from different perspectives is provided in [27], and a clarifying explanation of the relationship between distortion risk measures and distortion expectation theory is provided. Distortion risk measures are also studied in [4,17]. The definition of a distortion risk measure contains two key elements: first, the associated distortion function; and, second, the concept of the Choquet integral [15].

Definition 5. *Distortion Function.* *Let $g : [0,1] \to [0,1]$ be a function such that $g(0) = 0$, $g(1) = 1$ and g is injective and non-decreasing. Then g is called a distortion function.*

Definition 6. *Choquet Integral.* *The Choquet Integral with respect to a set function μ of a μ-measurable function $X : \Omega \to \overline{R}^{+} \cup \{0\}$ is denoted as $\int X d\mu$ and is equal to*

$$\int X d\mu = \int_0^{+\infty} S_{\mu,X}(x) dx,$$

if $\mu(\Omega) < \infty$, where $S_{\mu,X}(x) = \mu(\{X > x\})$ denotes the survival function of X with respect to μ. See [16] for more details.

Definition 7. *Distortion Risk Measure for Non-negative Random Variable.* *Let g be a distortion function. Consider a non-negative random variable X and its survival function $S_X(x) = P(X > x)$. Function ρ_g defined by*

$$\rho_g(X) = \int_0^{+\infty} g(S_X(x)) \, dx$$

is called a distortion risk measure.

3 GlueVaR Risk Measures

A new family of risk measures known as GlueVaR was introduced by Belles-Sampera et al. [5]. A GlueVaR risk measure is defined by a distortion function. Given confidence levels α and β, $\alpha \leq \beta$, the distortion function for a GlueVaR is:

$$\kappa_{\beta,\alpha}^{h_1,h_2}(u) = \begin{cases} \dfrac{h_1}{1-\beta} \cdot u, & \text{if } \ 0 \leq u < 1 - \beta \\[2mm] h_1 + \dfrac{h_2 - h_1}{\beta - \alpha} \cdot [u - (1-\beta)], \\ \quad \text{if } \ 1 - \beta \leq u < 1 - \alpha \\[2mm] 1, & \text{if } \ 1 - \alpha \leq u \leq 1 \end{cases} \tag{1}$$

where $\alpha, \beta \in [0,1]$ such that $\alpha \leq \beta$, $h_1 \in [0,1]$ and $h_2 \in [h_1, 1]$. Parameter β is the additional confidence level besides α. The shape of the GlueVaR distortion function is determined by the distorted survival probabilities h_1 and h_2 at levels $1 - \beta$ and $1 - \alpha$, respectively. Parameters h_1 and h_2 are referred to as the heights of the distortion function.

The GlueVaR family has been studied by [5–7,9], who showed that the associated distortion function $\kappa_{\beta,\alpha}^{h_1,h_2}$ can be defined as being concave in $[0,1]$. The concavity of the distortion risk measure is essential to guarantee tail-subadditivity.

Theorem 1. *Concave and continuous distortion risk measures are subadditive.*

Proof. A proof can be derived from [16]. $\qquad\blacksquare$

Corollary 1. *If a distortion risk measure is subadditive, it is also tail-subadditive in the restricted domain.*

Theorem 2. *GlueVaR risk measures are tail-subadditive if they are concave in the interval $[0, (1 - \alpha))$.*

Proof. For a GlueVaR risk measure, it suffices to check that its corresponding distortion function $\kappa_{\beta,\alpha}^{h_1,h_2}(u)$ is concave for $0 \leq u < (1 - \alpha)$. Note that by definition the distortion function is also continuous in that interval. Then it suffices to restrict the domain so that the variable only takes values that are larger than the VaR of the sum of losses and apply the previous theorem. Note also that the VaR of the sum of losses is always larger or equal than the VaR of each individual loss, since we consider that all losses are non-negative. [5] also provide a proof. $\qquad\blacksquare$

Let us comment on the practical application of the above results. Given two random variables, X_1 and X_2. Let us denote by $m_\alpha = VaR_\alpha(X_1 + X_2)$. Then, we define the truncated variables $X_1 | X_1 > m_\alpha$ and $X_2 | X_2 > m_\alpha$. Likewise, we consider the truncated random variable $(X_1 + X_2) | (X_1 + X_2) > m_\alpha$, then tail-subadditivity holds whenever

$$\rho_g\left[(X_1 + X_2) | (X_1 + X_2) > m_\alpha\right] \leq \rho_g\left[X_1 | X_1 > m_\alpha\right] + \rho_g\left[X_2 | X_2 > m_\alpha\right]. \quad (2)$$

Put simply, expression (2) means that the risk of the sum of the losses of two contracts that exceed the value-at-risk of the sum is less than or equal to the sum of the risks of losses from each contract above the risk of the sum.

The algorithm to calculate the rescaled GlueVaR risk measure in the tail that we implement below in Sect. 5 is as follows. We have restricted our data set to all values greater than m_{α_0} for a given confidence level α_0. For these data, we subtract m_{α_0} from each data point and redefine the tolerance parameters, so that $\alpha = 0$ and $\beta = 1 - (1 - \beta_0)/(1 - \alpha_0)$, where α_0 and β_0 are the original levels of confidence. Once the GlueVaR has been calculated for this set of data and parameters, we add $\alpha_0 m_{\alpha_0}$ to return to the original scale.

4 Nonparametric Estimation of Standard Risk Measures

Let $T(\cdot)$ be a concave transformation where $Y = T(X)$ is the transformed random variable and $Y_i = T(X_i)$, $i = 1 \ldots n$ are the transformed observed losses and n is the total number of observed data. Then the kernel estimator of the transformed cumulative distribution function of variable X is:

$$\widehat{F}_Y(y) = \frac{1}{n} \sum_{i=1}^{n} K\left(\frac{y - Y_i}{b}\right) = \frac{1}{n} \sum_{i=1}^{n} K\left(\frac{T(x) - T(X_i)}{b}\right), \tag{3}$$

The transformed kernel estimation of $F_X(x)$ is:

$$\widehat{F}_X(x) = \widehat{F}_{T(X)}(T(x)).$$

where b and $K\left(\frac{x - X_i}{b}\right)$ are defined as the bandwidth and the integral of the kernel function $k(\cdot)$, respectively (see [10] for more details).

In order to obtain the transformed kernel estimate, we need to determine which transformation should be used. Several authors have analysed the transformed kernel estimation of the density function ([10,14,28]).

A double transformation kernel estimation method was proposed by Bolancé et al. [11]. This requires an initial transformation of the data $T(X_i) = Z_i$, where the transformed variable distribution is close to a Uniform $(0, 1)$ distribution. Afterwards, the data are transformed again using the inverse of the distribution function of a *Beta* distribution. The resulting variable, with corresponding data values $M^{-1}(Z_i) = Y_i$, after the double transformation is close to a *Beta* (see, [10,12]) distribution, so it is quite symmetrical and the choice of the smoothing parameter can be optimized.

Following the double transformation of the original data, VaR_α is calculated with the Newton-Raphson method to solve the expression:

$$\widehat{F}_{T(X)}(T(x)) = \alpha$$

and once the result is obtained, the inverse of the transformations is applied in order to recover the original scale. The optimality properties and performance, even in small samples are studied by Alemany et al. [2].

When calculating the empirical $TVaR_\alpha$ a first moment of the data above VaR_α is used, but other numerical approximations based on the non-parametric estimate of the distribution function are also possible.

In general, a non-parametric estimation of distortion risk measures can be directly achieved in the transformed scale, which guarantees that the transformed variable is defined in a bounded domain. So, the non-parametric approach can simply be obtained by integrating the distorted estimate of the survival function of the transformed (or double transformed) variable $T(X)$, so:

$$\hat{\rho}_g(T(X)) = \sum_{i>1}^{n} g(1 - \widehat{F}_{T(X)}(T(X_{(i)})))(T(X_{(i)}) - T(X_{(i-1)})),$$

where subscript (i) indicates the ordered location. Once the result is obtained, the inverse of the transformations is applied in order to recover the original scale. The properties of this method have not yet been studied.

5   Example

Here we provide an example of the implementation of risk measurement and aggregation. The data have been provided by a Spanish Insurer and they contain information on two types of costs associated with the car accidents reported to the company. The first variable (X_1) is the cost of the medical expenses paid out to the insurance policy holder and the second variable (X_2) is the amount paid by the insurer corresponding to property damage. Medical expenses may contain medical costs related to a third person injured in the accident. More information on the data can be found in [13, 21, 22]. The sample size is 518 cases. The minimum, maximum and mean values of X_1 (in parentheses X_2) are 13 (1), 137936 (11855) and 1827.6 (283.9), respectively.

The empirical risk measures for different levels of tolerance are shown in Table 1. Risk in the tail region is shown in Table 2. The results in Table 1 confirm that VaR is not subadditive; nor is the GlueVaR example chosen here. However, tail-subadditivity holds in the tail, as shown in Table 2.

Table 1. Distortion risk measures (ρ) for car insurance cost data and subadditivity

ρ	α	$\rho(X_1)$	$\rho(X_2)$	$\rho(X_1 + X_2)$	$\rho(X_1) + \rho(X_2)$	Subadditivity
VaR_α	95.0 %	6450.00	1060.00	7926.00	7510.00	No
	99.0 %	20235.00	4582.00	25409.00	24817.00	No
$TVaR_\alpha$	95.0 %	18711.78	3057.88	20886.81	21769.66	Yes
	99.0 %	48739.25	7237.02	53259.39	55976.27	Yes
$GlueVaR^*$	95.0 %	10253.39	1558.42	11996.87	11811.81	No
	99.0 %	24817.56	4988.43	29992.21	29805.99	No

*The GlueVaR parameters are $h_1 = 1/20$, $h_2 = 1/8$ and $\beta = 99.5$ %.

Nonparametric estimates of VaR are shown in Table 3. The results also indicate that subadditivity is not fulfilled for a level of $\alpha = 95$ %. Note also that compared to the empirical results, the non-parametric approximation produces higher values for larger tolerance levels because the shape of the distribution in the extremes is smoothed and extrapolated. So, in this case, subadditivity is found for $\alpha = 99$ % and $\alpha = 99.5$ %.

Table 2. Distortion risk measures (ρ) for car insurance cost data and rescaled tail-measure

ρ	α	$\rho(X_1)$	$\rho(X_2)$	$\rho(X_1 + X_2)$	$\rho(X_1) + \rho(X_2)$	Tail-subadd.*
VaR_α	95.0%	7603.70	18978.70	7529.7	26582,40	Yes
	99.0%	36603.91	-	25409.00	-	-
$TVaR_\alpha$	95.0%	20380.47	69517.70	20440.66	89898.17	Yes
	99.0%	87142.91	-	49453.17	-	-
$GlueVaR^{**}$	95.0%	11740.70	27401.87	11588.64	39142.57	Yes
	99.0%	45027.08	-	29398.90	-	-

*Only values above the corresponding $VaR_\alpha(X_1 + X_2)$ are considered. For $\alpha = 99\%$, no values of X_2 are larger than this level.
**The GlueVaR parameters are $h_1 = 1/20$, $h_2 = 1/8$ and $\beta = 99.5\%$.

Table 3. Nonparametric estimates of Value-at-Risk (ρ) for car insurance cost data and subadditivity

α	$\rho(X_1)$	$\rho(X_2)$	$\rho(X_1 + X_2)$	$\rho(X_1) + \rho(X_2)$	Subadditivity
95.0%	6357.58	1049.77	7415.80	7407.35	No
99.0%	23316.56	4693.33	26606.16	28009.89	Yes
99.5%	36967.12	7921.23	36968.11	44888.35	Yes

6   Conclusion

We highlight the importance of transformations in the analysis of insurance data that present many extreme values. Distortion risk measures transform the survival function to focus on extreme losses, while advanced non-parametric kernel methods benefit from the transformation of the original data to eliminate asymmetry.

Extreme value theory plays an important methodological role in risk management for the insurance, reinsurance, and finance sectors, but many challenges remain with regards how best to measure and aggregate risk in these cases. Tails of loss severity distributions are essential for pricing [19] and creating the high-excess loss layers in reinsurance.

Distortion risk measures constitute a tool for increasing the probability density in those regions where there is more information available on extreme cases. Yet, the selection of the distortion function is not subject to an optimization procedure. Regulators have imposed the use of some easy-to-calculate measures, for example, in Solvency II the central risk measure is the VaR, while in the Swiss Solvency Test, TVaR is the standard approach. Non-parametric methods for risk measurement are flexible and do not require any assumptions regarding the statistical distribution that needs to be implemented. As such, they certainly impose fewer assumptions than when using a given parametric statistical distribution. We believe that distortion risk measures could optimize an objective function that reflects attitude towards risk. The relationship between risk measures and risk attitude was initially studied by [8]. The analysis of the attitudinal position and the risk aversion shown by the risk quantifier have not been addressed here and remain matters for future study.

References

1. Acerbi, C., Tasche, D.: On the coherence of expected shortfall. J. Bank. Financ. **26**(7), 1487–1503 (2002)
2. Alemany, R., Bolancé, C., Guillen, M.: A nonparametric approach to calculating value-at-risk. Insur. Math. Econ. **52**(2), 255–262 (2013)
3. Artzner, P., Delbaen, F., Eber, J.M., Heath, D.: Coherent measures of risk. Math. Financ. **9**(3), 203–228 (1999)
4. Balbás, A., Garrido, J., Mayoral, S.: Properties of distortion risk measures. Methodol. Comput. Appl. Probab. **11**(3), 385–399 (2009)
5. Belles-Sampera, J., Guillen, M., Santolino, M.: Beyond value-at-risk: GlueVaR distortion risk measures. Risk Anal. **34**(1), 121–134 (2014)
6. Belles-Sampera, J., Guillen, M., Santolino, M.: GlueVaR risk measures in capital allocation applications. Insur. Math. Econ. **58**, 132–137 (2014)
7. Belles-Sampera, J., Guillen, M., Santolino, M.: The use of flexible quantile-based measures in risk assessment. Commun. Stat. Theory Methods **45**(6), 1670–1681 (2016)
8. Belles-Sampera, J., Guillen, M., Santolino, M.: What attitudes to risk underlie distortion risk measure choices? Insur. Math. Econ. **68**, 101–109 (2016)
9. Belles-Sampera, J., Merigo, J.M., Guillen, M., Santolino, M.: The connection between distortion risk measures and ordered weighted averaging operators. Insur. Math. Econ. **52**(2), 411–420 (2013)
10. Bolancé, C., Guillen, M., Nielsen, J.P.: Kernel density estimation of actuarial loss functions. Insur. Math. Econ. **32**(1), 19–36 (2003)
11. Bolancé, C., Guillen, M., Nielsen, J.P.: Inverse Beta transformation in kernel density estimation. Stat. Probab. Lett. **78**(13), 1757–1764 (2008)
12. Bolancé, C., Guillén, M., Nielsen, J.P.: Transformation kernel estimation of insurance claim cost distributions. In: Corazza, M., Pizzi, C. (eds.) Mathematical and Statistical Methods for Actuarial Sciences and Finance, pp. 43–51. Springer, Milan (2010)
13. Bolancé, C., Guillen, M., Pelican, E., Vernic, R.: Skewed bivariate models and nonparametric estimation for the CTE risk measure. Insur. Math. Econ. **43**(3), 386–393 (2008)
14. Buch-Larsen, T., Nielsen, J.P., Guillen, M., Bolance, C.: Kernel density estimation for heavy-tailed distributions using the Champernowne transformation. Statistics **39**(6), 503–516 (2005)
15. Choquet, G.: Theory of capacities. Ann. de l'Inst. Fourier **5**, 131–295 (1954). Institut Fourier
16. Denneberg, D.: Non-additive Measure and Integral, vol. 27. Springer Science and Business Media, Netherlands (1994)
17. Denuit, M., Dhaene, J., Goovaerts, M., Kaas, R.: Actuarial Theory for Dependent Risks: Measures, Orders and Models. Wiley, Hoboken (2006)
18. Dhaene, J., Laeven, R.J., Vanduffel, S., Darkiewicz, G., Goovaerts, M.J.: Can a coherent risk measure be too subadditive? J. Risk Insur. **75**(2), 365–386 (2008)

19. Guelman, L., Guillen, M., Pérez-Marín, A.M.: A decision support framework to implement optimal personalized marketing interventions. Decis. Support Syst. **72**, 24–32 (2015)
20. Guillen, M.: Riesgo y seguro en economia. Discurso de ingreso en la Real Academia de Ciencias Economicas y Financieras. Barcelona (2015)
21. Guillen, M., Prieto, F., Sarabia, J.M.: Modelling losses and locating the tail with the Pareto Positive Stable distribution. Insur. Math. Econ. **49**(3), 454–461 (2011)
22. Guillen, M., Sarabia, J.M., Prieto, F.: Simple risk measure calculations for sums of positive random variables. Insur. Math. Econ. **53**(1), 273–280 (2013)
23. Jorion, P.: Value at Risk: The New Benchmark for Managing Financial Risk, vol. 3. McGraw-Hill, New York (2007)
24. MacKenzie, C.A.: Summarizing risk using risk measures and risk indices. Risk Anal. **34**(12), 2143–2162 (2014)
25. McNeil, A.J., Frey, R., Embrechts, P.: Quantitative Risk Management: Concepts, Techniques and Tools. Princeton University Press, Princeton (2015)
26. Szeg, G.: Measures of risk. J. Bank. Financ. **26**(7), 1253–1272 (2002)
27. Tsanakas, A., Desli, E.: Measurement and pricing of risk in insurance markets. Risk Anal. **25**(6), 1653–1668 (2005)
28. Wand, M.P., Marron, J.S., Ruppert, D.: Transformations in density estimation. J. Am. Stat. Assoc. **86**(414), 343–353 (1991)
29. Wang, S.: Insurance pricing and increased limits ratemaking by proportional hazards transforms. Insur. Math. Econ. **17**(1), 43–54 (1995)
30. Wang, S.: Premium calculation by transforming the layer premium density. Astin Bull. **26**(01), 71–92 (1996)
31. Yaari, M.E.: The dual theory of choice under risk. Econom. J. Econom. Soc. **55**, 95–115 (1987)

Privacy in Bitcoin Transactions: New Challenges from Blockchain Scalability Solutions

Jordi Herrera-Joancomartí[(✉)] and Cristina Pérez-Solà

Dept. d'Enginyeria de la Informació i les Comunicacions,
Universitat Autònoma de Barcelona, 08193 Bellaterra, Catalonia, Spain
jordi.herrera@uab.cat, cperez@deic.uab.cat

Abstract. Bitcoin has emerged as the most successful cryptocurrency since its appearance back in 2009. However, its main drawback to become a truly global payment system is its low capacity in transaction throughput. At present time, some ideas have been proposed to increase the transaction throughput, with different impact on the scalability of the system. Some of these ideas propose to decouple standard transactions from the blockchain core and to manage them through a parallel payment network, relegating the usage of the bitcoin blockchain only to transactions which consolidate multiple of those off-chain movements. Such mechanisms generate new actors in the bitcoin payment scenario, the Payment Service Providers, and new privacy issues arise regarding bitcoin users. In this paper, we provide a comprehensive description of the most relevant scalability solutions proposed for the bitcoin network and we outline its impact on users' privacy based on the early stage proposals published so far.

Keywords: Bitcoin · Scalability · Off-chain transactions · Lightning network · Duplex micropayment channels

1 Introduction

Bitcoin is an online virtual currency based on public key cryptography, proposed in 2008 in a paper authored by someone behind the Satoshi Nakamoto pseudonym [1]. It became fully functional on January 2009 and its broad adoption, facilitated by the availability of exchange markets allowing easy conversion with traditional currencies (EUR or USD), has brought it to be the most successful virtual currency.

The success of bitcoin has evidenced its weak design in terms of scalability since the number of transactions per second that the system may handle is orders of magnitude lower than standard globally used systems, like VISA.

In order to allow bitcoin to scale and to have a chance to be a global payment system, different solutions have been proposed. Although some of them are still in development, in this paper we point out the most relevant ones, that is, proposals that have a large acceptance degree in the community, focusing on those

© Springer International Publishing Switzerland 2016
V. Torra et al. (Eds.): MDAI 2016, LNAI 9880, pp. 26–44, 2016.
DOI: 10.1007/978-3-319-45656-0_3

that present an important shift in the bitcoin development: off-chain payment channels. Besides the general paradigm change that off-chain payment channels may suppose, we are mainly interested on how bitcoin users' privacy could be affected by such proposals.

The organization of the paper is as follows. In Sect. 2 we provide a general background of the bitcoin system and outline its scalability problems. Section 3 points out the main proposals to scale the bitcoin system, focusing in the off-chain payment channel solution. How bitcoin users' privacy will be affected by off-chain payment channels is discussed in Sect. 4, mainly analysing how actual techniques used to attack/protect users' privacy will be affected. Finally, Sect. 5 concludes the paper and gives some guidelines for further research in this field.

2   The Bitcoin System

In this section, we point out the main ideas to understand the basic functionality of the bitcoin cryptocurrency. Such background is needed to understand the scalability problems the system faces and the solutions that have been proposed. However, the complexity of bitcoin makes impossible to provide a full description of the system in this review, so interested readers can refer to Antonopoulos's book [2] for a detailed and more extended explanation on the bitcoin system.

Bitcoin is a cryptocurrency based on accounting entries. For that reason, it is not correct to look at bitcoins as digital tokens since bitcoins are represented as a balance in a bitcoin account. A **bitcoin account** is defined by an Elliptic Curve Cryptography key pair. The bitcoin account is publicly identified by its **bitcoin address**, obtained from its public key using a unidirectional function. Using this public information users can send bitcoins to that address[1]. Then, the corresponding private key is needed to spend the bitcoins of the account.

2.1   Bitcoin Payments

Payments in the bitcoin system are performed through transactions between bitcoin accounts. A **bitcoin transaction** indicates a bitcoin movement from source addresses to destination addresses. Source addresses are known as **input addresses** in a transaction and destination addresses are named **output addresses**. As it can be seen in Fig. 1, a single transaction can have one or multiple input addresses and one or multiple output addresses.

A transaction details the exact amount of bitcoins to be transferred from each input address. The same applies to the output addresses, indicating the total amount of bitcoins that would be transferred at each account. For consistency, the total amount of the input addresses (source of the money) must be greater or equal than the total amount of the output addresses (destination of the money). Furthermore, the bitcoin protocol forces that input addresses must spend the

---

[1] Notice that the terms public key, address or bitcoin account refer to the same concept.

Inputs

Previous output (index)	Amount	From address	Type	ScriptSig
c631567f352f...:1	3.02887912	1CGVyAgAx9gg1va5pGNVJtF6gdKpPUVTSf	Address	304402201700305a3d79a[....]2b985b15daa0ab9c50cd61449ca037dc9f0
c284ec14325f...:0	3.04042789	1GY84QPLfM9d4KqTjTbbHsb9BX9FF1kYQx	Address	3045022100e724004f2d3[....]91d95b56ad29f817f3e3259daffbd72f2a98
0fbec1d29b8e...0	2.99934316	1CGVyAgAx9gg1va5pGNVJtF6gdKpPUVTSf	Address	304402200f6c9b4281cb0[....]2b985b15daa0ab9c50cd61449ca037dc9f0
232715b3c51a...:1	3.00515088	17ALqzZFPbSqXz9aQhzgK6ts9htZfV8Mwu	Address	304402207311495478c1d[....]8d4656bf7613d47dd4e6a5b062d9fb6a34

Outputs

Index	Amount	To address	Type	ScriptPubKey
0	0.51682435	1LUHXNTsHPUGVJJeefPdb2rpdxtWoHrcKv	Address	OP_DUP OP_HASH160 d59336a017660c48be2adaa9a77153eccfdb8b0b8 OP_EQUALVERIFY OP_CHECKSIG
1	11.5569767	1HzAb4E1kZH4pDKoxML4KXBLPPyUootw4s	Address	OP_DUP OP_HASH160 ba51b9ace7595c72a2cbc1d4e3e90e356f77804 OP_EQUALVERIFY OP_CHECKSIG

Fig. 1. Bitcoin transaction example: four input addresses and two output addresses (data from blockexplorer.com).

exact amount of a previously received transaction[2] and, for that reason, in a transaction each input must unambiguously indicate the transaction[3] and the index of the output from which the bitcoins were received (the field *Previous output (index)* in Fig. 1).

Finally, the owner of the input addresses should perform a digital signature using his private keys, proving that he is the real owner of such accounts.[4]

Before accepting a payment from a standard transaction, the receiver should:

- Validate that the bitcoins of the input addresses are not previously spent.
- Validate that the digital signature is correct.

The first validation prevents doublespending in the bitcoin system and it is performed through a ledger where all previous transactions are annotated. Before accepting the payment, the receiver needs to be sure that there is no other transaction already in the ledger that has an input with the same *Previous output (Index)*. For that reason, the integrity of the system is based on the fact that this ledger is not modifiable, although it should be possible to add new transactions. In the bitcoin system, this append-only ledger is called blockchain.[5] The second validation can be performed with the information included in the transaction itself (field *ScriptSig*) together with the information of the transaction identified in the *Previous output (Index)* (field *ScriptPubKey*).

---

[2] Notice that in Fig. 1, there are two input addresses that are exactly the same which indicates that bitcoins have arrived to this bitcoin account in two separate transactions.

[3] A transaction is identified in the bitcoin system by its hash value.

[4] Although this is the standard form of bitcoin verification for regular bitcoin transfer transactions, the verification of a transaction can be much more complex and is based on the execution of a stack-based scripting language (more details can be found in Chap. 5 of [2]).

[5] Note that the non-modifiable property of the blockchain implies that bitcoin payments are non reversible.

2.2   The Blockchain and the Mining Process

The **blockchain** is a general append-only ledger containing all bitcoin transactions performed since the system started to operate, back in 2009. Such approach implies that the size of the blockchain is constantly increasing and, for that reason, scalability is probably the biggest challenge that the system faces. The blockchain is freely replicated and stored in different nodes of the bitcoin network, making the bitcoin a completely distributed system.

Transactions are included in the blockchain at time intervals, rather than in a flow fashion, and such addition is performed by collecting all new transactions of the system, compiling them together in a data structure called block, and including the block at the top of the blockchain. Every time that a block containing a specific transaction is included in the blockchain such transaction is said to be a **confirmed transaction** since it has been already included in the blockchain and can be checked for doublespending prevention.

Blocks are data structures that mainly contain a set of transactions that have been performed in the system (see Fig. 2). To achieve the append-only property, the inclusion of a block in the blockchain is a hard problem, so adding blocks to the blockchain is time and work consuming. Furthermore, every block is indexed using its hash value and every new block contains the hash value of the previous one (see the field *Previous block* in Fig. 2). Such mechanism ensures that the modification of a block from the middle of the chain would imply to modify all remaining blocks of the chain from that point to the top in order to match all hash values.

Block 125552

Hash: 00000000000000001e8d6829a8a21adc5d38d0a473b144b6765798e61f98bd1d
Previous block: 00000000000008a3a41b85b8b29ad444def299fee21793cd8b9e567eab02cd81
Time: 2011-05-21 17:26:31
Difficulty: 244 112.487774
Transactions: 4
Total BTC: 84.52
Size: 1.496 kilobytes
Merkle root: 2b12fcf1b09288fcaff797d71e950e71ae42b91e8bdb2304758dfcffc2b620e3
Nonce: 2504433986

Transactions

Transaction	Fee	Size (kB)	From (amount)	To (amount)
51d37bdd87...	0	0.135	Generation: 50 + 0.01 total fees	15nNvBTUdMaiZ6d3GWCeXFu2MagXL3XM1q: 50.01
60c25dda8d...	0	0.259	1HuppjXz7dPrt2a67LqacDW5T4VanFrpqC: 29.5	1B8vkT58i8KUPVJvvyQfrbc8Wjwu3vEarQ: 0.5 1BQbxzgRSLEsmv1JNc8MG76wdUgMwbsaww: 29
01f314cdd8...	0.01	0.617	1NdzSE6sHubscXJrv7jJn2gd4fL9L3ai6E: 0.03 1Jjv9m5VrRUE7VoktCsj18KUSqkqchhbum: 0.02 1HsYJJPqTn34DEjMnTb3VfKckX7ZcWPibm: 4.82	175FNxcLc1YrTwwG6TcsywcsHYdVqyhbwC: 0.01 1MueNMRJmcqVQeqE7v4dqogpNbhyxqq8R6: 4.85
b519286a10...	0	0.404	12DCoCVvDCkQShZ5RTh9bysgCkmkRMNQbT: 0.14 13CJwnnXJPwkzY4Xnaoqf8dnyNBwrHG9fe: 0.01	1Mos7p8fqJKBcYNRG1TdT5hBRxdMP6YHPy: 0.15

Fig. 2. Example of a bitcoin block (data from blockexplorer.com).

Adding a block to the blockchain is known as the **mining process**, a process that is also distributed and that can be performed by any user of the bitcoin network using specific-purpose software (and hardware). The mining process uses a hashcash proof-of-work system, first proposed by Adam Back as an anti-spam mechanism [3]. The proof-of-work consists in finding a hash of the new block with a value lower than a predefined target[6]. This process is performed by brute force varying the nonce value of the block. Once the value has been found, the new block becomes the top block of the blockchain and all miners discard their work on that block and move to the next one.

Mining new blocks is a structural task in the bitcoin system since it helps to confirm the transactions of the system. For that reason, and also assuming that mining implies a hard work, miners have to be properly rewarded. In the bitcoin system, miners are rewarded with two mechanisms. The first one provides them with newly created bitcoins. Every new block includes a special transaction, called **generation transaction** or coinbase transaction (see the first transaction in Fig. 2), in which it does not appear any input address and the output address is determined by the miner who creates the block, who obviously indicates one of its own addresses.[7] The second rewarding mechanism is the fees that each transaction pays to the miner. The fee for each transaction is calculated by computing the difference between the total input amount and the total output amount of the transaction (notice that in example block of Fig. 2 the first transaction does not provide any fee while the second one generates a 0.01 fee). All fees collected from transactions in a block are included in the generation transaction.

2.3   The Bitcoin Network

The bitcoin system needs to disseminate different kinds of information, essentially, transactions and blocks. Since both data are generated in a distributed way, the system transmits such information over the Internet through a distributed peer to peer (P2P) network. Such distributed network is created by bitcoin users in a dynamic way, and nodes of the bitcoin P2P network [4] are computers running the software of the bitcoin network node. This software is included by default into bitcoin's full-client wallets, but it is not usually incorporated in light wallet versions, such as those running in mobile devices. It is important to stress such distinction in case to perform network analysis, because when discovering nodes in the P2P bitcoin network, depending on the scanning techniques, not all bitcoin users are identified, but only those running a full-client and those running a special purpose bitcoin P2P node. Furthermore, online bitcoin accounts,

---

[6] Notice that the value of the target determines the difficulty of the mining process. Bitcoin adjusts the target value depending on the hash power of the miners in order to set the throughput of new blocks to 1 every 10 min (in mean).

[7] The amount of a generation transaction is not constant and it is determined by the bitcoin system. Such value, started in 50 bitcoins, is halved every four years, fixing asymptotically to 21 millions the total number of bitcoins that will ever be created.

provided by major bitcoin Internet sites, can also be considered as light weight clients, so they do not represent a full bitcoin P2P node neither.

2.4   Bitcoin Scalability Issues

Recently, the increase of both the popularity and the usage of the bitcoin system has shown its bounds regarding its ability to scale with the number of users. It is obvious that a system with a unique (although replicated) register containing all system transactions (i.e. the blockchain) may present a bottleneck.

Scalability issues can be measured in different ways as it is pointed out in [5]. From latency (the time for a transaction to be confirmed) to bootstrap time (the time it takes a new node to download and process the history necessary to validate payments) through cost per confirmed transactions, different measures can be used to evaluate the efficiency of a payment system. Croman et al. [5] give approximations of all of these metrics for the Bitcoin network. However, probably the easiest measure to compare Bitcoin with existing global payment systems and the one that has a direct impact on scalability is the system transaction throughput. The transaction throughput can be measured by the maximum number of transactions per second that a system may deal with and it is often chosen to evaluate systems because it is objective, easy to compute, and can be used to compare different payment systems easily. For instance, Visa reported to allow around 2,000 transactions per second in normal situation [5] while reaching a peak of 56,000 in a stress test [6]. Paypal manages lower values, providing 136 transactions per second as mean throughput on his payment network.[8]

Bitcoin throughput can be measured taking into account different parameters, from network communication latency to processing power of the nodes performing transaction validation. However, the restriction that limits the most the throughput of the system is the maximum size of blocks. Currently (June 2016), block size is fixed at a maximum value of 1 MB.[9] Yet this limit has not always been in place: the initial release of the code in February 2009 did not explicitly contain a block size limit and it was not until late 2010 when the 1 MB limit was enforced. The procedure for activating the block size limit was gradual: first, the core was changed so that no large blocks were mined;[10] second, the consensus rules were updated to reject blocks larger than 1 MB;[11] finally, the new rules started to be enforced on block height higher than 79,400 (which was reached[12] in September 12nd, 2010). From that moment on, the block size limit has been kept to 1 MB.

---

[8] PayPal Q1 2016 Results [7] reported handling 1.41B payment transactions, which leads to an estimated $1.41B/4/30/24/60/60 = 136$ transactions per second.

[9] https://github.com/bitcoin/bitcoin/blob/a6a860796a44a2805a58391a009ba22752f64 e32/src/consensus/consensus.h#L9.

[10] https://github.com/bitcoin/bitcoin/commit/a30b56ebe76ffff9f9cc8a6667186179413c c6349.

[11] https://github.com/bitcoin/bitcoin/commit/8c9479c6bbbc38b897dc97de9d04e4d5a 5a36730#diff-118fcbaaba162ba17933c7893247df3aR1421.

[12] https://blockchain.info/block-height/79400.

Limiting the size of blocks to 1 MB implies a maximum throughput of 7 transactions per second [5]. The 7 transactions per second limit is an approximation obtained by dividing the maximum size of blocks by the average size of Bitcoin transactions (250 bytes) and the inter-block time (10 min). Therefore, a block of maximum size may contain 1,000,000/250 = 100 average sized transactions, thus giving a throughput of 100/600 = 6.6 transactions per second. Notice that such value is very far from the numbers that other payment systems, like Visa or PayPal, may deal with.

3   Bitcoin Scalability Proposals

Modification proposals in the Bitcoin core protocol, even those of utter importance like the ones affecting the scalability of the system, are often difficult to tackle since they have to be deployed with extreme precaution and maximum consensus. Furthermore, if changes affect the consensus mechanisms of the protocol, their implications may cause a blockchain fork and that could have a big impact in a cryptocurrency with a market capitalization of more than 11.5 billion dollars.[13] Moreover, the collateral implications of changes need to be also considered beforehand to prevent unexpected consequences, specially those related to security and decentralization.

Changes in the Bitcoin consensus rules may be introduced by soft (protocol) forks or hard (protocol) forks. A soft fork is produced when the protocol rules are changed so that the new rules are more strict than the old rules. In this case, all blocks accepted by the new rules will also be recognized as valid by the old rules. On the contrary, hard forks make the protocol rules less strict. Therefore, all blocks accepted by the old rules will also be valid by the new rules but there may be blocks that are valid with the new rules that were invalid with the older rules. Soft forks are preferred for updating rules because they do not break compatibility with previous versions and they do no require all participants to upgrade [8].

The effects of hard and soft forks on the network are also different. As an example, let's consider the case where 95 % of the mining power of the network upgrades to a new set of rules. In a hard fork, the upgraded 95 % will eventually create a block which is valid under the new rules but invalid following the old rules. From that moment on, a (blockchain) fork will remain in the network: the upgraded clients will consider the block valid and will keep mining on top of it, whereas the non-upgraded 5 % will recognize the block as invalid and discard it together with all subsequent blocks. The non-upgraded 5 % will always consider the block invalid, and thus will create an alternative branch of the chain and will remain in that branch, making the (blockchain) fork persistent. On the contrary, in a soft fork, most of the blocks will be mined by the upgraded nodes (since they have 95 % of the mining power) and they will be accepted by all nodes (regardless of their upgrading status). Sooner or later, one of the non-upgraded miners will create a new block which will be seen as valid by the 5 % of the

---

[13] Information from http://coinmarketcap.com/ on June 17th, 2016.

miners but invalid for the rest. As a consequence, 5 % of the mining power will start mining on top of that new block, and the 95 % left will keep mining at the same height. Since the upgraded nodes have the majority of the mining power, their branch will soon be longer than the branch created by the non-upgraded miner. Seeing that this branch is longer and valid, the non upgraded nodes will change to the upgraded branch, and thus all the network will be mining again on the same branch.

In the next subsections, we review the techniques that have been proposed to boost the scalability of bitcoin. It is worth mention that since in this paper we are focused on the bitcoin system, we do not consider those scalability proposals that are envisaged for general decentralized blockchain systems, like the ones proposed Croman et al. [5], but could not be applied to the bitcoin system due to the impractical solutions to redefine some primitives, like modify the proof-of-work protocol.

3.1 Tuning Bitcoin Protocol Parameters

Tuning protocol parameters may allow Bitcoin to improve its scalability, although previous studies have concluded that high scalability in the longer term requires a protocol redesign [5].

The parameter that has been most discussed by the Bitcoin community in order to improve system scalability is the block size limit. Some proposals suggest to increase the limit following different strategies or even propose to remove the limit. Jeff Garzik's BIP 100 [9] proposed to change the 1 MB fixed limit to a new floating block size limit, where miners may increase the block size by consensus. Gavin Andresen BIP 101 [10] proposal (currently withdrawn) consisted in initially increasing block size to 8 MB and doubling the size every two years for 20 years, after which the block size remains fixed. Jeff Garzik's BIP 102 [11] proposes to simply increase block size to 2 MB. Pieter Wuille's BIP 103 [12] proposed to increase the maximum block size by 4.4 % every 97 days until 2063, implying a 17.7 % block size increase per year. Gavin Andresen's BIP 109 [13] propose a fixed block size increase to 2 MB with a new limit on the amount of data that can be hashed to compute signature hashes and a change on how the maximum number of signatures is counted. All of these proposals have to be deployed via a hard fork, since all blocks bigger than 1 MB will be seen as invalid by the current version nodes.

However, the increase on the block size limit can not be done arbitrarily. Recent studies [5] argued that with the current 10 min average block interval and taking into account block propagation times in the network, block size limit should not be increased to more than 4 MB.

Segregated Witness [14] is another proposal that does not increase the block size limit, but reduces the amount of information stored per transaction, thus effectively allowing more transactions per block. Additionally, segregated witness solves the malleability problem (refer to Sect. 3.2 for a description) and introduces many other benefits.

3.2 Off-Chain Payment Channels

It is unclear whether tuning the protocol parameters alone will provide enough scaling benefits to satisfy bitcoin needs in the future. For that reason, one of the proposals that has been broadly accepted in the bitcoin community as a relevant bitcoin scalability solution is an improvement that has been enumerated in the previous section: the segregated witness approach. As a single proposal, segregated witness only provides, in the best case, a 4x increase in throughput of the bitcoin network, falling in the buy-time-now solution for the bitcoin scalability problem. But the segregated witness ability to resolve the transaction malleability problem allows to develop new mechanisms that could provide a much more powerful tool for bitcoin scalability issues: off-chain payment channels.

Transaction Malleability Problem. As we pointed out in Sect. 2.1 Bitcoin transactions are identified with its hash value, a value computed using a double SHA256 function over the raw data that defines the transaction. However, for space considerations, this identifier is not stored in the blockchain. Since signatures are not performed over all transaction data, after its creation a transaction can be modified adding some irrelevant data, resulting in a slightly different transaction but with a completely different identifier. Notice that, in this case, we will have two different valid transactions with different identifiers, and only after the transaction is included in the blockchain, the final identifier of the transaction will become unique. It is important to mention that such a modification, that provides malleability, does not affect the ability of an attacker to spend/steal the bitcoins present in the transaction inputs (since the attacker cannot perform the digital signature of the owner). The attacker is only able to modify the identifier of the transaction in a value that differs from the one its real owner has established. For that reason, although transaction malleability is known back from 2011, it has never been considered as a security issue.

Nonetheless, transaction malleability supposes a problem for smart contracts when a child transaction wants to spend a parent output before the parent transaction appears on the blockchain. In case that a malleabled parent transaction is finally included in the blockchain, then all pre-signed child transactions would be invalid.

Basic Off-Chain Payment Channel Ideas. Off-chain Payment Channels are mechanisms that allow payments between two parties, A and B, payments that can be performed without including a transaction for the payment itself in the blockchain.

The first proposal of such a mechanism was first targeted at micropayments from one payer to one payee. Its main goal was to avoid the fees that transactions in the blockchain imply and that are not affordable for micropayment transactions [15]. To set up the payment channel, a transaction is included in the blockchain as a deposit of the money that will be used in the payment channel. A refund transaction is also created, allowing the payer to recover the deposited funds if the payee does not cooperate. The refund transaction can not be included in the blockchain until a certain point in the future, and thus the channel may remain open until that moment arrives. Between the set up and the closing of the payment channel, the payer can perform multiple payments to the payee through transactions that, although formatted in standard bitcoin format, would be transferred privately between A and B without using the standard bitcoin P2P network. Furthermore, the individual payment transactions will not appear in the blockchain: only the set up transaction that opens the channel and the last transaction that closes the channel will be broadcast through the bitcoin P2P network and will be included in the blockchain.

The channel can be closed at any time by B by signing and broadcasting the last transaction received from A. If A has never sent a transaction to B or B does not cooperate, A can get back her funds using the refund transaction, but she will have to wait until the transaction is valid as specified by the time lock. Moreover, if all the funds deposited in the channel by A have already been transferred to B, the channel is exhausted and can no longer be used. In that case, B can sign and broadcast the last transaction received from A, which transfers the whole amount of the channel to B and closes the channel.

In order to create the described unidirectional micropayment channel two bitcoin features are used: multisignature outputs and transactions with lock time.

Multisignature outputs are transaction outputs that may require more than one signature to unlock. For instance, two signatures may be required to unlock a single output. Multisignature outputs are used in the set up transaction of the basic micropayment channel explained above in order to lock the funds that are being used by the channel. In the set up transaction, the payer deposits a certain amount of bitcoins in the channel by sending that amount of bitcoins to a multisignature output controlled by both the payer and the payee.

Bitcoin transactions may have a time lock specifying either a unix timestamp or a blockchain block height. Transactions with a time lock can not be included in any block until the specified time has arrived. Time locks are used in micropayment channels as a mechanism that allows to replace transactions. A certain transaction with a time lock can be replaced by creating a new transaction with a smaller time lock spending the same outputs (or some of those outputs). The new transaction can then be broadcast sooner (because it has a smaller time lock) and thus replaces the old transaction. Note that, in order to effectively replace the transaction, the interested party must broadcast the new transaction to the network before the older one becomes valid. In the basic micropayment channel described above, a time lock is placed on the refund transaction to ensure that it can not be used to return all the money to the payer before the channel extinguishes. Payment transactions spend the same outputs and do not have a time lock, so they can replace the refund transaction.

Such basic approach is restricted to a unidirectional channel between A and B, allowing A to perform off-chain payments to B but without the ability for B to pay to A. The straightforward approach to generate bidirectional channels is to create two unidirectional channels, one from A to B and another from B to A. The problem with such approach is that both channels are independent and if one of the channels runs out of money (suppose the channel $A \to B$) no more payments can be performed from A to B even if in the other payment channel ($B \to A$) A has a positive balance with B.

To construct bidirectional off-chain payment channels without this restriction, two different schemes have been proposed: duplex micropayment channels and lightning channels. In the following paragraphs, we provide a high level overview of both proposals.

Duplex Micropayment Channels. Duplex micropayment channels (DMC) are proposed by Decker and Wattenhofer [16]. DMC are able to provide bidirectional payments between two entities within a finite time frame. The main idea behind DMC is indeed to create two unidirectional channels between the two parties A and B as described before, but using a technique that allows to reset the channels when needed and thus effectively overcoming the problem of exhausting the funds in one of the channels while having a positive amount of bitcoins in the other. Therefore, the main contribution of the proposal is the technique that allows to reset the unidirectional channels: the invalidation tree.

The invalidation tree is a tree structure of depth d made of transactions with multisignature outputs. Each transaction in the tree has a time lock such that any two transactions spending coins from the same output have different time locks and the time lock of a children transaction is at least the same of the time lock of the parent transaction. A branch of the tree is thus a set of d transactions of (non-strictly) increasing time lock. At any given moment, only one branch of the tree is valid, while the other branches are effectively replaced because of the time lock.

The unidirectional channels are then build on top of a leaf of the invalidation tree and are operated in the same way than the basic channels explained previously in this section. User A pays to user B using the $A \to B$ channel while user B pays to A using the $B \to A$ channel independently. However, when one of the channels is exhausted and the sender has funds in the other channel, a reset of the channels can be triggered so that these funds become available to spend. In order to do so, A and B create a new branch of the invalidation tree that replaces the currently active branch, and new unidirectional channels are appended to the leaf of the new branch.

DMC have a finite time frame defined by the time locks used in the invalidation tree. Moreover, DMC also have a finite number of available channel resets, which can be freely determined by the depth of the invalidation tree and the time locks used.

Lightning Channels. Lightning channels are proposed by Poon and Dryja [17] and are able to create bidirectional channels without any time limitations, that is, channels that can remain open until any of the parties decides to close them.

Unlike DMC, lightning channels do not create unidirectional channels: they create a unique channel that allows to send bitcoins in both directions.

In order to do so, for each new payment the two users agree on the current balances of the channel and create two transactions that represent these new balances (both transactions represent the same balances). One transaction is then kept by each user, allowing her to sign it and broadcast it to the network and thus closing the channel in its current state. These transactions have some particularities. The transaction that A can broadcast to the network sends immediately to B his amount, and prevents A from getting her share until after some blocks have been created on top of the block including the transaction. During this time frame, B can also claim A's amount if he reveals a secret that only A knows. Similarly, the transaction that B keeps sends bitcoins immediately to A and prevents B from getting his amount until some blocks, and during this time A can claim B's amount if she reveals a secret than only B knows.

Whenever a new payment has to be made in the channel, both users update the balances and agree to the new state of the channel: they create two transactions with the new balances, keeping again each one of the transactions. Now, both users can broadcast their transaction to the network and thus secure their balances. However, at this point there is nothing preventing the users from broadcasting their transactions from the previous state of the channel. In order to ensure that none of the users cheat by broadcasting old transactions, at every new transaction the users exchange their respective secrets for the previous transaction. Now, if one of the users tries to cheat by broadcasting an old transaction, the other party can claim all the funds of the channel by revealing the secret.

Off-Chain Payment Networks. Duplex micropayment channels and lightning networks as described above provide a mechanism to stablish bidirectional channels between two different users. However, it is impractical that users will open a new off-chain payment channel with a counterpart unless the number of payments between both parties is high. To overcome such problem, both proposals allow an improvement by which two-side off-chain payment channels can be somehow concatenated in order to allow users to perform payment through multiple established off-chain channels without the parts needing to trust the intermediary ones. The idea to implement this feature uses the ability to spend a transaction once a secret value is known. In Fig. 3, a single hop example is showed. C, who receives the payment, generates a random value, x, and computes its hash, $h(x)$, that will send to A. A creates a transaction $Tx_1(B, h(x))$ and sends it to B. B can charge the transaction providing the value x, that he does not have. To obtain x, B creates another transaction $Tx_2(C, h(x))$ that can be charged by C when he provides x. Since C does indeed know the value x, when he reveals it he can charge the second transaction and B can charge the first one.

Notice that with this scenario, the payment between users A and C can be performed not only in case A and B have a direct payment channel but also when there is a path through multiple payment channels that link them. Based on this

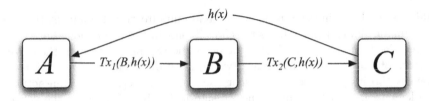

Fig. 3. One hop off-chain payment channel.

principle, it is straight forward to envisage the appearance of such intermediary nodes, the payment service providers, nodes that will create a highly connected network that will route and perform such off-chain payments.

4   Privacy Implications of Scalability Solutions

Anonymity is probably one of the properties that has contributed to the success of bitcoin deployment. Anonymity in the bitcoin network is based on the fact that users can create any number of anonymous bitcoin addresses that will be used in their bitcoin transactions. This basic approach is a good starting point, but the underlaying non-anonymous Internet infrastructure, together with the availability of all bitcoin transactions in the blockchain, has proven to be an anonymity threat. In [18], research performed on bitcoin privacy is categorized in three main areas: Blockchain analysis, traffic analysis and mixing networks. Next, we review how the different ideas proposed in those areas would be affected by the implementation and adoption of off-chain payment channels.

4.1   Blockchain Analysis

A direct approach to analyze the anonymity offered by the bitcoin system is to dig information out of the blockchain. A simple analysis provides information about the movements of bitcoins: from which bitcoin addresses the money comes and to which bitcoin addresses it goes. However, such basic approach has two main drawbacks:

- Users with multiple addresses: since users in the bitcoin system can create any number of addresses, in order to obtain insightful information from the blockchain researchers try to cluster all addresses that belong to the same user. As we will see, authors apply different techniques to perform such clustering.
- Blockchain data volume: at present time, the size of the blockchain data is 72 GB. However, such data only include the raw blockchain data, which is not a database and it is not suitable for data queries. Storing such information in a searchable database expands the data size and produces a much larger database which is more difficult to deal with. For that reason, research papers on privacy issues regarding blockchain analysis have drastically decreased in the last years.

Furthermore, a basic assumption is made on the blockchain information when doing such analysis: the blockchain includes all transactions of the system. But, in the light of off-chain payment channels, such assumption is no longer valid since only a fraction of transactions are finally stored in the blockchain.

Address clustering has been one of the blockchain analysis techniques used to deanonymize users in the bitcoin networks. The idea is to cluster different bitcoin addresses belonging to the same user in order to trace his economic movements. Different heuristics have been used to perform such clustering. A common assumption is to consider that all addresses included as inputs in a transaction belong to the same user [19–21]. Another technique to cluster addresses is to consider that in a transaction with multiple outputs, in case that one of them goes to a not previously used addresses (and the others use addresses already appeared in the blockchain) the new address can be clustered with all input addresses of such transaction [20, 22].

Notice that those techniques cannot be effectively used when off-chain payments channels would become common use. First of all, the assumption that all inputs from a single transaction belong to the same user no longer holds since set up transactions for payments channels include two inputs from exactly two different users. Furthermore, a portion of transactions in the blockchain with exactly two outputs would come from closing payment channels, for which no assumption can be performed between input and output addresses. At most, transactions closing payment channels can be linked with set up transactions since both share the same address, but since input addresses from set up channels will be different from output addresses from closing channels it will not be possible to infer how much money each address have been spent/earned as a consolidate balance of the payment channel.

On the other hand, with the adoption of payment channels, the size of the address cluster for a typical user will hardly be reduced. Notice that without payment channels, users are free to use a new address for every single operation performed (paying, cashing or taking the change). Such amount of addresses hardens the possibility to obtain a single cluster for each user. However, once a payment channel is opened the user performs all payments through such channel without involving any new address. Nevertheless, all those payments are performed off-chain so they cannot be traced by blockchain analysis techniques.

User anonymity has been also analyzed using k-anonymity measures. Ober et al. [23] indicate that to estimate the level of k-anonymity provided by bitcoin it is necessary to estimate the number of active entities since, for instance, dormant coins (those included in an address not active for a long time) reduce the anonymity set. Furthermore, they also indicate that to better estimate the k-anonymity at a certain point of time, active entities should be defined based on a window time around this period (hours, days, weeks, ...). Then, an active entity is the one that has performed a payment within this time window. With off-chain payment channels, entities activity is very hard to estimate since once a channel is opened, it cannot be determined by blockchain analysis if such channel is an active one or not for the obvious mechanism of off-chain communication.

Finally, tools like BitIodine [24] are based on mining the blockchain information assuming that all transactions performed by the system are included. With the off-chain payment channels, such assumption is no longer valid and, depending on the degree of its adoption, such a tool will not be able to provide significant information.

4.2 Traffic Analysis

As we already mentioned, the anonymity degree of users in the bitcoin system is also bounded by the underlying technologies used. Transactions in the bitcoin system are transmitted through a P2P network, so the TCP/IP information obtained from that network can be used to reduce the anonymity of the system, as it is pointed out in [19]. Although it is true that most wallets are able to work over TOR anonymous network,[14] a high number of bitcoin users do not use such services, and then, there is still room for network analysis. Moreover, using TOR to obtain anonymity while using bitcoin may not be the best choice [25].

Koshy et al. [26] performed an anonymity study based on real-time transaction traffic collected during 5 months. For that purpose, authors develop Coin-Seer, a bitcoin client designed exclusively for data collection. For more than 5 million transactions, they collected information on the IP address from where the CoinSeer received such transaction and, in the general case, they assigned as the IP corresponding to the transaction the one that broadcast the transaction for the first time. In order to perform a pure network analysis, authors do not apply any address clustering process, so only single input transactions (almost four million) are taken into account in the analyzed data set. Then, to match an IP with a bitcoin address, they consider a vote on the link between IP_i and $address_j$ if a transaction first broadcasted form an IP_i contains the bitcoin $address_j$ as input address. Although Koshy et al. could not provide positive results for deanonymizing users, the techniques they propose were interesting and could be a threat to user privacy when more data is available for the analysis. Note that there already exist some actors in the current bitcoin scenario that are able to collect and process this kind of data. For instance, some blockchain explorers keep and publicly show information on the IP address from which they first receive transactions and blocks.

However, new off-chain payment channels present a new scenario since, as we already explained in Sect. 3.2, off-chain payments use a separate network. Although nodes of the channel need to be connected to the bitcoin network for channel set up and also to monitor the correctness of channel counter-party, communication of the channel will pass through a different network, the one that connects users from different off-chain payment channels through their payment service providers. Users will access the off-chain payment networks through a single node[15] for which the user will maintain an open payment channel. So a

---

[14] https://www.torproject.org/.

[15] It is difficult to predict at present time whether users will maintain multiple payment channels with multiple payment service providers but multiple channels could be not viable depending on the fees needed to open and close those channels.

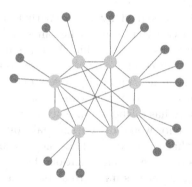

Fig. 4. Possible off-chain payment channel topology.

possible network topology for the network would be similar to the one showed in Fig. 4, with a highly connected component which include the payment service provider nodes and final users connected to those nodes.

This architecture differs from the distributed topology of the bitcoin network, where every user is able to maintain multiple simultaneous network connections with different nodes of the network to obtain some security features.

Notice that with this type of topology, transaction anonymity is lost when the payment between two users is performed through off-chain payment channels that only include a single payment service provider, since the entity providing such off-chain channels may know both, source and destination of the payment.

4.3   Mixing

In the bitcoin environment, mixing is used to anonymize bitcoins. Mix services shuffle the inputs and outputs of a transaction in order to hinder the relation between them. The goal is to allow bitcoin users to send bitcoins from one address to a mix service and receive from the mix service the bitcoins to another address that can not be linked with the original one. Therefore, the process detaches the link between a source (the input address) and a destination (the output address). Although such mixing service can be implemented straightforward using a central authority which receives payments and pays back to different addresses, the trusted level of such central authority would be too high. For that reason, different proposals that avoid or reduce the trusted role of the central authority have been presented.

A basic mix service can be implemented using a multiple-input and multiple-output transaction, as it is described in CoinJoin [27]. The idea is that multiple users can jointly create a transaction with multiple input addresses and multiple output addresses. To be a valid transaction, the transaction should be signed by all users participating in the mixing. One of the problems of this proposal (and to some extent of the majority of mixing proposals) is that one of the anonymous users of the mix service can perform a DoS attack. Since the final valid transactions should be signed by all users that include bitcoins in the transactions, each mixing transaction never becomes valid in case the attacker simply does not sign any transaction in which he takes part.

In [28], Bonneau *et al.* present Mixcoin, a more sophisticated centralized mixing system that relies on accountability. Users of the system obtain, prior the mixing phase, a signed warranty that can be used to prove, in case of the event, that the mixer entity has misbehaved. Authors point out that such public verifiable proof of misbehavior would discourage malicious mixing. Furthermore, to reduce the possibility that the mixer could deanonymize users using his stored information, the authors propose a concatenation of several mixer services, thus reducing the strategy of a malicious mixer to a collusion with the other mixers.

Mixing services as described so far can be applied to standard bitcoin addresses and transactions but when off-chain payment channels are used, such mixing techniques cannot be applied in the same form. The reason is that in the standard bitcoin model, payments can bee seen as one hope transactions that are visible by all participants (they appear in the blockchain). Fortunately, in this scenario every user can create multiple addresses without any cost. For that reason, standard mixing services use multiple new addresses to hinder identities since there is no way to allow payments through secret multiple hops, because each hope (a transaction, in fact) must be recorded publicly in the blockchain. Conversely, in off-chain payments, on one hand, users may be more restricted on the number of payment channels that they create (due to fee costs) but on the other hand, payments are processed with multiple hops through different payment service providers, and such hops could remain secret since they take place in the off-chain payment network and there is no need to store them. So the natural idea to detach the link between source and destination in this scenario is to perform payments through secret multiple hop routes.

At that point it is worth notice that, in off-chain payment networks, payment anonymity highly resembles standard communication anonymity where a path between source and destination is hidden to protect communication identification. For that reason, common onion routing techniques [29] could be applied to allow anonymous payments in a similar way than TOR network provides anonymous browsing. Nevertheless, in the same way that single TOR utilization does not guarantee that you are browsing the www anonymously (since, for instance, the browser configuration may reveal some details about your identity), details on the protocol for multi-hop off-chain payment channels will have to be carefully analyzed (when they are available) in order not to disclose the link between the source and the destination.

5   Conclusions

Bitcoin scalability is one of the relevant topics in the broad crytocurrency field since some limitations Bitcoin faces are common to all blockchain based cryptocurrencies. Different ideas have been proposed so far, being the segregated witness approach the one that most support has received. Segregated witness has the potential to solve transaction malleability and once solved, bitcoin will have the ability to work with unsigned transactions and off-chain payment channels will be able to start working in practice. From that moment on, payment networks will be able to grow and flourish. Moreover, payment networks will have a life of its own, being able to operate on top of the existing Bitcoin protocol but with the freedom that off-chain transacting will offer. This is potentially one of the biggest changes the bitcoin ecosystem has ever seen and, as such, will have an impact on many bitcoin properties, among which users' privacy is included.

Once off-chain payment networks become main use, some Bitcoin premises on decentralization and openness of its payment routing, multiple address generation and full transaction disclosure through the blockchain may be modified. Research performed so far has proven that the way the system uses payment addresses may unveil some information from their owners, when all transactions performed by the system were freely available in the blockchain for analysis and transactions were published through open P2P networks. However, if most of the transactions occur off-chain, this kind of analysis will no longer be as effective as before. Depending on the final payment channels implementations, some of these techniques may be able to adapt to extract information about the existing channels. Nonetheless, we will have to wait until the payment networks are deployed to evaluate to what extent this kind of analysis remains feasible.

Acknowledgments. This work was partially supported by the Spanish Ministerio funds under grant MINECO/TIN2014-55243-P and FPU-AP2010-0078, and through the Catalan Government funded project AGAUR/2014-SGR-691.

References

1. Nakamoto, S.: Bitcoin: a peer-to-peer electronic cash system (2008)
2. Antonopoulos, A.M.: Mastering Bitcoins. O'Reilly, Media (2014)
3. Back, A.: A partial hash collision based postage scheme (1997). http://www.hashcash.org/papers/announce.txt. Accessed June 2016
4. Donet Donet, J.A., Pérez-Solà, C., Herrera-Joancomartí, J.: The bitcoin P2P network. In: Böhme, R., Brenner, M., Moore, T., Smith, M. (eds.) FC 2014 Workshops. LNCS, vol. 8438, pp. 87–102. Springer, Heidelberg (2014)
5. Croman, K., Decker, C., Eyal, I., Gencer, A.E., Juels, A., Kosba, A., Miller, A., Saxena, P., Shi, E., Gün, E.: On scaling decentralized blockchains. In: Proceedings of 3rd Workshop on Bitcoin and Blockchain Research (2016)
6. Visa: 56,582 transaction messages per second!, July 2014. http://visatechmatters.tumblr.com/post/108952718025/56582-transaction-messages-per-second. Accessed June 2016
7. Paypal: Paypal q1 2016 fast facts, June 2016. Accessed June 2016
8. Core, B.: Bitcoin core statement, January 2016. https://bitcoincore.org/en/2016/01/07/statement/. Accessed June 2016
9. Garzik, J.: BIP 100: making decentralized economic policy (2015). Accessed June 2016
10. Andresen, G.: BIP 101: increase maximum block size (2015). Accessed June 2016
11. Garzik, J.: BIP 102: block size increase to 2MB (2015). Accessed June 2016

12. Wuille, P.: BIP 103: block size following technological growth (2015). Accessed June 2016
13. Andresen, G.: BIP 109: two million byte size limit with sigop and sighash limits (2016). Accessed June 2016
14. Lombrozo, E., Lau, J., Wuille, P.: BIP 141: segregated witness (consensus layer) (2015). Accessed June 2016
15. Hearn, M., Spilman, J.: Bitcoin contracts. https://en.bitcoin.it/wiki/Contract. Accessed June 2016
16. Decker, C., Wattenhofer, R.: A fast and scalable payment network with bitcoin duplex micropayment channels. In: Pelc, A., Schwarzmann, A.A. (eds.) SSS 2015. LNCS, vol. 9212, pp. 3–18. Springer, Heidelberg (2015)
17. Poon, J., Dryja, T.: The bitcoin lightning network: scalable off-chain instant payments. Technical report (draft). https://lightning.network (2015)
18. Herrera-Joancomartí, J.: Research and challenges on bitcoin anonymity. In: Garcia-Alfaro, J., Herrera-Joancomartí, J., Lupu, E., Posegga, J., Aldini, A., Martinelli, F., Suri, N. (eds.) DPM/SETOP/QASA 2014. LNCS, vol. 8872, pp. 3–16. Springer, Heidelberg (2015)
19. Reid, F., Harrigan, M.: An analysis of anonymity in the bitcoin system. In: Altshuler, Y., Elovici, Y., Cremers, A.B., Aharony, N., Pentland, A. (eds.) Security and Privacy in Social Networks, pp. 197–223. Springer, New York (2013)
20. Androulaki, E., Karame, G.O., Roeschlin, M., Scherer, T., Capkun, S.: Evaluating user privacy in bitcoin. In: Sadeghi, A.-R. (ed.) FC 2013. LNCS, vol. 7859, pp. 34–51. Springer, Heidelberg (2013)
21. Ron, D., Shamir, A.: Quantitative analysis of the full bitcoin transaction graph. In: Sadeghi, A.-R. (ed.) FC 2013. LNCS, vol. 7859, pp. 6–24. Springer, Heidelberg (2013)
22. Meiklejohn, S., Pomarole, M., Jordan, G., Levchenko, K., McCoy, D., Voelker, G.M., Savage, S.: A fistful of bitcoins: characterizing payments among men with no names. In: Proceedings of the 2013 Conference on Internet Measurement Conference, IMC 2013, pp. 127–140. ACM, New York (2013)
23. Ober, M., Katzenbeisser, S., Hamacher, K.: Structure and anonymity of the bitcoin transaction graph. Future Internet 5(2), 237–250 (2013)
24. Spagnuolo, M., Maggi, F., Zanero, S.: BitIodine: extracting intelligence from the bitcoin network. In: Christin, N., Safavi-Naini, R. (eds.) FC 2014. LNCS, vol. 8437, pp. 457–468. Springer, Heidelberg (2014)
25. Biryukov, A., Pustogarov, I.: Bitcoin over tor isn't a good idea. In: 2015 IEEE Symposium on Security and Privacy (SP), pp. 122–134. IEEE (2015)
26. Koshy, P., Koshy, D., McDaniel, P.: An analysis of anonymity in bitcoin using P2P network traffic. In: Christin, N., Safavi-Naini, R. (eds.) FC 2014. LNCS, vol. 8437, pp. 469–485. Springer, Heidelberg (2014)
27. Maxwell, G.: Coinjoin: bitcoin privacy for the real world. Post on Bitcoin Forum
28. Bonneau, J., Narayanan, A., Miller, A., Clark, J., Kroll, J.A., Felten, E.W.: Mixcoin: anonymity for bitcoin with accountable mixes. In: Christin, N., Safavi-Naini, R. (eds.) FC 2014. LNCS, vol. 8437, pp. 486–504. Springer, Heidelberg (2014)
29. Chaum, D.L.: Untraceable electronic mail, return addresses, and digital pseudonyms. Commun. ACM 24(2), 84–90 (1981)

Aggregation Operators and Decision Making

On k–additive Aggregation Functions

Anna Kolesárová[1], Jun Li[2], and Radko Mesiar[3(✉)]

[1] Faculty of Chemical and Food Technology, Slovak University of Technology,
Radlinského 9, 812 37 Bratislava, Slovakia
anna.kolesarova@stuba.sk
[2] School of Science, Communication University of China,
100024 Beijing, People's Republic of China
lijun@cuc.edu.cn
[3] Faculty of Civil Engineering, Slovak University of Technology,
Radlinského 11, 810 05 Bratislava, Slovakia
radko.mesiar@stuba.sk

Abstract. Inspired by the Grabisch idea of k–additive measures, we introduce and study k–additive aggregation functions. The Owen multilinear extension of a k–additive capacity is shown to be a particular k–additive aggregation function. We clarify the relation between k–additive aggregation functions and polynomials of a degree not exceeding k. We also describe $n^2 + 2n$ basic 2–additive n–ary aggregation functions whose convex closure forms the class of all 2–additive n–ary aggregation functions.

Keywords: Aggregation function · k–additive aggregation function · k–additive capacity

1 Introduction

Consider a fixed finite space $X = \{1, \ldots, n\}$, $n \in \mathbb{N}$. Recall that a set function $m : 2^X \to [0, 1]$ is called a *capacity* if it is monotone, i.e., $m(A) \leq m(B)$ whenever $A \subseteq B$, and if $m(\emptyset) = 0$, $m(X) = 1$. The additivity of a capacity m makes m a probability measure, determined by n values of singletons, $w_1 = m(\{1\}), \ldots, w_n = m(\{n\})$, constrained by the condition $w_1 + \cdots + w_n = 1$. The additivity of a probability measure excludes the possibility of interactions between single subsets of X. On the other hand, a general capacity m allows to model interaction of any group of subsets of X, but it requires the knowledge of $2^n - 2$ values $m(A)$, $\emptyset \neq A \neq X$. To reduce the complexity of a general capacity m, but to allow to model interaction of some groups of subsets of X, Grabisch [5] introduced the notion of k–additive capacities, $k \in \{1, \ldots, n\}$. Note that 1–additive capacities are just probability measures on X, and n–additive capacities are general capacities on X. Though k–additive capacities were originally defined by means of the Möbius transform, supposing that the Möbius transform of each subset with cardinality exceeding k is equal to zero [5], they can also be characterized as follows (see [11,12]).

© Springer International Publishing Switzerland 2016
V. Torra et al. (Eds.): MDAI 2016, LNAI 9880, pp. 47–55, 2016.
DOI: 10.1007/978-3-319-45656-0_4

A capacity $m\colon 2^X \to [0,1]$ is said to be $k\text{-}additive$ if for any system $(A_i)_{i=1}^{k+1}$ of pairwise disjoint subsets of X we have

$$\sum_{i=1}^{k+1}(-1)^{k+1-i}\left(\sum_{I\subseteq\{1,\ldots,k+1\},|I|=i} m\left(\bigcup_{j\in I} A_j\right)\right) = 0. \tag{1}$$

So, for example, a capacity $m\colon 2^X \to [0,1]$ is 2–additive, if for any pairwise disjoint sets $A, B, C \subseteq X$ we have

$$m(A\cup B\cup C)-(m(A\cup B)+m(B\cup C)+m(A\cup C))+m(A)+m(B)+m(C) = 0. \tag{2}$$

Note that each 2–additive capacity m is determined by $\binom{n}{2} + n = \frac{n(n+1)}{2}$ values $w_{ij} = m(\{i,j\})$, $1 \le i < j \le n$ and $w_i = m(\{i\})$, $i \in X$, constrained by the condition $\sum_{i<j} w_{ij} + \sum_i w_i = 1$, and some monotonicity conditions.

Based on the representation of subsets $A \in 2^X$ by means of the characteristic functions $1_A\colon X \to \{0,1\}$, one can represent a capacity $m\colon 2^X \to [0,1]$ as a pseudo-Boolean function $H\colon \{0,1\}^n \to [0,1]$, whose value at any $\mathbf{x} = (x_1,\ldots,x_n) \in \{0,1\}^n$ is given by

$$H(\mathbf{x}) = m(\{i \in X \mid x_i = 1\}).$$

Then the Eq. (1) for k-additivity of m can be changed into the equation

$$\sum_{i=1}^{k+1}(-1)^{k+1-i}\left(\sum_{I\subseteq\{1,\ldots,k+1\},|I|=i} H\left(\sum_{j\in I}\mathbf{x}_j\right)\right) = 0, \tag{3}$$

that is satisfied for any $\mathbf{x}_1,\ldots,\mathbf{x}_{k+1} \in \{0,1\}^n$ such that also $\mathbf{x}_1 + \cdots + \mathbf{x}_{k+1} \in \{0,1\}^n$.

Similarly, the 2-additivity of a capacity m, see (2), can be represented in the form

$$H(\mathbf{x}+\mathbf{y}+\mathbf{z}) - (H(\mathbf{x}+\mathbf{y}) + H(\mathbf{y}+\mathbf{z}) + H(\mathbf{x}+\mathbf{z})) + H(\mathbf{x}) + H(\mathbf{y}) + H(\mathbf{z}) = 0, \tag{4}$$

for any $\mathbf{x}, \mathbf{y}, \mathbf{z} \in \{0,1\}^n$ such that also $\mathbf{x} + \mathbf{y} + \mathbf{z} \in \{0,1\}^n$.

In what follows, inspired by formulas (3), (4), and several extension methods extending pseudo-Boolean functions H representing capacities m, such as the Lovász and Owen extensions, see [8] and [10], respectively, and also other extensions characterized in [7], we propose and study k–additive aggregation functions.

The paper is organized as follows. After the introduction and exemplification of k–additive aggregation functions in the next section, in Sect. 3, we derive a complete characterization of 2–additive n–ary aggregation functions and also provide a characterization of k–additive aggregation functions for $k \in \mathbb{N}$, independently of $n = |X|$. Finally, some concluding remarks are added.

2 k–additive Aggregation Functions

We start by recalling the definition of an aggregation function.

Definition 2.1. *Let n be a fixed natural number. A mapping $F\colon [0,1]^n \to [0,1]$ is called an (n–ary) aggregation function if it is monotone and satisfies the boundary conditions $F(\mathbf{0}) = F(0,\dots,0) = 0$ and $F(\mathbf{1}) = F(1,\dots,1) = 1$.*

Note that we also have adopted this definition for $n = 1$. For more details concerning aggregation functions we recommend recent monographs [1,2,6].

Definition 2.2. *Let $k,n \in \mathbb{N}$ and let $F\colon [0,1]^n \to [0,1]$ be an aggregation function. Then F is called k–additive whenever for all collections $\mathbf{x}_1,\dots,\mathbf{x}_{k+1} \in [0,1]^n$ such that also $\sum_{i=1}^{k+1} \mathbf{x}_i \in [0,1]^n$ we have*

$$\sum_{i=1}^{k+1}(-1)^{k+1-i}\left(\sum_{I\subseteq\{1,\dots,k+1\},|I|=i} F\left(\sum_{j\in I}\mathbf{x}_j\right)\right) = 0. \tag{5}$$

In particular, F is 2–additive if and only if for all $\mathbf{x},\mathbf{y},\mathbf{z} \in [0,1]^n$ such that $\mathbf{x}+\mathbf{y}+\mathbf{z} \in [0,1]^n$ it holds that

$$F(\mathbf{x}+\mathbf{y}+\mathbf{z}) - (F(\mathbf{x}+\mathbf{y}) + F(\mathbf{y}+\mathbf{z}) + F(\mathbf{x}+\mathbf{z})) + F(\mathbf{x}) + F(\mathbf{y}) + F(\mathbf{z}) = 0. \tag{6}$$

Observe that 1–additivity is just the standard additivity of aggregation functions. Moreover, it is not difficult to check that each k–additive aggregation function F is also p–additive for any integer $p > k$.

Example 2.1.

(i) Consider the standard product $P\colon [0,1]^n \to [0,1]$, $P(\mathbf{x}) = \prod_{i=1}^{n} x_i$. Then P is an n–additive aggregation function that is not $(n-1)$–additive. In particular, for $n = 2$, P is 2–additive, but not additive.

(ii) Let $F\colon [0,1]^n \to [0,1]$, $F(\mathbf{x}) = \frac{1}{n}\sum_{i=1}^{n} x_i^k$, $k \geq 2$. Then F is a k–additive aggregation function, but it is not $(k-1)$–additive.

(iii) Recall that the Choquet integral [4,6] based on an additive capacity is an additive aggregation function, i.e., a weighted arithmetic mean. However, this link between k–additive capacities and k–additivity of the related Choquet integrals is no more valid if $n > 1$, $k > 1$. Consider, e.g., $n = 2$ and the smallest capacity m_* on $X = \{1,2\}$, given by $m_*(A) = 1$ if $A = X$, and otherwise, $m_*(A) = 0$. Then the corresponding Choquet integral is just the minimum, $Ch_{m_*}(x_1,x_2) = \min\{x_1,x_2\}$. The aggregation function $Ch_{m_*} = Min\colon [0,1]^2 \to [0,1]$ is not k–additive for any $k \in \mathbb{N}$. We illustrate the violation of 3–additivity only:

Let $\mathbf{x}_1 = \left(\frac{1}{3},0\right)$, $\mathbf{x}_2 = \left(\frac{2}{3},0\right)$, $\mathbf{x}_3 = \left(0,\frac{1}{3}\right)$ and $\mathbf{x}_4 = \left(0,\frac{2}{3}\right)$. Applying (5) to Min, we have

$$\sum_{i=1}^{4}(-1)^{4-i}\left(\sum_{I\subseteq\{1,\dots,4\},|I|=i} Min\left(\sum_{j\in I}\mathbf{x}_j\right)\right) = 1 - \left(\frac{1}{3}+\frac{2}{3}+\frac{1}{3}+\frac{2}{3}\right)$$

$$+ \left(0+\frac{1}{3}+\frac{1}{3}+\frac{1}{3}+\frac{2}{3}+0\right) - (0+0+0+0) = \frac{2}{3} \neq 0.$$

It is known that the discrete Choquet integral with respect to a capacity m can be seen as the Lovász extension of the pseudo-Boolean function representing m [3,8]. For the Owen extension [10], that is also called a multilinear extension [6], we have the next connection between k-additive capacities and the k-additivity of the corresponding Owen extension.

Definition 2.3. ([10]) *Let* $m\colon 2^X \to [0,1]$ *be a capacity. Its Möbius transform* $M_m\colon 2^X \to [0,1]$ *is given by*

$$M_m(A) = \sum_{B\subseteq A}(-1)^{|A\setminus B|}m(B),$$

and the Owen extension $O_m\colon [0,1]^n \to [0,1]$ *of the pseudo-Boolean function* $H\colon \{0,1\}^n \to [0,1]$ *corresponding to* m *(or simply the Owen extension of* m*) is given by*

$$O_m(\mathbf{x}) = \sum_{\emptyset\neq A\subseteq X} M_m(A)\left(\prod_{j\in A} x_j\right). \tag{7}$$

Theorem 2.1. *Let* $m\colon 2^X \to [0,1]$ *be a capacity and* $O_m\colon [0,1]^n \to [0,1]$ *the Owen extension of* m*. Then the following are equivalent.*

(i) O_m *is a* k-*additive aggregation function.*
(ii) m *is a* k-*additive capacity.*

Proof. The necessity (i) $\Rightarrow$ (ii) follows from the fact that if O_m is a k-additive aggregation function, then $H\colon \{0,1\}^n \to [0,1]$, $H(\mathbf{1}_I) = O_m(\mathbf{1}_I) = m(I)$, is also k-additive.

To show the sufficiency (ii) $\Rightarrow$ (i), suppose that m is a k-additive capacity, i.e., $M_m(I) = 0$ whenever $|I| > k$. Let $\mathbf{x}_1,\dots,\mathbf{x}_{k+1}, \sum_{i=1}^{k+1}\mathbf{x}_i \in [0,1]^n$. For $\emptyset \neq I \subseteq \{1,\dots,k+1\}$ put $\sum_{j\in I}\mathbf{x}_j = \mathbf{x}_I$. Evaluating the expression in (5) for O_m, we obtain

$$\sum_{i=1}^{k+1}(-1)^{k+1-i}\left(\sum_{I\subseteq\{1,\dots,k+1\},|I|=i} O_m\left(\mathbf{x}_I\right)\right)$$

$$= \sum_{i=1}^{k+1}(-1)^{k+1-i}\left(\sum_{I\subseteq\{1,\dots,k+1\},|I|=i}\left(\sum_{\emptyset\neq A\subseteq X,|A|\leq k} M_m(A)\left(\prod_{j\in A} x_{I,j}\right)\right)\right)$$

$$= \sum_{\emptyset\neq A\subseteq X,|A|\leq k} M_m(A)\left(\sum_{i=1}^{k+1}(-1)^{k+1-i}\left(\sum_{I\subseteq\{1,\dots,k+1\},|I|=i}\prod_{j\in A} x_{I,j}\right)\right) = 0,$$

where we have utilized the fact that the product restricted to the components from the index set A, is a $|A|$–additive aggregation function.   □

Observe that the k–additive aggregation function introduced in Example 2.1 (ii) cannot be obtained by means of the Owen extension, i.e., Owen extensions of k–additive capacities form a proper subset of all k–additive aggregation functions.

3   Characterization of k–additive Aggregation Functions

In this section we describe all k–additive n–ary aggregation functions.

As already mentioned, 1–additive, i.e., additive aggregation functions $F\colon [0,1]^n \to [0,1]$ are completely characterized by a weighting vector $\mathbf{w} = (w_1, \ldots, w_n) \in [0,1]^n$ such that $\sum_{i=1}^{n} w_i = 1$, and it holds that $F(\mathbf{x}) = \sum_{i=1}^{n} w_i x_i$.

Denote by $\mathcal{A}_{k,n}$ the set of all k–additive n–ary aggregation functions. Our aim is to provide a description of $\mathcal{A}_{k,n}$, in general, but because of transparency, we give complete proofs for 2–additive aggregation functions only. The proof for $k = 2$ will be divided into 3 steps. We start by characterizing the set $\mathcal{A}_{2,1}$.

For $k = 2$ and $n = 1$, we have

$$\mathcal{A}_{2,1} = \{F\colon [0,1] \to [0,1] \mid F(0) = 0, F(1) = 1, F \text{ is increasing and 2–additive }\}.$$

Note that here, 2–additivity means that for all $x, y, z, x + y + z \in [0,1]$,

$$F(x+y+z) - F(x+y) - F(y+z) - F(x+z) + F(x) + F(y) + F(z) = 0. \quad (8)$$

Proposition 3.1. *A function $F\colon [0,1] \to [0,1]$ is a 2–additive aggregation function if and only if*

$$F(x) = (a+1)x - ax^2, \ a \in [-1,1]. \quad (9)$$

Proof. It is only a matter of computation to show that each function F of the form (9) is an aggregation function satisfying (8), i.e., $F \in \mathcal{A}_{2,1}$.

Now, suppose that $F \in \mathcal{A}_{2,1}$. Putting $F\left(\frac{1}{2}\right) = b$ and $a = 4b - 2$, we can write $F\left(\frac{1}{2}\right)$ in the form $F\left(\frac{1}{2}\right) = (a+1)\frac{1}{2} - a\frac{1}{4}$. Next, using (8), for $x = y = z = \frac{1}{4}$ we obtain

$$F\left(\frac{3}{4}\right) - 3b + 3F\left(\frac{1}{4}\right) = 0,$$

and similarly, for $x = y = \frac{1}{4}$ and $z = \frac{1}{2}$ we have

$$1 - 2F\left(\frac{3}{4}\right) - F\left(\frac{1}{2}\right) + 2F\left(\frac{1}{4}\right) + F\left(\frac{1}{2}\right) = 0, \ i.e., \ 1 - 2F\left(\frac{3}{4}\right) + 2F\left(\frac{1}{4}\right) = 0.$$

Thus we can derive

$$F\left(\frac{1}{4}\right) = \frac{6b-1}{8} = (a+1)\frac{1}{4} - a\frac{1}{16},$$

$$F\left(\frac{3}{4}\right) = \frac{6b+3}{8} = (a+1)\frac{3}{4} - a\frac{9}{16}.$$

Proceeding by induction, we can show that F is of the form $F(r) = (a+1)r - ar^2$ for any dyadic rational $r \in [0,1]$ (recall that $r \in [0,1]$ is a dyadic rational number if and only if there is a $k \in \mathbb{N}$ such that also $r \cdot 2^k \in \mathbb{N}$), and due to the monotonicity of F, we can conclude that

$$F(x) = (a+1)x - ax^2, \text{ for all } x \in [0,1].$$

Clearly, F is an aggregation function only if $a \in [-1,1]$.   □

The next result gives a complete characterization of the set $\mathcal{A}_{2,2}$.

Proposition 3.2. *A function $F \colon [0,1]^2 \to [0,1]$ is a 2–additive n–ary aggregation function if and only if*

$$F(x_1, x_2) = ax_1 + bx_2 + cx_1^2 + dx_2^2 + ex_1x_2, \tag{10}$$

where

$$a + b + c + d + e = 1, \quad a \geq 0, \quad b \geq 0, \quad a+e \geq 0, \quad b+e \geq 0,$$

$$a + 2c \geq 0, \quad b + 2d \geq 0, \quad a + 2c + e \geq 0, \quad a + 2d + e \geq 0.$$

Proof. The necessity can be proved by similar arguments as in Proposition 3.1. On the other hand, one can show by a direct computation that functions of the form (10) satisfying the given conditions, are 2–additive aggregation functions. □

Now, we are ready to describe the set $\mathcal{A}_{2,n}$ for any $n \in \mathbb{N}$.

Theorem 3.1. *A function $F \colon [0,1]^n \to [0,1]$ is a 2–additive aggregation function if and only if*

$$F(x_1, \ldots, x_n) = \sum_{i,j \in \{1,\ldots,n\}, i \leq j} a_{ij}x_ix_j + \sum_{i=1}^{n} b_ix_i, \tag{11}$$

and the coefficients a_{ij} and b_i satisfy the conditions:

$$\sum_{i,j \in \{1,\ldots,n\}, i \leq j} a_{ij} + \sum_{i=1}^{n} b_i = 1,$$

and for each fixed $i \in \{1, \ldots, n\}$,

$$0 \leq b_i \leq - \sum_{a_{pj} < 0} a_{pj} \left(\mathbf{1}_{\{p\}}(i) + \mathbf{1}_{\{j\}}(i)\right).$$

Proof. It remains to prove the claim for $n \geq 3$ only. The proof of (11) is by induction on n. For $n = 3$, the necessity can be proved as a consequence of Propositions 3.1 and 3.2, using the following equality valid for functions $F \in \mathcal{A}_{23}$ and all $(x_1, x_2, x_3) \in [0,1]^3$,

$$F(x_1, x_2, x_3) = F(x_1, x_2, 0) + F(0, x_2, x_3) + F(x_1, 0, x_3)$$
$$- F(x_1, 0, 0) - F(0, x_2, 0) - F(0, 0, x_3).$$

The restrictions put on the coefficients a_{ij} and b_i follow from the boundary condition $F(1, \ldots, 1) = 1$ and the monotonicity of F.

The sufficiency is a matter of an easy computation. It follows from the 2–additivity of functions $F_i, F_{ij}\colon [0,1]^n \to [0,1]$ given by $F_i(\mathbf{x}) = x_i$, $F_{ij}(\mathbf{x}) = x_i x_j$, where $i, j \in \{1, \ldots, n\}$, $i \leq j$. $\quad\square$

Using similar arguments as in the case of 2–additive n–ary aggregation functions, we get the following characterization of k–additive n–ary aggregation functions.

Theorem 3.2. *A function $F\colon [0,1]^n \to [0,1]$ is a k–additive n–ary aggregation function, i.e., $F \in \mathcal{A}_{k,n}$, if and only if there are appropriate constants (ensuring the boundary condition $F(1, \ldots, 1) = 1$ and the monotonicity of F) such that for all $(x_1, \ldots, x_n) \in [0,1]^n$,*

$$F(x_1, \ldots, x_n) = \sum_{i=1}^{k} \left(\sum_{1 \leq j_{1,i} \leq \cdots \leq j_{i,i} \leq n} a_{j_{1,i}, \ldots, j_{i,i}} \left(\prod_{p=1}^{i} x_{j_{p,i}} \right) \right),$$

i.e., F is a polynomial of a degree not exceeding k.

The fact that $F \in \mathcal{A}_{k,n}$ only if F is a polynomial in n variables of degree at most k, is substantial for the following result.

Corollary 3.1. *Let $F \in \mathcal{A}_{k,n}$ and $G_1, \ldots, G_n \in \mathcal{A}_{p,m}$. Then the composite function $G\colon [0,1]^m \to [0,1]$ given by*

$$G(\mathbf{x}) = F(G_1(\mathbf{x}), \ldots, G_n(\mathbf{x}))$$

is a $(k \cdot p)$–additive aggregation function.

Proof. Under given assumptions, G is clearly an m–ary aggregation function, and so only its $(k \cdot p)$–additivity remains to be proved. As the compositions of polynomials is again a polynomial, G is a polynomial of a degree κ,

$$\kappa \leq \lambda \max\{\lambda_1, \ldots, \lambda_n\} \leq k \cdot p,$$

where λ is the degree of F and $\lambda_1, \ldots, \lambda_n$ are the degrees of $G_1, \ldots, G_n$, respectively. $\quad\square$

From Corollary 3.1, it follows that each class $\mathcal{A}_{k,n}$ is convex. We describe the set $\mathcal{V}_{2,n}$ of all vertices of $\mathcal{A}_{2,n}$, i.e., the minimal set whose convex closure gives just $\mathcal{A}_{2,n}$. For $\mathcal{A}_{k,n}$, $k > 2$, a similar result can be proved.

Theorem 3.3. *For a fixed $n \in \mathbb{N}$, let $\mathcal{V}_{2,n} = \{F_i, F_{ij}, G_{ij}\}_{1 \leq i \leq j \leq n}$, where, for each $\mathbf{x} \in [0,1]^n$,*

$$F_i(\mathbf{x}) = x_i, \quad F_{ij}(\mathbf{x}) = x_i x_j, \quad G_{ij}(\mathbf{x}) = x_i + x_j - x_i x_j.$$

Then neither F_i nor F_{ij} and G_{ij} can be expressed as a non–trivial convex combinations of 2–additive n–ary aggregation functions, and $F \in \mathcal{A}_{2,n}$ if and only if there are non-negative constants α_i, β_{ij} and γ_{ij} such that

$$\sum_1^n \alpha_i + \sum_{1 \leq i \leq j \leq n} (\beta_{ij} + \gamma_{ij}) = 1 \quad and \quad F = \sum_1^n \alpha_i F_i + \sum_{1 \leq i \leq j \leq n} (\beta_{ij} F_{ij} + \gamma_{ij} G_{ij}).$$

(12)

Proof. The proof of this theorem follows directly from Theorem 3.1.   □

Example 3.1. Let $n = 2$. Consider a 2–additive capacity $m \colon 2^{\{1,2\}} \to [0,1]$ with $m(\{1\}) = a$, $m(\{2\}) = b$, $a, b \in [0,1]$. Its Owen extension $O_m \colon [0,1]^2 \to [0,1]$ is a 2–additive aggregation function, given by

$$O_m(x_1, x_2) = ax_1 + bx_2 + (1 - a - b)x_1 x_2.$$

If $a + b \leq 1$, then immediately

$$O_m(\mathbf{x}) = aF_1(\mathbf{x}) + bF_2(\mathbf{x}) + (1 - a - b)F_{12}(\mathbf{x}),$$

and if $a + b \geq 1$, we have

$$O_m(\mathbf{x}) = (1 - b)F_1(\mathbf{x}) + (1 - a)F_2(\mathbf{x}) + (a + b - 1)G_{12}(\mathbf{x}),$$

where $F_1(\mathbf{x}) = x_1$, $F_2(\mathbf{x}) = x_2$, $F_{12}(\mathbf{x}) = x_1 x_2$, and $G_{12}(\mathbf{x}) = x_1 + x_2 - x_1 x_2$.

As mentioned above, the Owen extensions of k–additive capacities form a proper subset of all k–additive aggregation functions. For $k = 2$ we have the following corollary of Theorems 2.1 and 3.3.

Corollary 3.2. *Let $F \in \mathcal{A}_{2,n}$. Then F is the Owen extension of some 2–additive capacity m, i.e., $F = O_m$, if and only if in the convex decomposition (12) of F there are coefficients $\alpha_{ii} = \beta_{ii}$ for each $i \in \{1, \ldots, n\}$.*

4   Conclusion

We have introduced and discussed k–additive aggregation functions. A special attention was paid to 2–additive n–ary aggregation functions. We have clarified the relation between k–additive capacities, their Owen (multilinear) extension and k–additive aggregation functions. The convex structure of k–additive aggregation functions, and related Owen extensions, as well, have also been clarified.

We expect applications of k–additive aggregation functions in multicriteria decision problems, where interaction of single criteria is expected. Another development of our ideas is expected in generalizations of aggregation functions where some kind of additivity can be replaced by k–additivity. In particular, the notion of k–OWA operators is a promising object of our next study. As an interesting example with expected applications consider a generalization of the arithmetic mean $M\colon [0,1]^2 \to [0,1]$, $M(x,y) = \frac{x+y}{2}$. Recall that M is the only idempotent symmetric additive aggregation function. Replacing additivity by 2–additivity, a one-parametric family $(M_\alpha)_{\alpha \in [-1/4, 1/4]}$ of aggregation functions is obtained, where $M_\alpha(x,y) = \frac{x+y}{2} + \alpha(x-y)^2$.

Acknowledgement. A. Kolesárová and R. Mesiar kindly acknowledge the support of the project of Science and Technology Assistance Agency under the contract No. APVV–14–0013. J. Li acknowledges the support of the National Natural Science Foundation of China (Grants No. 11371332 and No. 11571106).

References

1. Beliakov, G., Pradera, A., Calvo, T.: Aggregation Functions: A Guide for Practitioners. Springer, Heidelberg (2007)
2. Beliakov, G., Bustince, H., Calvo, T.: A Practical Guide to Averaging Functions. Springer, Heidelberg (2016)
3. Chateauneuf, A., Jaffray, J.Y.: Some characterizations of lower probabilities and other monotone capacities through the use of Möbius inversion. Math. Soc. Sci. **17**, 263–283 (1989)
4. Choquet, G.: Theory of capacities. Ann. Inst. Fourier **5**, 131–295 (1953)
5. Grabisch, M.: k-order additive discrete fuzzy measures and their representation. Fuzzy Sets Syst. **92**, 167–189 (1997)
6. Grabisch, M., Marichal, J.-L., Mesiar, R., Pap, E.: Aggregation Functions. Cambridge University Press, Cambridge (2009)
7. Kolesárová, A., Stupňanová, A., Beganová, J.: Aggregation-based extensions of fuzzy measures. Fuzzy Sets Syst. **194**, 1–14 (2012)
8. Lovász, L.: Submodular function and convexity. In: Bachem, A., Korte, B., Grötschel, M. (eds.) Mathematical Programming: The state of the art, pp. 235–257. Springer, Berlin (1983)
9. Marichal, J.-L.: Aggregation of interacting criteria by means of the discrete Choquet integral. In: Calvo, T., Mayor, G., Mesiar, R. (eds.) Aggregation Operators. New Trends and Applications, pp. 224–24. Physica-Verlag, Heidelberg (2002)
10. Owen, G.: Multilinear extensions of games. In: Shapley, S., Roth, A.E. (eds.) The Shapley value. Essays in Honour of Lloyd, pp. 139–151. Cambridge University Press, Cambridge (1988)
11. Valášková, L.: A note to the 2-order additivity. In: Proceedings of MAGIA, Kočovce, pp. 53–55 (2001)
12. Valášková, L.: Non-additive measures and integrals. Ph.D. thesis, STU Bratislava, (2007)

A Representative in Group Decision by Means of the Extended Set of Hesitant Fuzzy Linguistic Term Sets

Jordi Montserrat-Adell[1,2]([✉]), Núria Agell[2], Mónica Sánchez[1],
and Francisco Javier Ruiz[1]

[1] UPC-BarcelonaTech, Barcelona, Spain
{jordi.montserrat-adell,monica.sanchez,francisco.javier.ruiz}@upc.edu
[2] Esade - Universitat Ramon Llull, Barcelona, Spain
nuria.agell@esade.edu

Abstract. Hesitant fuzzy linguistic term sets were introduced to grasp
the uncertainty existing in human reasoning when expressing preferences.
In this paper, an extension of the set of hesitant fuzzy linguistic term sets
is presented to capture differences between non-compatible preferences.
In addition, an order relation and two closed operation over this set
are also introduced to provide a lattice structure to the extended set
of hesitant fuzzy linguistic term sets. Based on this lattice structure a
distance between hesitant fuzzy linguistic descriptions is defined. This
distance enables differences between decision makers to be quantified.
Finally, a representative of a decision making group is presented as the
centroid of the group based on the introduced distance.

Keywords: Linguistic modeling · Group decision making · Uncertainty
and fuzzy reasoning · Hesitant fuzzy linguistic term sets

Introduction

Different approaches involving linguistic assessments have been introduced in
the fuzzy set literature to deal with the impreciseness and uncertainty connate
with human preference reasoning [2,4,5,7,9]. Additionally, different extensions
of fuzzy sets have been presented to give more realistic assessments when uncertainty increases [1,3,8]. In particular, Hesitant Fuzzy Sets were introduced in [10],
to capture this kind of uncertainty and hesitance. Following this idea, Hesitant
Fuzzy Linguistic Term Sets (HFLTSs) were introduced in [8] to deal with situations in which linguistic assessments involving different levels of precision are used.
In addition, a lattice structure was provided to the set of HFLTSs in [6].

In this paper, we present an extension of the set of HFLTSs, $\overline{\mathcal{H}_S}$, based on
an equivalence relation on the usual set of HFLTSs. This enables us to establish
differences between non-compatible HFLTSs. An order relation and two closed
operation over this set are also introduced to define a new lattice structure
in $\overline{\mathcal{H}_S}$.

© Springer International Publishing Switzerland 2016
V. Torra et al. (Eds.): MDAI 2016, LNAI 9880, pp. 56–67, 2016.
DOI: 10.1007/978-3-319-45656-0_5

In order to describe group decision situations in which Decision Makers (DMs) are evaluating different alternatives, Hesitant Fuzzy Linguistic Descriptions (HFLDs) were presented in [6]. A distance between HFLTSs is defined based on the lattice of $\overline{\mathcal{H}_S}$. This allows us to present a distance between HFLDs that we can use to quantify differences among assessments of different DMs. Taking into consideration this distance, a group representative is suggested to describe the whole group assessment. Due to this representative is the HFLD that minimizes distances with the assessments of all the DMs, it is called the centroid of the group.

The rest of this paper is organized as follows: first, Sect. 1 presents a brief review of HFLTSs and its lattice structure. The lattice of the extended set of HFLTSs is introduced in Sect. 2. In Sect. 3, the distances between HFLTSs and HFLDs are defined and the centroid of the group is presented in Sect. 4. Lastly, Sect. 5 contains the main conclusions and lines of future research.

1   The Lattice of Hesitant Fuzzy Linguistic Term Sets

In this section we present a brief review of some concepts about HFLTSs already presented in the literature that are used throughout this paper [6,8].

From here on, let S denote a finite total ordered set of linguistic terms, $S = \{a_1, \ldots, a_n\}$ with $a_1 < \cdots < a_n$.

Definition 1. [8] A *hesitant fuzzy linguistic term set (HFLTS)* over S is a subset of consecutive linguistic terms of S, i.e. $\{x \in S \mid a_i \leq x \leq a_j\}$, for some $i, j \in \{1, \ldots, n\}$ with $i \leq j$.

The HFLTS S is called the *full HFLTS*. Moreover, the empty set $\{\} = \emptyset$ is also considered as a HFLTS and it is called the *empty HFLTS*.

For the rest of this paper, the non-empty HFLTS, $H = \{x \in S \mid a_i \leq x \leq a_j\}$, is denoted by $[a_i, a_j]$. Note that, if $j = i$, the HFLTS $[a_i, a_i]$ is expressed as the singleton $\{a_i\}$.

The set of all the possible HFLTSs over S is denoted by $\mathcal{H}_S$, being $\mathcal{H}_S^* = \mathcal{H}_S - \{\emptyset\}$ the set of all the non-empty HFLTSs. This set is provided with a lattice structure in [6] with the two following operations: on the one hand, the *connected union of two HFLTSs*, $\sqcup$, which is defined as the least element of $\mathcal{H}_S$, based on the subset inclusion relation $\subseteq$, that contains both HFLTSs, and on the other hand, the intersection of HFLTSs, $\cap$, which is defined as the usual intersection of sets. The reason of including the empty HFLTS in $\mathcal{H}_S$ is to make the intersection of HFLTSs a closed operation in $\mathcal{H}_S$.

For the sake of comprehensiveness, let us introduce the following example that is used throughout all this paper to depict all the concepts defined.

Example 1. Given the set of linguistic terms $S = \{a_1, a_2, a_3, a_4, a_5\}$, being $a_1 =$ *very bad*, $a_2 =$ *bad*, $a_3 =$ *regular*, $a_4 =$ *good*, $a_5 =$ *very good*, possible linguistic assessments and their corresponding HFLTSs by means of S would be:

Assessments	HFLTSs
A = "between bad and regular"	$H_A = [a_2, a_3]$
B = "bad"	$H_B = \{a_2\}$
C = "above regular"	$H_C = [a_4, a_5]$
D = "below regular"	$H_D = [a_1, a_2]$
E = "not very good"	$H_E = [a_1, a_4]$

2 The Extended Lattice of Hesitant Fuzzy Linguistic Term Sets

With the aim of describing differences between couples of HFLTSs with empty intersections, an extension of the intersection of HFLTSs is presented in this section, resulting their intersection if it is not empty or a new element that we will call *negative HFLTS* related to the rift, or gap, between them if their intersection is empty. In order to present said extension of the intersection between HFLTSs, we first need to introduce the mathematical structure that allows us to define it as a closed operation. To this end, we define the extended set of HFLTSs in an analogous way to how integer numbers are defined based on an equivalence relation on the natural numbers. To do so, we first present some needed concepts:

Definition 2. Given two non-empty HFLTSs, $H_1, H_2 \in \mathcal{H}_S^*$, we define:

(a) The *gap between H_1 and H_2* as:

$$gap(H_1, H_2) = (H_1 \sqcup H_2) \cap \overline{H_1} \cap \overline{H_2}.$$

(b) H_1 and H_2 are *consecutive* if and only if $H_1 \cap H_2 = \emptyset$ and $gap(H_1, H_2) = \emptyset$.

Proposition 1. Given two non-empty HFLTSs, $H_1, H_2 \in \mathcal{H}_S^*$, the following properties are met:

1. $gap(H_1, H_2) = gap(H_2, H_1)$.
2. If $H_1 \subseteq H_2$, $gap(H_1, H_2) = \emptyset$.
3. If $H_1 \cap H_2 \neq \emptyset$, $gap(H_1, H_2) = \emptyset$.
4. If $H_1 \cap H_2 = \emptyset$, $gap(H_1, H_2) \neq \emptyset$ or H_1 and H_2 are consecutive.
5. If H_1 and H_2 are consecutive, there exist $j \in \{2, \ldots, n-1\}$, $i \in \{1, \ldots, j\}$ and $k \in \{j+1, \ldots, n\}$, such that $H_1 = [a_i, a_j]$ and $H_2 = [a_{j+1}, a_k]$ or $H_2 = [a_i, a_j]$ and $H_2 = [a_{j+1}, a_k]$.

Proof. The proof is straightforward.   □

Note that neither $[a_1, a_j]$ nor $[a_i, a_n]$ can ever be the result of the *gap* between two HFLTSs for any i and for any j.

Notation. Given two consecutive HFLTSs, $H_1 = [a_i, a_j]$ and $H_2 = [a_{j+1}, a_k]$, then $\{a_j\}$ and $\{a_{j+1}\}$ are named as the linguistic terms that provide the consecutiveness of H_1, H_2.

Example 2. Following Example 1, $gap(H_B, H_C) = \{a_3\}$, while the HFLTSs H_A and H_C are consecutive and their consecutiveness is given by $\{a_3\}$ and $\{a_4\}$.

Definition 3. Given two pairs of non-empty HFLTSs, (H_1, H_2) and (H_3, H_4), the *equivalence relation* $\sim$, is defined as:

$$(H_1, H_2) \sim (H_3, H_4) \iff \begin{cases} H_1 \cap H_2 = H_3 \cap H_4 \neq \emptyset \\ \vee \\ gap(H_1, H_2) = gap(H_3, H_4) \neq \emptyset \\ \vee \\ \text{both pairs are consecutive and} \\ \text{their consecutiveness is provided} \\ \text{by the same linguistic terms} \end{cases}$$

It can be easily seen that $\sim$ relates couples of non-empty HFLTSs with the same intersection if they are compatible, with consecutiveness provided by the same linguistic terms if they are consecutive and with the same *gap* between them in the case that they are neither compatible nor consecutive.

Example 3. Following Example 1, the pairs of HFLTSs (H_A, H_B) and (H_A, H_D) are related according to $\sim$ given that they have the same intersection, $\{a_2\}$. Additionally, $(H_C, H_B) \sim (H_C, H_D)$ since they have the same *gap* between them, $\{a_3\}$.

Applying this equivalence relation over the set of all the pairs of non-empty HFLTSs, we get the quotient set $(\mathcal{H}_S^*)^2 / \sim$, whose equivalence classes can be labeled as:

- $[a_i, a_j]$ for the class of all pairs of compatible non-empty HFLTSs with intersection $[a_i, a_j]$, for all $i, j = 1, \ldots, n$ with $i \leq j$.
- $-[a_i, a_j]$ for the class of all pairs of incompatible non-empty HFLTSs whose *gap* is $[a_i, a_j]$, for all $i, j = 2, \ldots, n-1$ with $i \leq j$.
- α_i for the class of all pairs of consecutive non-empty HFLTSs whose consecutiveness is provided by $\{a_i\}$ and $\{a_{i+1}\}$, for all $i = 1, \ldots, n-1$.

For completeness and symmetry reasons, $(\mathcal{H}_S^*)^2 / \sim$ is represented as shown in Fig. 1 and stated in the next definition.

Example 4. Subsequent to this labeling, and following Example 1, the pair (H_C, H_B) belongs to the class $-\{a_3\}$ and so does the pair (H_C, H_D). The pair (H_C, H_A) belongs to the class α_3 and the pair (H_C, H_E) belongs to the class $\{a_4\}$.

Definition 4. Given a set of ordered linguistic term sets $\mathcal{S} = \{a_1, \ldots, a_n\}$, the *extended set of HFLTSs*, $\overline{\mathcal{H}_S}$, is defined as:

$$\overline{\mathcal{H}_S} = (-\mathcal{H}_S^*) \cup \mathcal{A} \cup \mathcal{H}_S^*,$$

where $-\mathcal{H}_S^* = \{-H \mid H \in \mathcal{H}_S^*\}$ and $\mathcal{A} = \{\alpha_0, \ldots, \alpha_n\}$.

In addition, by analogy with real numbers $-\mathcal{H}_S^*$ is called the *set of negative HFLTSs*, $\mathcal{A}$ is called the *set of zero HFLTSs*, and, from now on, $\mathcal{H}_S^*$ is called the *set positive HFLTSs*.

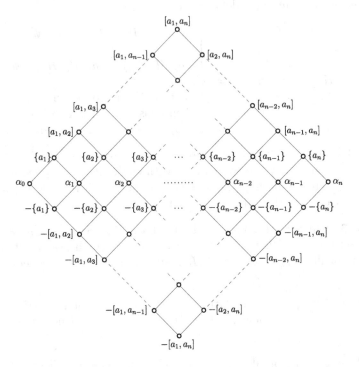

Fig. 1. Graph of the extended set of HFLTSs.

Note that HFLTSs can be characterized by couples of zero HFLTSs. This leads us to introduce a new notation for HFLTSs:

Notation. Given a HFLTS, $H \in \overline{\mathcal{H}_S}$, it can be expressed as $H = \langle \alpha_i, \alpha_j \rangle$, where the first zero HFLTS identifies the bottom left to top right diagonal and the second one identifies the top left to bottom right diagonal. Thus, $\langle \alpha_i, \alpha_j \rangle$ corresponds with $[a_{i+1}, a_j]$ if $i < j$, with $-[a_{i+1}, a_j]$ if $i > j$ and α_i if $i = j$.

This notation is used in the following definition that we present in order to latter introduce an order relation within $\overline{\mathcal{H}_S}$.

Definition 5. Given $H \in \overline{\mathcal{H}_S}$ described by $\langle \alpha_i, \alpha_j \rangle$ the coverage of H is defined as:

$$cov(H) = \{ \langle \alpha_{i'}, \alpha_{j'} \rangle \in \overline{\mathcal{H}_S} \mid i' \geq i \wedge j' \leq j \}.$$

Example 5. The coverage of H_A from Example 1 can be seen in Fig. 2.

The concept of coverage of a HFLTS enables us to define the *extended inclusion relation* between elements of $\overline{\mathcal{H}_S}$.

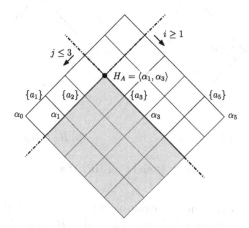

Fig. 2. Coverage of H_A.

Definition 6. The *extended inclusion relation in* $\overline{\mathcal{H}_S}$, $\preccurlyeq$, is defined as:

$$\forall H_1, H_2 \in \overline{\mathcal{H}_S}, \quad H_1 \preccurlyeq H_2 \iff H_1 \in cov(H_2).$$

Note that, restricting to only the positive HFLTSs, the extended inclusion relation coincides with the usual subset inclusion relation. According to this relation in $\overline{\mathcal{H}_S}$, we can define the *extended connected union* and the *extended intersection* as closed operations within the set $\overline{\mathcal{H}_S}$ as follows:

Definition 7. Given $H_1, H_2 \in \overline{\mathcal{H}_S}$, the *extended connected union of* H_1 *and* H_2, $H_1 \sqcup H_2$, is defined as the least element that contains H_1 and H_2, according to the extended inclusion relation.

Definition 8. Given $H_1, H_2 \in \overline{\mathcal{H}_S}$, the *extended intersection of* H_1 *and* H_2, $H_1 \sqcap H_2$, is defined as the largest element being contained in H_1 and H_2, according to the extended inclusion relation.

It is straightforward to see that the extended connected union of two positive HFLTSs coincides with the connected union presented in [6]. This justifies the use of the same symbol. About the extended intersection of two positive HFLTSs, it results the usual intersection of sets if they overlap and the *gap* between them if they do not overlap. Notice that the empty HFLTS is not needed to make the extended intersection a closed operation in $\overline{\mathcal{H}_S}$.

Proposition 2. Given two non-empty HFLTSs, $H_1, H_2 \in \mathcal{H}_S^*$, if $H_1 \preccurlyeq H_2$, then $H_1 \sqcup H_2 = H_2$ and $H_1 \sqcap H_2 = H_1$.

Proof. The proof is straightforward. $\qquad\qquad\qquad\qquad\qquad\qquad\qquad\qquad\qquad\qquad$ $\square$

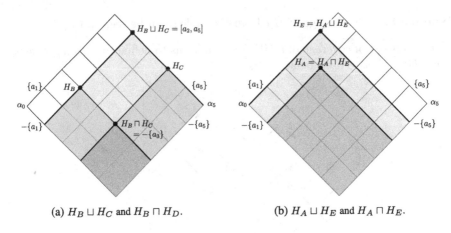

(a) $H_B \sqcup H_C$ and $H_B \sqcap H_D$.   (b) $H_A \sqcup H_E$ and $H_A \sqcap H_E$.

Fig. 3. $\sqcup$ and $\sqcap$ of HFLTSs.

Example 6. Figure 3 provides an example with the extended connected union and the extended intersection of H_B and H_C and of H_A and H_E from Example 1: $H_B \sqcup H_C = [a_2, a_5]$, $H_B \sqcap H_C = -\{a_3\}$, $H_A \sqcup H_E = H_E$ and $H_A \sqcap H_E = H_A$.

Proposition 3. $(\overline{\mathcal{H}_S}, \sqcup, \sqcap)$ is a distributive lattice.

Proof. According to their respective definitions, both operations, $\sqcup$ and $\sqcap$, are trivially commutative and idempotent.

The associative property of $\sqcup$ is met since $(H_1 \sqcup H_2) \sqcup H_3 = H_1 \sqcup (H_2 \sqcup H_3)$ given that both parts equal the least element that contains H_1, H_2 and H_3. About the associativeness of $\sqcap$, $(H_1 \sqcap H_2) \sqcap H_3 = H_1 \sqcap (H_2 \sqcap H_3)$ given that in both cases it results the largest element contained in H_1, H_2 and H_3.

Finally, the absorption laws are satisfied given that: on the one hand $H_1 \sqcup (H_1 \sqcap H_2) = H_1$ given that $H_1 \sqcap H_2 \preccurlyeq H_1$ and on the other hand $H_1 \sqcap (H_1 \sqcup H_2) = H_1$ given that $H_1 \preccurlyeq H_1 \sqcup H_2$.

Furthermore, the lattice $(\overline{\mathcal{H}_S}, \sqcup, \sqcap)$ is distributive given that none of its sublattices are isomorphic to the diamond lattice, M_3, or the pentagon lattice, N_5. □

3   A Distance Between Hesitant Fuzzy Linguistic Term Sets

In order to define a distance between HFLTSs, we introduce a generalization of the concept of cardinal of a positive HFLTS to all the elements of the extended set of HFLTSs.

Definition 9. Given $H \in \overline{\mathcal{H}_S}$, the *width of* H is defined as:

$$W(H) = \begin{cases} card(H) & \text{if } H \in \mathcal{H}_S^*, \\ 0 & \text{if } H \in \mathcal{A}, \\ -card(-H) & \text{if } H \in (-\mathcal{H}_S^*). \end{cases}$$

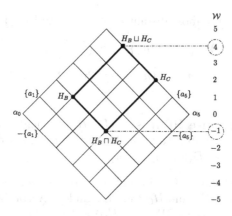

Fig. 4. Distance between HFLTSs.

Note that the width of a HFLTS could be related as well with the height on the graph of $\overline{\mathcal{H}_S}$, associating the zero HFLTSs with height 0, the positive HFLTSs with positive heights and the negative HFLTSs with negative values of heights as shown in Fig. 4.

Proposition 4. $D(H_1, H_2) = \mathcal{W}(H_1 \sqcup H_2) - \mathcal{W}(H_1 \sqcap H_2)$ provides a distance in the lattice $(\overline{\mathcal{H}_S}, \sqcup, \sqcap)$.

Proof. $D(H_1, H_2)$ defines a distance given that it is equivalent to the geodesic distance in the graph $\overline{\mathcal{H}_S}$. The geodesic distance between H_1 and H_2 is the length of the shortest path to go from H_1 to H_2. Due to the fact that $H_1 \sqcap H_2 \preccurlyeq H_1 \sqcup H_2$, $\mathcal{W}(H_1 \sqcup H_2) - \mathcal{W}(H_1 \sqcap H_2)$ is the length of the minimum path between $H_1 \sqcup H_2$ and $H_1 \sqcap H_2$. Thus, we have to check that the length of the shortest path between $H_1 \sqcup H_2$ and $H_1 \sqcap H_2$ coincides with the length of the shortest path between H_1 and H_2.

If one of them belong to the coverage of the other one, let us suppose that $H_1 \preccurlyeq H_2$, then $H_1 \sqcup H_2 = H_2$ and $H_1 \sqcap H_2 = H_1$ and the foregoing assertion becomes obvious. If not, H_1, $H_1 \sqcup H_2$, H_2 and $H_1 \sqcap H_2$ define a parallelogram on the graph. Two consecutive sides of this parallelogram define the shortest path between $H_1 \sqcup H_2$ and $H_1 \sqcap H_2$ while two other consecutive sides of the same parallelogram define the shortest path between H_1 and H_2. Thus, the assertion becomes true as well.

Proposition 5. Given two HFLTSs, $H_1, H_2 \in \overline{\mathcal{H}_S}$, then $D(H_1, H_2) \leq 2n$. If, in addition, $H_1, H_2 \in \mathcal{H}_S^*$, then $D(H_1, H_2) \leq 2n - 2$.

Proof. If $H_1, H_2 \in \overline{\mathcal{H}_S}$, then, the most distant pair is α_0 and α_n. Then,

$$\mathcal{W}(\alpha_0 \sqcup \alpha_n) - \mathcal{W}(\alpha_0 \sqcap \alpha_n) = \mathcal{W}([a_1, a_n]) - \mathcal{W}(-[a_1, a_n]) =$$

$$n - (-n) = 2n.$$

If $H_1, H_2 \in \mathcal{H}_\mathcal{S}^*$, then, the most distant pair is $\{a_1\}$ and $\{a_n\}$. Then,

$$\mathcal{W}(\{a_1\} \sqcup \{a_n\}) - \mathcal{W}(\{a_1\} \sqcap \{a_n\}) = \mathcal{W}([a_1, a_n]) - \mathcal{W}(-[a_2, a_{n-1}]) =$$

$$n - (-(n-2)) = 2n - 2.$$

$$\square$$

Notice that for positive HFLTSs, $D(H_1, H_2)$ coincides with the distance $D_2(H_1, H_2)$ introduced in [6]. Additionally, in this case, the distance presented can also be calculated as $D([a_i, a_j], [a_{i'}, a_{j'}]) = |i - i'| + |j - j'|$.

Example 7. Figure 4 shows the width of the extended connected union and the extended intersection of H_B and H_C from Example 1. According to these results, $D(H_B, H_C) = \mathcal{W}(H_B \sqcup H_C) - \mathcal{W}(H_B \sqcap H_C) = 4 - (-1) = 5$.

4   A Representative of a Group Assessment

The aim of this section is to model the assessments given by a group of Decision Makers (DMs) that are evaluating a set of alternatives $\Lambda = \{\lambda_1, \ldots, \lambda_r\}$ by means of positive HFLTSs over $\mathcal{S} = \{a_1, \ldots, a_n\}$. To do so, we use the definition of Hesitant Fuzzy Linguistic Description (HFLD) introduced in [6].

Definition 10. A *Hesitant fuzzy linguistic description of the set* Λ *by* $\mathcal{H}_\mathcal{S} - \{\emptyset\}$ is a function F_H on Λ such that for all $\lambda \in \Lambda$, $F_H(\lambda)$ is a non-empty HFLTS, i.e., $F_H(\lambda) \in \mathcal{H}_\mathcal{S} - \{\emptyset\}$.

According to this definition, we can extend the distance between HFLTSs presented in Sect. 3 to a distance between HFLDs as follows:

Definition 11. Let us consider F_H^1 and F_H^2 two HFLDs of a set $\Lambda = \{\lambda_1, \ldots, \lambda_r\}$ by means of $\mathcal{H}_\mathcal{S}$, with $F_H^1(\lambda_i) = H_i^1$ and $F_H^2(\lambda_i) = H_i^2$, for all $i \in \{1, \ldots, r\}$. Then, the *distance* $D^\mathcal{F}$ *between these two HFLDs* is defined as:

$$D^\mathcal{F}(F_H^1, F_H^2) = \sum_{t=1}^{r} D(H_t^1, H_t^2).$$

Thus, given a set ok k DMs, we have k different HFLDs of the set of alternatives Λ. In order to summarize this k different assessments, we propose a HFLD that serves as a group representative.

Definition 12. Let Λ be a set of r alternatives, G a group of k DMs and $F_H^1, \ldots, F_H^k$ the HFLDs of Λ provided by the DMs in G, then, the*centroid of the group*is:

$$F_H^C = arg \min_{F_H^x \in (\mathcal{H}_\mathcal{S}^*)^r} \sum_{t=1}^{k} D^\mathcal{F}(F_H^x, F_H^t),$$

identifying each HFLD F_H with the vector $(H_1, \ldots, H_r) \in (\mathcal{H}_\mathcal{S}^*)^r$, where $F_H(\lambda_i) = H_i$, for all $i = 1, \ldots, r$.

Note that the HFLD of the centroid of the group does not have to coincide with any of the HFLDs given by the DMs. In addition, there can be more than one HFLDs minimizing the addition of distances to the assessments given by the DMs, so the centroid of the group is not necessarily unique. Consequently, we proceed with a further study of the possible unicity of the centroid of the group.

Proposition 6. For a specific alternative λ, let $F_H^1(\lambda), \ldots, F_H^k(\lambda)$ be the HFLTSs given as assessments of λ by a group of k DMs. Then, if $F_H^p(\lambda) = [a_{i_p}, a_{j_p}], \forall p \in \{1, \ldots, k\}$, the set of all the HFLTSs associated to the centroid of the group for λ is:

$$\{[a_i, a_j] \in \mathcal{H}_S^* \mid i \in med(i_1, \ldots, i_k), j \in med(j_1, \ldots, j_k)\},$$

where $med(\ )$ contains the median of the values sorted from smallest to largest if k is odd or any integer number between the two central values sorted in the same order if k is even.

Proof. It is straightforward to check that the distance D between HFLTSs is equivalent to the Manhattan distance, also known as taxicab distance, because the graph of $\overline{\mathcal{H}_S}$ can be seen as a grid. Thus, finding the HFLTSs that corresponds to the centroid of the group is reduced to finding the HFLTSs in the grid that minimizes the addition of distances to the other HFLTSs given by the DMs.

The advantage of the taxicab metric is that it works with two independent components, in this case, initial linguistic term and ending linguistic term. Therefore, we can solve the problem for each component separately. For each component, we have a list of natural numbers and we want to find the one minimizing distances. It is well known that the median is the number satisfying a minimum addition of distances to all the points, generalizing the median to all the numbers between the two central ones if there is an even amount of numbers.

Thus, all the HFLTSs satisfying a minimum addition of distances are:

$$\{[a_i, a_j] \in \overline{\mathcal{H}_S} \mid i \in med(i_1, \ldots, i_k), j \in med(j_1, \ldots, j_k)\}.$$

Finally, we have to check that the HFLTSs associated to the centroid are positive HFLTSs for the F_H^C to be a HFLD. If $F_H^p(\lambda) = [a_{i_p}, a_{j_p}] \in \mathcal{H}_S^*, \forall p \in \{1, \ldots, k\}$, that means $i_p \leq j_p, \forall p \in \{1, \ldots, k\}$. Therefore, if k is odd, the median of $i_1, \ldots, i_k$ is less than or equal to the median of $j_1, \ldots, j_k$, and if k is even, the minimum value of $med(i_1, \ldots, i_k)$ is less than or equal than the maximum value of $med(j_1, \ldots, j_k)$. Accordingly, there is always at least one HFLTS associated to the centroid which is a positive HFLTS. Thus,

$$\{[a_i, a_j] \in \mathcal{H}_S^* \mid i \in med(i_1, \ldots, i_k), j \in med(j_1, \ldots, j_k)\}.$$

$\square$

Example 8. Let us assume that H_A, H_B, H_C, H_D, H_E from Example 1 are the assessments given by 5 DMs about the same alternative. In such case, $med(2, 2, 4, 1, 1) = 2$ and $med(3, 2, 5, 2, 4) = 3$, and, therefore, the central assessment for this alternative is $[a_2, a_3]$.

Corollary 1. For a group of k DMs, if k is odd, the centroid of the group is unique.

Proof. If k is odd, both medians are from a set with an odd amount of numbers, so both medians are unique. Therefore, the corresponding HFLTS minimizing the addition of distances is also unique.   □

Corollary 2. For each alternative in Λ, the set of all the HFLTSs corresponding to any centroid of the group is a connected set in the graph of $\overline{\mathcal{H}_S}$.

Proof. If k is odd, by Corollary 1, the proof results obvious. If k is even, by the definition of $med(\ )$, the set of possible results is also connected.   □

Example 9. Let G be a group of 5 DMs assessing a set of alternatives $\Lambda = \{\lambda_1, \ldots, \lambda_4\}$ by means of HFLTSs over the set $\mathcal{S} = \{a_1, a_2, a_3, a_4, a_5\}$ from Example 1, and let $F_H^1, F_H^2, F_H^3, F_H^4, F_H^5$ the HFLDs describing their corresponding assessments shown in the following table together with the HFLD corresponding to the centroid of the group:

	F_H^1	F_H^2	F_H^3	F_H^4	F_H^5	F_H^C
λ_1	$[a_2, a_3]$	$\{a_2\}$	$[a_4, a_5]$	$[a_1, a_2]$	$[a_1, a_4]$	$[a_2, a_3]$
λ_2	$[a_1, a_2]$	$\{a_1\}$	$[a_2, a_3]$	$[a_1, a_2]$	$\{a_2\}$	$[a_1, a_2]$
λ_3	$[a_3, a_5]$	$\{a_3\}$	$\{a_4\}$	$[a_1, a_4$	$[a_2, a_4]$	$[a_3, a_4]$
λ_4	$[a_4, a_5]$	$\{a_5\}$	$\{a_5\}$	$\{a_5\}$	$[a_1, a_2]$	$\{a_5\}$

As the last alternative shows, the centroid of the group is not sensible to outliers, due to the fact that is based on the calculation of two medians.

5   Conclusions and Future Research

This paper presents an extension of the set of Hesitant Fuzzy Linguistic Term Sets by introducing the concepts of negative and zero HFLTSs to capture differences between pair of non-compatible HFLTSs. This extension enables the introduction of a new operation studying the intersection and the gap between HFLTSs at the same time. This operation is used to define a distance between HFLTSs that allows us to analyze differences between the assessments given by a group of decision makers. Based on the study of these differences, a centroid of the group has been proposed.

Future research is focused in two main directions. First, the study of the consensus level of the total group assessments to analyze the agreement or disagreement within the group. And secondly, a real case study will be performed in the marketing research area to examine consensus and heterogeneities in consumers' preferences.

References

1. Deschrijver, G., Kerres, E.E.: On the relationship between some extensions of fuzzy set theory. Fuzzy Sets Syst. **133**, 227–235 (2003)
2. Espinilla, M., Liu, J., Martínez, L.L.: An extended hierarchical linguistic model for decision-making problems. Comput. Intell. **27**(3), 489–512 (2011)
3. Greenfield, S., Chiclana, F.: Accuracy and complexity evaluation of defuzzification strategies for the discretised interval type-2 fuzzy set. Int. J. Approximate Reasoning **54**(8), 1013–1033 (2013)
4. Herrera, F., Herrera-Viedma, E., Martínez, L.: A fuzzy linguistic methodology to deal with unbalanced linguistic term sets. IEEE Trans. Fuzzy Syst. **16**(2), 354–370 (2008)
5. Herrera-Viedma, E., Herrera, F., Chiclana, F.: A consensus model for multiperson decision making with different preference structures. IEEE Trans. Syst. Man Cybern. Part A: Syst. Hum. **32**, 394–402 (2002)
6. Montserrat-Adell, J., Agell, N., Sánchez, M., Prats, F., Ruiz, F.J.: Modeling group assessments by means of hesitant fuzzy linguistic term sets. To appear in J. Appl. Logic
7. Parreiras, R.O., Ekel, P.Y., Martini, J.S.C., Palhares, R.M.: A flexible consensus scheme for multicriteria group decision making under linguistic assessments. Inf. Sci. **180**(7), 1075–1089 (2010)
8. Rodríguez, R.M., Martínez, L., Herrera, F.: Hesitant fuzzy linguistic terms sets for decision making. IEEE Trans. Fuzzy Syst. **20**(1), 109–119 (2012)
9. Tang, Y., Zheng, J.: Linguistic modelling based on semantic similarity relation among linguistic labels. Fuzzy Sets Syst. **157**(12), 1662–1673 (2006)
10. Torra, V.: Hesitant fuzzy sets. Int. J. Intell. Syst. **25**(6), 529–539 (2010)

Suitable Aggregation Operator for a Realistic Supplier Selection Model Based on Risk Preference of Decision Maker

Sirin Suprasongsin[1,2(✉)], Pisal Yenradee[2], and Van Nam Huynh[1]

[1] Japan Advanced Institute of Science and Technology, Nomi, Ishikawa, Japan
{sirin,huynh}@jaist.ac.jp
[2] Sirindhorn International Institute of Technology, Thammasat University,
Pathumthani, Thailand
pisal@siit.tu.ac.th

Abstract. In this paper, we propose (a) the realistic models for Supplier Selection and Order Allocation (SSOA) problem under fuzzy demand and volume/quantity discount constraints, and (b) how to select the suitable aggregation operator based on risk preference of the decision makers (DMs). The aggregation operators under consideration are additive, maximin, and augmented operators while the risk preferences are classified as risk-averse, risk-taking, and risk-neutral ones. The fitness of aggregation operators and risk preferences of DMs is determined by statistical analysis. The analysis shows that the additive, maximin, and augmented aggregation operators are consistently suitable for risk-taking, risk-averse, and risk-neutral DMs, respectively.

1 Introduction

Selecting appropriate suppliers is one of the critical business decisions faced by purchasing managers, and it has a long term impact on a whole supply chain. For most firms, raw material costs account for up to 70 % of product cost as observed in Ghodspour and O'Brien (2001). Thus, a supplier selection process is an important issue in strategic procurement to enhance the competitiveness of the firm [1]. Effective selection of appropriate suppliers involves not only scanning the price list, but also requirements of organization which are increasingly important due to a high competition in a business market. Typically, Dickson (1996) indicated that major requirements are meeting customer demand, reducing cost, increasing product quality and on time delivery performance [2]. Hence, supplier selection is a Multi-Criteria-Decision-Making (MCDM) problem which includes both qualitative and quantitative data, and some of which may be conflicting. In a case of conflicting criteria, DMs need to compromise among criteria. To do so, decision criteria are transformed to objective functions or constraints. The relative importance (weight) of each criterion may be also applied to the model.

Essentially, to prevent a monopolistic supply base as well as to meet all the requirements of firms, most firms have multiple sources which lead to the problem of how many units of each product should be allocated to each of suppliers. Thus, it becomes a Supplier Selection and Order Allocation (SSOA) problem.

© Springer International Publishing Switzerland 2016
V. Torra et al. (Eds.): MDAI 2016, LNAI 9880, pp. 68–81, 2016.
DOI: 10.1007/978-3-319-45656-0_6

Interestingly, to attract large order quantities, suppliers frequently offer trade discounts. Commonly, volume and quantity discounts are popular trade-discount strategies. The quantity-discount policy aims to reduce a unit cost, while the volume-discount encourages firms to reduce the total purchasing cost. Both discounts are triggered at a certain purchasing level. For example, buyers purchase at $20 per unit from $25 per unit when they purchase more than 100 units or receive a 10 % discount when the total purchase cost of all products is greater than $1000. It is interesting here to observe that the trade discount complicates the allocation of order quantities placed to suppliers. Thus, determining the joint consideration of different pricing conditions is a crucial task of DMs to make the most beneficial buying decision.

Practically, firms try to place an order at the level of predicted demand to avoid excess inventory. However, when trade discounts are offered, firms usually purchase more than predicted demand to receive a lower price. Hence, to flexibly optimize the benefit, fuzzy demand is incorporated in models. Note that the satisfaction of demand criteria decreases whenever the order quantity is greater or less than predicted demand. Regarding the issue of uncertainty (fuzziness), fuzzy set theory (FST) developed by Zadeh (1965) has been extensively used to deal with uncertain data, like in this case [3].

During the last decades, we have witnessed many decision techniques for handling MCDM problem. Among several techniques suggested in Ho et al. (2010) [4], linear weighting programming model proposed by Wind and Robinson (1968) [5], is widely applied to assess the performances of suppliers. The model is relatively easy to understand and implement. Later, with the use of pairwise comparisons, an analytical hierarchy process (AHP) allows more accurate scoring method [6]. Generally, this technique decomposes the complex problem into multiple levels of hierarchical structure. Similarly, Analytic Network Process (ANP), Goal Programming(GP), Neural Network (NN), etc., are also introduced to deal with the MCDM problem.

Although several advanced techniques have been proposed to deal with the MCDM problem, little attention has been addressed to determine which aggregation operator is suitable for a specific risk preference of DMs. Basically, the risk preference of DMs can be distinguished into three types, namely, risk-taking, risk-averse, and risk-neutral. Another concerning issue is that previous research works related to the SSOA problem have been conducted based on either volume or quantity discount, not both of them at the same time.

Based on these motivations, this paper proposes realistic models with important practical constraints, especially volume and quantity discount constraints under fuzzy demand. Interestingly, three types of aggregation operators are applied to the models to determine which operator is suitable for risk-taking, risk-averse, and risk-neutral DMs. The aggregation operators are (1) additive, (2) maximin, and (3) augmented operators. The models are developed from Amid et al. [7], Amid et al. [8], and Feyzan [9], accordingly. In addition, to test the sensitivity of the models as well as the effect of aggregation operators, statistical analysis is conducted based on two performance indicators, namely, the average and the lowest satisfaction levels.

The rest of this paper is organized as follows. In Sect. 2, related works are mentioned. Then, six developed models are presented in Sect. 3. In Sect. 4, statistical experiments are designed to analyze the performances of the aggregation operators using MINITAB software. Results are discussed in Sect. 5 and some concluding remarks are presented in Sect. 6.

2   Related Work

To aggregate multiple criteria, many advanced aggregation operators have been proposed in decades. However, in this paper, three basic types of operators are investigated with relative importance of criteria.

Additive Aggregation Operator. The weighted additive technique is probably the best known and widely used method for calculating the total score when multiple criteria are considered. In [7], the objective function is

$$\text{Max} \quad \Sigma_{i=1}^{I} w_i \lambda_i$$

where w_i is the relative importance of criteria i and λ_i is the satisfaction level (SL) of criteria i. Note that to deal with multiple criteria, dimensions of criteria are transformed to SLs which are dimensionless.

Maximin Aggregation Operator. In [8], this operator is looking for SL that meets the need of all criteria. Therefore, s is the smallest SL of all criteria.

$$\text{Max} \quad s$$

Augmented Aggregation Operator. In [9], the author propose this operator in order to keep both advantages of additive and maximin operators. The objective function is developed as follows.

$$\text{Max} \quad s + \Sigma_{i=1}^{I} w_i \lambda_i$$

3   Model Development

There are six proposed models for SSOA problem under fuzzy demand and volume/quantity discount constraints. These models are based on risk preference of DMs which are risk-taking, risk-averse, and risk-neutral. Models under consideration are shown in Fig. 1.

3.1   Problem Description

In this study, DMs must properly allocate the order quantities to each supplier so that the maximum satisfaction is achieved. They have four criteria in mind: (1) the total cost, (2) the quality of product, (3) the on time delivery performance, and (4) the preciseness of demand, where relative importances of criteria (weights) are given. We reduce dominant effects among criteria by transforming them into satisfaction levels (SLs) in a range from 0.0 to 1.0. Demand of each product is allowed to be fuzzy. As multiple products are considered, the overall demand SL is the least SL of all products. In addition, the price-discount models were developed from Xia and Wu (2007) [10], Wang and Yang (2009) [11], and Suprasongsin et al. (2014) [12].

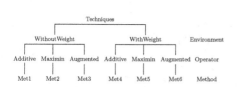

Fig. 1. A combined model diagram

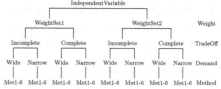

Fig. 2. Experimental's factor of each data set

3.2 Notations

Let us assume that there are five products and five suppliers under consideration. Supplier k $(k = 1, ..., K)$ offers either volume discount or quantity discount when product j $(j = 1, ..., J)$ is purchased at a discount level c $(c = 1, ..., C)$. It is also assumed that supplier 3 offers a volume discount policy, while other suppliers offer a quantity discount policy.

Indices

i	index of criteria		$i = 1...I$
j	index of products		$j = 1...J$
k	index of suppliers		$k = 1...K$
c	index of business volume breaks and price breaks levels		$c = 1...C$
m	index of fuzzy demand		$m = 1...M$
n	index of demand(d) levels		$n = 1$ if $d \leq M$
			$n = 2$ if $d \geq M$

Input parameters

dc_j constant (crisp) demand of product j (unit)
h_{jk} capacity for product j from supplier k (unit)
u_j maximum number of supplier that can supply product j (supplier)
l_j minimum number of supplier that can supply product j (supplier)
o_{jk} minimum order quantity of product j from supplier k (unit)
sr_{jk} 1 if supplier k supplies product j ; 0 otherwise (unitless)
r_{jk} minimum fraction of total demand of product j purchased from supplier k (unitless)
p_{cjk} price of product j offered from supplier k at discount level c ($)
$z1_{jk}$ unit price of product j from supplier k ($)
$z2_{jk}$ quality score of product j from supplier k (scores)
$z3_{jk}$ delivery lateness of product j from supplier k (days)

e_{cjk} break point of quantity discount at level c of product j from supplier k (unit)

g_{ck} discount fraction of volume discount from supplier k at discount level c (unitless)

b_{ck} break point of volume discount at level c from supplier k ($)

f_k 1 if supplier k offers quantity discount; 0 otherwise (unitless)

w_i weight of criteria i (unitless)

σ weight of fuzzy demand (unitless)

mn_i minimum value of criteria i ($, scores, days)

md_i moderate value of criteria i ($, scores, days)

mx_i maximum value of criteria i ($, scores, days)

bo_{mj} boundary of demand level m of product j (unit)

Decision variables

x_{cjkn} purchased quantity at discount level c of product j from supplier k at demand level n (unit)

v_{cjk} purchased quantity at discount level c of product j from supplier k (unit) at constant demand

π_{jk} 1 if supplier k supplies product j; 0 otherwise (unitless)

t_{cjk} total purchasing cost j from supplier k at level c for quantity discount ($)

a_{ck} total purchasing cost j from supplier k at level c for volume discount ($)

α_{ck} 1 if quantity discount level c is selected for supplier k; 0 otherwise (unitless)

β_{ck} 1 if volume discount level c is selected for supplier k; 0 otherwise (unitless)

λ_i satisfaction level of criteria i; cost, quality and delivery lateness (unitless)

s overall satisfaction level formulated by weighted maximin model (unitless)

sl the minimum of satisfaction levels of all criteria (unitless)

γ achievement level of fuzzy demand from all products (unitless)

z_{jn} 1 if demand level n is selected for product j; 0 otherwise (unitless)

sld_j satisfaction level of fuzzy demand of each product j (unitless)

d_{jn} total demand of product j at level n (unit)

3.3 Mathematical Formulation

In this section, six models are presented as the following.

Additive Model. In this model, we assume that all criteria are equally important. The model aims to maximize the average SLs of all criteria including the achievement level of fuzzy demand as shown in (1).

Maximize

$$(\Sigma_i \lambda_i + \gamma)/(I + 1) \tag{1}$$

Price Discount. In quantity discount constraints (2–4), the purchasing quantity x_{cjkn} must be corresponding to a suitable discount level. Similarly, in volume discount constraints (5–7), the business volume a_{ck} from supplier k should be in a suitable discount level c.

$$\Sigma_c t_{cjk} \cdot f_k = \Sigma_c \Sigma_n p_{cjk} \cdot x_{cjkn} \cdot f_k \qquad \forall j, k \tag{2}$$

$$e_{c-1,jk} \cdot \alpha_{ck} \cdot f_k \leq \Sigma_j x_{cjkn} \cdot f_k < e_{cjk} \cdot \alpha_{ck} \cdot f_k \qquad \forall c, k, n \qquad (3)$$

$$\Sigma_c \alpha_{ck} \cdot f_k \leq 1 \qquad \forall k \qquad (4)$$

$$\Sigma_c a_{ck} \cdot (1 - f_k) = \Sigma_c \Sigma_j \Sigma_n z1_{jk} \cdot x_{cjkn} \cdot (1 - f_k) \qquad \forall k \qquad (5)$$

$$b_{c-1,k} \cdot \beta_{ck} \cdot (1 - f_k) \leq a_{ck} \cdot (1 - f_k) < b_{ck} \cdot \beta_{ck} \cdot (1 - f_k) \qquad \forall c, j, k \qquad (6)$$

$$\Sigma_c \beta_{ck} \cdot (1 - f_k) \leq 1 \qquad \forall k \qquad (7)$$

Available Supplier. A supplier may supply only some products but not all of the products.

$$\pi_{jk} \leq sr_{jk} \quad \forall j, k \qquad (8)$$

Capacity. The total purchasing quantity x_{cjkn} must be less than the supply capacity h_{jk} and it is active only if the assigned π_{jk} is equal to 1.

$$\Sigma_c \Sigma_n x_{cjkn} \leq h_{jk} \cdot \pi_{jk} \quad \forall j, k \qquad (9)$$

Limited Number of Supplier. The number of suppliers cannot exceed the available suppliers.

$$l_j \leq \Sigma_k \pi_{jk} < u_j \quad \forall j \qquad (10)$$

Minimum Order Quantity. The total purchasing quantity x_{cjkn} must be greater than the required minimum order quantity of product j from supplier k

$$o_{jk} \cdot \pi_{jk} \leq \Sigma_c \Sigma_n x_{cjkn} \quad \forall j, k \qquad (11)$$

Relationship. The agreement with a supplier k that a firm will purchase the product j at least some percentage of the total demand from this supplier k.

$$r_{jk} \cdot \Sigma_n d_{jn} \leq \Sigma_c \Sigma_n x_{cjkn} \quad \forall j, k \qquad (12)$$

Fuzzy Demand. Total purchasing quantity x_{cjkn} must be in a range of minimum $bo_{m,j}$ and maximum $bo_{m+1,j}$ demand levels and only one demand level z_{jn} must be selected.

$$bo_{mj} \cdot z_{jn} \leq d_{jn} < bo_{m+1,j} \cdot z_{jn} \qquad \forall j, m, n \qquad (13)$$

$$\Sigma_c \Sigma_k x_{cjkn} = d_{jn} \qquad \forall j, n \qquad (14)$$

$$\Sigma_n z_{jn} = 1 \qquad \forall j \qquad (15)$$

Satisfaction Level. Constraints (16–18) describe the SLs of cost, quality, and delivery lateness criteria. Constraints (19–21) calculate the SL (called achievement level) of the fuzzy demand.

$$\lambda_1 \leq \frac{mx_1 - \Sigma_c \Sigma_j \Sigma_k t_{cjk} \cdot f_k + \Sigma_c \Sigma_k a_{ck} \cdot (1 - g_{ck}) \cdot (1 - f_k)}{mx_1 - md_1} \qquad (16)$$

$$\lambda_2 \leq \frac{\Sigma_c \Sigma_j \Sigma_k \Sigma_n z2_{jk} \cdot x_{cjkn} - mn_2}{md_2 - mn_2} \qquad (17)$$

$$\lambda_3 \leq \frac{mx_3 - \Sigma_c \Sigma_j \Sigma_k \Sigma_n z3_{jk} \cdot x_{cjkn}}{mx_3 - md_3} \tag{18}$$

$$sld_j \leq \frac{bo_{3j} - \Sigma_n d_{jn}}{bo_{3j} - bo_{2j}} \qquad \forall j \tag{19}$$

$$sld_j \leq \frac{\Sigma_n d_{jn} - bo_{1j}}{bo_{2j} - bo_{1j}} \qquad \forall j \tag{20}$$

$$\gamma \leq sld_j \qquad \forall j \tag{21}$$

Non-negativity Conditions and the Range of Values. Constraints (22–24) are non-negativity conditions and the range of values.

$$0 \leq \lambda_i < 1 \qquad \forall i \tag{22}$$

$$0 \leq sld_j < 1 \qquad \forall j \tag{23}$$

$$0 \leq \gamma < 1 \tag{24}$$

Weighted Additive Model. A basic concept of this model is to use a single utility function representing the overall preference of DMs corresponding to the relative importance of each criterion.

Maximize

$$(\Sigma_i w_i \cdot \lambda_i) + (\sigma \cdot \gamma) \tag{25}$$

All constraints are the same as those of the additive model (2–24).

Maximin Model. Different from the additive model, the maximin model attempts to maximize the minimum SLs of all criteria, rather than maximize the average value of all SLs. In this model, all criteria are equally important.

Maximize

$$sl \tag{26}$$

Constraints (2–24) are used and three non-negativity constraints are added.

$$sl \leq \gamma \tag{27}$$

$$sl \leq \lambda_i \qquad \forall i \tag{28}$$

$$0 \leq sl < 1 \tag{29}$$

Weighted Maximin Model. It is similar to the maximin model but weights are considered. Constraints (31–36) are adapted from constraints (16–21).

Maximize

$$s \tag{30}$$

The constraints are subjected to (2–15), (23) and the following constraints.

$$w_1 \cdot s \leq \frac{mx_1 - \Sigma_c \Sigma_j \Sigma_k t_{cjk} \cdot f_k + \Sigma_c \Sigma_k a_{ck} \cdot (1 - g_{ck}) \cdot (1 - f_k)}{mx_1 - md_1} \tag{31}$$

$$w_2 \cdot s \leq \frac{\Sigma_c \Sigma_j \Sigma_k \Sigma_n z2_{jk} \cdot x_{cjkn} - mn_2}{md_2 - mn_2} \tag{32}$$

$$w_3 \cdot s \leq \frac{mx_3 - \Sigma_c \Sigma_j \Sigma_k \Sigma_n z_{3jk} \cdot x_{cjkn}}{mx_3 - md_3} \tag{33}$$

$$\sigma \cdot sld_j \leq \frac{bo_{3j} - \Sigma_n d_{jn}}{bo_{3j} - bo_{2j}} \tag{34}$$

$$\sigma \cdot sld_j \leq \frac{\Sigma_n d_{jn} - bo_{1j}}{bo_{2j} - bo_{1j}} \quad \forall j \tag{35}$$

$$s \leq sld_j \quad \forall j \tag{36}$$

$$0 \leq s < 1 \tag{37}$$

Augmented Model. Technically, to maximize the average SLs and the minimum SLs of all criteria simultaneously, the objective function is changed to (38).

Maximize

$$(sl + (\Sigma_i \lambda_i + \gamma))/(I+1) \tag{38}$$

All constraints are drawn from the maximin model (2–24) and (27–29).

Weighted Augmented Model. Weighted augmented model is developed from augmented model by taking weights into account. All constraints are the same as augmented model (Tables 1, 2, 3, 5, 6, 7, 8, 9 and 10).

Maximize

$$sl + (\Sigma_i w_i \cdot \lambda_i + \sigma \cdot \gamma) \tag{39}$$

Table 1. Weight sets (w_i, σ)

Factor/weight	Weight set 1	Weight set 2
Cost	31 %	38 %
Quality	24 %	28 %
Delivery lateness	13 %	11 %
Demand	32 %	23 %

Table 2. Crisp demand of each product (dc_j)

Product	Predicted demand
1	500
2	30
3	100
4	700
5	2500

Table 3. Narrow(N) and wide(W) demand range (bo_{mj})

Level/product	P1		P2		P3		P4		P5	
	N	W	N	W	N	W	N	W	N	W
Minimum variation	450	100	25	10	50	20	650	200	2300	1500
Predicted demand	500	500	30	30	100	100	700	700	2500	2500
Maximum variation	550	1000	32	80	160	500	720	1500	3000	5000

Table 4. Unit (LIST) price, quality score and delivery lateness for Incomplete trade-off(I) and Complete trade-off(C); $(z1_{jk}),(z2_{jk})$ and $(z3_{jk})$

Data	P/S	S1		S2		S3		S4		S5	
		I	C	I	C	I	C	I	C	I	C
Unit (list) price	P1	50	50	40	40	55	55	50	50	45	45
	P2	0	0	200	200	0	0	230	230	0	0
	P3	70	70	75	75	72	*69*	0	0	0	0
	P4	0	0	0	0	8	8	10	10	5	5
	P5	0	0	0	0	0	0	20	20	20	20
Quality score	P1	3	3	5	8	6	6	2	2	4	4
	P2	0	0	6	6	0	0	7	7	0	0
	P3	5	5	7	7	6	*8*	0	0	0	0
	P4	0	0	0	0	8	8	10	10	5	5
	P5	0	0	0	0	0	0	8	8	9	9
Delivery lateness	P1	3	3	1	1	2	2	4	4	3	3
	P2	0	0	4	4	0	0	3	3	0	0
	P3	2	2	2	2	1	1	0	0	0	0
	P4	0	0	0	0	3	3	5	5	4	4
	P5	0	0	0	0	0	0	5	5	3	3

Table 5. Limited number of supplier (u_j, l_j)

No. of supplier	P1	P2	P3	P4	P5
Maximum	2	5	3	4	3
Minimum	1	1	1	1	1

Table 6. Break point of volume discount (b_{ck}) and volume discount percentage (g_{ck})

Level	Supplier 3	
	b_{ck}	g_{ck}
1	0	0
2	10000	0.05
3	50000	0.1

Table 7. Available supplier for each product (sr_{jk})

P/S	S1	S2	S3	S4	S5
1	1	1	1	1	1
2	0	1	0	1	0
3	1	1	1	0	0
4	0	0	1	1	1
5	0	0	0	1	1

Table 8. Price of each product for quantity discount levels (p_{cjk})

Level/sup.	S1			S2				S3					S5		
	P1	P3	P2,4,5	P1	P2	P3	P4-5	P1	P2	P3	P4	P5	P1	P2-3	P4	P5
Level 1	50	70	0	40	200	75	0	50	230	0	32	20	45	0	29	20
Level 2	45	68	0	39	180	74	0	48	220	0	30	18	43	0	28	17
Level 3	43	65	0	38	170	73	0	46	210	0	28	16	42	0	25	14

Table 9. Break point of quantity discount at level (e_{cjk})

Level/	S1	S2		S4		S5
supplier	P1-5	P1,3,4,5	P2	P1,3,4,5	P2	P1-5
Level 1	0	0	0	0	0	0
Level 2	100	100	50	100	20	100
Level 3	500	500	60	500	30	500

Table 10. Boundaries of each criterion (mn_i, md_i, mx_i)

Criteria i	mn_i	md_i	mx_i	Units
z1(Cost)	-	87574	94096	$
z2(Quality score)	28891	32798	-	Score
z3(Delivery lateness)	-	12101	13298	Day

4   Design of Experiment to Statistically Analyze Effects of Each Aggregation Operator

To statistically analyze the sensitivities of optimal solutions and the advantages of aggregation operators, we generate five data sets by varying randomly the capacity, the number of supplier, the minimum order quantity, and the relationships to suppliers. In designing the experiment, independent and dependent variables are required. Models investigate how independent variables have significant effects on dependent variables. The experimental results are analyzed by MINITAB software (Table 11).

Independent Variable. Four independent variables are considered in this study: (1) two sets of weights, (2) two types of demand ranges(wide and narrow demand ranges), (3) six models, and (4) two types of trade-offs (Incomplete and Complete trade-off). Incomplete trade-off means that there are some dominant

Table 11. Capacity (h_{jk}), Minimum order quantity (MOQ) (o_{jk}) and Min % of demand to be purchased (%Demand)(r_{jk})

Data	P/S	S1	S2	S3	S4	S5
Capacity (h_{jk})	P1	1000	500	400	1500	700
	P2	0	50	0	40	0
	P3	300	1000	100	0	0
	P4	0	0	500	2000	600
	P5	0	0	0	3000	2000
MOQ (o_{jk})	P1	0	0	0	0	0
	P2	0	0	0	0	0
	P3	0	10	0	0	0
	P4	0	0	0	0	0
	P5	0	0	0	100	0
%Demand (r_{jk})	P1	0	0	0	0	0
	P2	0	0	0	0	0
	P3	0	0.1	0	0	0
	P4	0	0	0	0	0
	P5	0	0	0	0	0.05

Table 12. Optimal purchasing quantity of weighted additive technique: weight set1, complete trade-off, narrow demand range

P/S	S1	S2	S3	S4	S5
P1	-	50	-	-	450
P2	-	-	-	30	-
P3	-	10	90	-	-
P4	-	-	179	471	50
P5	-	-	-	500	2000

suppliers. For example, supplier 1 is considered as a dominant supplier if supplier 1 provides the lowest cost, the highest quality and the lowest delivery lateness. Each data set consists of 48 combinations as illustrated in Fig. 2.

Dependent Variable. Dependent variables are the performance indicators and are used as responses in MINITAB software. The average SL and the lowest SL are two responses in this study.

5   Results and Discussion

Results are evaluated in four aspects, namely, verification of reasonable results, average SL, lowest SL, dominated solution, and how to select the aggregation operator to match the risk preferences of DMs.

5.1   Reasonable Result Verification

From Table 12, it can be seen that the model yields reasonable results as follows. For Product 4 (P4), it is supplied by 3 suppliers. Unquestionably, if there is only cost criterion, all units must be ordered from S5 due to the lowest price offered. As multiple criteria are concerned, the model is required to make trade-offs among criteria with respect to assigned weights from DMs. As we have seen from Table 4, the quality score of S4 is greater than S5 (10:5) and the delivery lateness of S5 is less than S4 (4:5). Thus, to achieve the highest satisfaction of DMs, DMs purchase P4 at a bit higher price and gain a much better quality and a bit worse delivery lateness. In addition, as the fuzzy demand has the highest weight (32 %), DMs prefer to purchase at the amount closed to the predicted demand. Hence, the total demand of P4 in this model is exactly 700 units.

5.2   Level of Average Satisfaction

By means of statistical analysis, a two-level full factorial design of experiment is applied and each insignificant factor is gradually deleted each time beginning with the highest p-value of interaction factors, until only significant factors are left. The results show that the method and demand range have significant interaction effects. Using Tukey test presented in Fig. 3 and interaction plot in Fig. 4, techniques with the additive operators (Tech.1 and 4) have significantly higher average SL than those of augmented operators (Tech.3 and 6) and max-imin operators (Tech.2 and 5) in both environments. Although, in Fig. 4, the demand range and method have significant interaction effect, conclusion can be concluded in the same way.

```
Grouping Information Using Tukey
Method and 95.0% Confidence

Method  N  Mean  Grouping
1       8  0.6   A
4       8  0.6   A B
6       8  0.6     B
3       8  0.6     B
2       8  0.5         C
5       8  0.5           D

Means that do not share a letter
are significantly different.
```

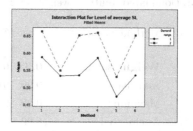

Fig. 3. Grouping for the average SL

Fig. 4. Interaction plot of method and demand range for the average SL

5.3 Level of the Lowest Satisfaction

The results show that an interaction between method and demand range is statistically significant. It is because the model has more ability to search for a better solution when demand range is wider. In Figs. 5 and 6, the maximin aggregation operator (Tech.2) has significantly higher lowest SL than techniques based on the additive operators (Tech.1 and 4). The benefit of the maximin operator is to avoid very bad performance in any aspect. Paradoxically, although the weighted maximin technique is developed using maximin operator, it provides the lowest SL (Lowest SL = 0.1), instead of the highest SL (Highest SL = 0.4) as presented in Fig. 6.

```
Grouping Information Using Tukey
Method and 95.0% Confidence

Method  N  Mean  Grouping
2       8  0.4   A
3       8  0.4   A
6       8  0.4   A
1       8  0.2       B
4       8  0.2         C
5       8  0.1           D

Means that do not share a letter
are significantly different.
```

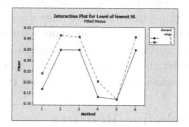

Fig. 5. Grouping for the lowest SL

Fig. 6. Interaction plot of method and demand range for the lowest SL

5.4 Dominated Solution

A solution is considered as a dominated solution whenever the SLs of all criteria are worse than or the same as those of other solutions. The results show that all techniques, except the weighted maximin technique, do not provide any dominated solution as shown in Table 13. We can see that every SLs of weighted maximin technique is lower than the weighted additive technique. This is because if SLs of all criteria are equal to their assigned weights, the weighted maximin technique will get the optimal solution (the sum of all SLs = 1.0) and it has no effort to strive for a better solution. Thus, there is high chance that it will be dominated by the others since the sum of their SLs can be greater than one.

Table 13. Dominated solution (weight set 2, complete trade-off, narrow demand range)

Method/criteria	Cost	Quality	Delivery lateness	Demand
Weight set 2	**0.38**	**0.28**	**0.11**	**0.23**
Additive	0.99	0.6	0.18	0.57
Maximin	0.99	0.46	0.34	0.34
Augmented	0.99	0.46	0.34	0.34
Weighted additive	1	0.63	0.11	0.6
Weighted maximin	0.99	0.33	0.11	0.36
Weighted augmented	0.99	0.46	0.34	0.34

5.5 How to Select the Aggregation Operator to Match the Risk Preferences of DMs

The risk-taking DM normally prefers the solution with relatively high value of average SLs of all criteria even some criteria may have very low or zero SL. The risk-taking DM will feel that scarifying a criterion for a betterment of many other criteria is worth to take a risk. In opposite, the risk-averse DM is very unhappy if a criterion has a very low or zero degree of satisfaction although many other criteria will have very high degree of satisfaction. The risk-neutral DM has a moderate opinion about risk which is somewhere between the risk-taking and risk-averse ones. This type of risk preference DM feels that the average SLs of all criteria is important but the lowest degree of satisfaction should not be too low. Based on the above mentioned characteristics of risk preference, most risk-taking DMs should prefer the additive aggregation operator while most risk-averse DMs should prefer the maximin operator. Similarly, most risk-neutral DMs will find that the augmented operator provides the most preferable solution for them.

6 Concluding Remarks

In this paper, we have proposed the realistic FMOLP models that involve volume and quantity discounts under fuzzy demand and how to select a proper aggregation operator based on risk preference of DMs. The effects of aggregation operator are statistically analyzed. The results reveal that solutions are reasonable with different sets of input parameters. The statistical results also show that the additive aggregation operator matches the preference of the risk-taking DMs since it offers relatively high average SL but a criterion may have very low SL. In opposite, the maximin aggregation operator is acceptable for the risk-averse DMs since it yields a solution with not too low degree of the lowest satisfaction. The augmented aggregation operator, which tries to combine the additive and maximin aggregation operators, provides the solution that is acceptable for the risk-neutral DMs. In addition, it also reveals that the weighted maximin technique should be applied with caution since it may generate a dominated solution.

References

1. Ghodsypour, S.H., OBrien, C.: The total cost of logistics in supplier selection, under conditions of multiple sourcing multiple criteria and capacity constraint. Int. J. Prod. Econ. **73**, 15–27 (2001)
2. Dickson, G.W.: An analysis of vendor selection systems and decisions. J. Purchasing **2**, 5–17 (1966)
3. Zadeh, L.A.: Fuzzy sets. Inf. Control **8**, 338–353 (1965)
4. Ho, W., Xu, W., Dey, P.K.: Multi-criteria decision making approaches for supplier evaluation and selection: a literature review. Eur. J. Oper. Res. **202**, 16–24 (2010)
5. Wind, Y., Robinson, P.J.: The determinants of vendor selection: evaluation function approach. J. Purchasing Mater. Manag., 29–41 (1968)
6. Narisimhan, R.: An analytic approach to supplier selection. J. Purchasing Supply Manag. **1**, 27–32 (1983)
7. Amid, A., Ghodsypour, S.H., OBrien, C.: A weighted additive fuzzy multiobjective model for supplier selection in a supply chain under price breaks. Int. J. Prod. Econ. **121**, 323–332 (2009)
8. Amid, A., Ghodsypour, S.H., OBrien, C.: A weighted max-min model for fuzzy multi-objective supplier selection in a supply chain. Int. J. Prod. Econ. **131**, 145–189 (2011)
9. Feyzan, A.: A fuzzy solution approach for multi objective supplier selection. Expert Syst. Appl. **40**, 947–952 (2013)
10. Xia, W., Wu, Z.: Supplier selection with multiple criteria in volume discount environments. Omega Int. J. Manag. Sci. **35**, 494–504 (2007)
11. Wang, T.Y., Yang, Y.H.: A fuzzy supplier selection in quantity discount environment. Expert Syst. Appl. **36**, 12179–12187 (2009)
12. Suprasongsin, S., Yenradee, P.: Supplier selection with multi criteria and multi products in volume discount and quantity discount environments. In: International Conference Image Processing, Computers and Industrial Engineering, Kuala Lumpur, Malaysia, 15–16 January, pp. 18–22 (2014)

Weighted Quasi-Arithmetic Means on Two-Dimensional Regions: An Independent Case

Yuji Yoshida[(✉)]

Faculty of Economics and Business Administration, University of Kitakyushu,
4-2-1 Kitagata, Kokuraminami, Kitakyushu 802-8577, Japan
yoshida@kitakyu-u.ac.jp

Abstract. Weighted quasi-arithmetic means on two-dimensional regions when weighting functions are independent and utility functions have independent forms are introduced, and some conditions on weighting functions are discussed to characterize the properties. The first-order stochastic dominance and risk premiums on two-dimensional regions are demonstrated. Several examples of two-dimensional utility functions are given by one-dimensional utility functions to explain main results.

1 Introduction

Weighted quasi-arithmetic means are important concepts not only for mathematical theory like the mean value theorems but also for application like subjective estimation of data in management science, artificial intelligence and so on. Weighted quasi-arithmetic means of an interval is derived mathematically by aggregation operations. (Kolmogorov [6], Nagumo [7] and Aczél [1]). In micro-economics, subjective estimations with preference relations are formulated as utility functions (Fishburn [4]). From the view point of utility functions, Yoshida [9,12] have studied the relations between weighted quasi-arithmetic means on an interval and decision maker's attitude regarding risks. For example, for a continuous strictly increasing function $\varphi : [a, b] \mapsto (-\infty, \infty)$ as a decision maker's *utility function* and a continuous function $\omega : [a, b] \mapsto (0, \infty)$ as a *weighting function*, a *weighted quasi-arithmetic mean* μ on a closed interval $[a, b]$ is defined by

$$\mu = \varphi^{-1} \left(\int_a^b \varphi(x)\, \omega(x)\, dx \Big/ \int_a^b \omega(x)\, dx \right). \tag{1.1}$$

Then μ is a *mean value* satisfying

$$\varphi(\mu) \int_a^b \omega(x)\, dx = \int_a^b \varphi(x)\, \omega(x)\, dx \tag{1.2}$$

in the *mean value theorem for integration*. As a special case,

$$\nu = \int_a^b x\, \omega(x)\, dx \Big/ \int_a^b \omega(x)\, dx \tag{1.3}$$

© Springer International Publishing Switzerland 2016
V. Torra et al. (Eds.): MDAI 2016, LNAI 9880, pp. 82–93, 2016.
DOI: 10.1007/978-3-319-45656-0_7

is called a risk neutral mean. Then some properties of utility functions were discussed when risk averse/risk neutral/risk loving conditions $\mu \lesseqgtr \nu$ hold. Bustince et al. [2] discussed aggregation operations on two-dimensional OWA operators, and Labreuche and Grabisch [5] demonstrated Choquet integral for aggregation in multicriteria decision making, and Torra and Godo [8] studied continuous WOWA operators for defuzzification. Yoshida [13] has introduced weighted quasi-arithmetic means on two-dimensional regions, which are related to multi-object decision making. In this paper, we investigate weighted quasi-arithmetic means on two-dimensional regions when utility functions have an independent form and weighting functions are independent, and we discuss some properties for weighting functions for decision maker's attitude. We also demonstrate the first-order stochastic dominance and risk premiums in micro-economics.

In Sect. 2 we discuss weighted quasi-arithmetic means on two-dimensional regions when weighting functions are independent and utility functions have independent forms. In this case we can give a special risk neutral function. We also give some characterizations of weighting functions for decision maker's risk averse/risk neutral/risk loving attitudes and the comparison between weighted quasi-arithmetic means.

In Sect. 3 we demonstrate weighted quasi-arithmetic means on two-dimensional regions and the first-order stochastic dominance and risk premiums on two-dimensional regions, which are important concepts for risk management in micro-economics. In Sect. 4 we investigate several examples for two-dimensional utility functions constructed from one-dimensional utility functions, and we explain the results in previous sections.

2 Weighted Quasi-Arithmetic Means on Two-Dimensional Regions

In this section, we discuss weighted quasi-arithmetic means on two-dimensional regions when utility functions have an independent form and weighting functions are independent, and we give their characterization regarding weighting functions. Let a domain D be a non-empty open subset of $(0, \infty)^2$ such that $D = D_1 \times D_2 \subset (0, \infty)^2$, where D_1 and D_2 are open intervals. Let (Ω, P) be a probability space, where P is a non-atomic probability measure on a sample space Ω. For $i = 1, 2$, let $\mathcal{X}_i$ be a family of real valued random variables X on Ω which have C^1-class density functions $w_i : D_i \mapsto (0, \infty)$ and which also satisfy the following tail condition:

$$\lim_{x \to \infty} xP(|X| \geq x) = 0. \tag{2.1}$$

Denote by $\mathcal{X}^2$ a family of random vectors (X_1, Y_1) on Ω such that $X_1 \in \mathcal{X}_1$, $Y_1 \in \mathcal{X}_2$, and X_1 and Y_1 are *independent*. Then a density function w of random vector $(X_1, Y_1) \in \mathcal{X}^2$ is given by

$$w(x, y) = w_1(x)w_2(y) \tag{2.2}$$

for $(x, y) \in D = D_1 \times D_2$, where w_1 and w_2 are C^1-class density functions of X_1 and Y_1 respectively. In this paper, these density functions are used as weights in quasi-arithmetic means, and they are also called *weighting functions*. Let Γ be a set of 4-tuples $(\alpha, \beta, \gamma, \delta) \in (-\infty, \infty)^4$ satisfying one of the following conditions $(\Gamma.\text{a})$ and $(\Gamma.\text{b})$:

$(\Gamma.\text{a})$ $\alpha > 0$, $\beta > 0$ and $\gamma = 0$.
$(\Gamma.\text{b})$ $\alpha \geq 0$, $\beta \geq 0$ and $\gamma > 0$.

For $i = 1, 2$, let $\mathcal{L}_i$ be a family of C^1-class strictly increasing functions $f_i : D_i \mapsto (0, \infty)$. Denote by $\mathcal{L}$ a family of utility functions $f : D \mapsto (0, \infty)$ given by

$$f(x, y) = \alpha f_1(x) + \beta f_2(y) + \gamma f_1(x) f_2(y) + \delta \qquad (2.3)$$

for $(x, y) \in D = D_1 \times D_2$ with coefficients $(\alpha, \beta, \gamma, \delta) \in \Gamma$ and functions $f_i \in \mathcal{L}_i$ ($i = 1, 2$). Then f are C^1-class strictly increasing functions on D because $f_x(x, y) = \alpha f_1'(x) + \gamma f_1'(x) f_2(y) > 0$ and $f_y(x, y) = \beta f_2'(y) + \gamma f_1(x) f_2'(y) > 0$ for $(x, y) \in D = D_1 \times D_2$. In this paper, (2.3) is called an *independent form* because in the right-hand side of (2.3) utility $f(x, y)$ is constructed from the sum and the scalar product of one-dimensional utilities $f_1(x)$ and $f_2(y)$. Denote a family of rectangle regions by $\mathcal{R}(D) = \{R = I \times J \mid I \text{ and } J \text{ are bounded closed intervals and } R \subset D\}$. For a rectangle region $R \in \mathcal{R}(D)$, *weighted quasi-arithmetic means* on region R with utility $f(\in \mathcal{L})$ and weighting w are given by a subset $M_w^f(R)$ of R as follows.

$$M_w^f(R) = \left\{ (\tilde{x}, \tilde{y}) \in R \mid f(\tilde{x}, \tilde{y}) \iint_R w(x, y) \, dx \, dy = \iint_R f(x, y) w(x, y) \, dx \, dy \right\}.$$

Then we have $M_w^f(R) \neq \emptyset$ because f is continuous on R and

$$\min_{(\tilde{x}, \tilde{y}) \in R} f(\tilde{x}, \tilde{y}) \leq \iint_R f(x, y) w(x, y) \, dx \, dy \Big/ \iint_R w(x, y) \, dx \, dy \leq \max_{(\tilde{x}, \tilde{y}) \in R} f(\tilde{x}, \tilde{y}).$$

We define a point $(\overline{x}_R, \overline{y}_R)$ on a rectangle region $R(\in \mathcal{R}(D))$ by the following weighted quasi-arithmetic means:

$$\overline{x}_R = \iint_R x \, w(x, y) \, dx \, dy \Big/ \iint_R w(x, y) \, dx \, dy, \qquad (2.4)$$

$$\overline{y}_R = \iint_R y \, w(x, y) \, dx \, dy \Big/ \iint_R w(x, y) \, dx \, dy. \qquad (2.5)$$

Hence, $(\overline{x}_R, \overline{y}_R)$ is called an *invariant risk neutral point* on R with weighting w (Yoshida [13]). For a random vector $(X_1, Y_1) \in \mathcal{X}^2$, from (2.2) we also have

$$\overline{x}_R = \int_I x \, w_1(x) \, dx \Big/ \int_I w_1(x) \, dx, \qquad (2.6)$$

$$\overline{y}_R = \int_J y \, w_2(y) \, dy \Big/ \int_J w_2(y) \, dy. \qquad (2.7)$$

Now we introduce the following relations between decision maker's attitude and his utility.

Definition 2.1. Let a utility function $f \in \mathcal{L}$ and let a rectangle region $R \in \mathcal{R}(D)$.

(i) Decision making with utility f is called *risk neutral on R* if

$$f(\overline{x}_R, \overline{y}_R) \iint_R w(x, y) \, dx \, dy = \iint_R f(x, y) w(x, y) \, dx \, dy \qquad (2.8)$$

for all density functions w of random vectors $(X_1, Y_1) \in \mathcal{X}^2$.

(ii) Decision making with utility f is called *risk averse on R* if

$$f(\overline{x}_R, \overline{y}_R) \iint_R w(x, y) \, dx \, dy \geq \iint_R f(x, y) w(x, y) \, dx \, dy \qquad (2.9)$$

for all density functions w of random vectors $(X_1, Y_1) \in \mathcal{X}^2$.

(iii) Decision making with utility f is called *risk loving on R* if

$$f(\overline{x}_R, \overline{y}_R) \iint_R w(x, y) \, dx \, dy \leq \iint_R f(x, y) w(x, y) \, dx \, dy \qquad (2.10)$$

for all density functions w of random vectors $(X_1, Y_1) \in \mathcal{X}^2$.

Definition 2.2. Let I and J be closed intervals satisfying $I \subset D_1$ and $J \subset D_2$, and let $f_i \in \mathcal{L}_i$ be utility functions for $i = 1, 2$.

(i) Decision making with utility f_1 (f_2) is called *risk neutral on I (J)* if

$$f_1(\overline{x}_R) \int_I w_1(x) \, dx = \int_I f_1(x) w_1(x) \, dx \ \left(f_2(\overline{y}_R) \int_J w_2(y) \, dy = \int_J f_2(y) w_2(y) \, dy \right)$$

$$(2.11)$$

for all density functions w_1 (w_2) of random variables $X_1 \in \mathcal{X}_1$ ($Y_1 \in \mathcal{X}_2$ resp.).

(ii) Decision making with utility f_1 (f_2) is called *risk averse on I (J)* if

$$f_1(\overline{x}_R) \int_I w_1(x) \, dx \geq \int_I f_1(x) w_1(x) \, dx \ \left(f_2(\overline{y}_R) \int_J w_2(y) \, dy \geq \int_J f_2(y) w_2(y) \, dy \right)$$

$$(2.12)$$

for all density functions w_1 (w_2) of random variables $X_1 \in \mathcal{X}_1$ ($Y_1 \in \mathcal{X}_2$ resp.).

(iii) Decision making with utility f_1 (f_2) is called *risk loving on I (J)* if

$$f_1(\overline{x}_R) \int_I w_1(x) \, dx \leq \int_I f_1(x) w_1(x) \, dx \ \left(f_2(\overline{y}_R) \int_J w_2(y) \, dy \leq \int_J f_2(y) w_2(y) \, dy \right)$$

$$(2.13)$$

for all density functions w_1 (w_2) of random variables $X_1 \in \mathcal{X}_1$ ($Y_1 \in \mathcal{X}_2$ resp.).

The following result, which can be checked directly by (2.2), (2.6) and (2.7), implies special utility function g is risk neutral.

Theorem 2.1. *Let* $g : D \mapsto (0, \infty)$ *be a utility function defined by*

$$g(x, y) = \alpha x + \beta y + \gamma xy + \delta \tag{2.14}$$

for $(x, y) \in D$ *with a coefficient* $(\alpha, \beta, \gamma, \delta) \in \Gamma$. *Then utility function* g *is risk neutral on any rectangle regions* $R \in \mathcal{R}(D)$.

Hence we investigate this special risk neutral utility function g given by 2.14. Firstly g is strictly increasing, i.e. it holds that $g_x(x, y) = \alpha + \gamma x > 0$ and $g_y(x, y) = \beta + \gamma y > 0$ for $(x, y) \in D$. Further Hessian matrix of g is

$$H^g = \begin{pmatrix} g_{xx} & g_{xy} \\ g_{yx} & g_{yy} \end{pmatrix} = \begin{pmatrix} 0 & \gamma \\ \gamma & 0 \end{pmatrix} \tag{2.15}$$

and its determinant is $|H^g| = -\gamma^2 \leq 0$. For example, we take a domain $D = (0, \infty)^2$, a region $R = [1, 9]^2$ and a weighting function $w(x, y) = 1$ for all $(x, y) \in D$. Then Fig. 1 illustrates weighted quasi-arithmetic means $M_w^g(R)$ for risk neutral functions g which are given respectively by three cases: (a) $g(x, y) = x + 3y$ and (b) $g(x, y) = x + 2y$ in $(\Gamma.a)$ and (c) $g(x, y) = xy$ in $(\Gamma.b)$. We find an *invariant risk neutral point* is $(\overline{x}_R, \overline{y}_R) = (5, 5)$. In Yoshida [13], weighted quasi-arithmetic means for risk neutral functions are given by only straight lines. However, in case (c), weighted quasi-arithmetic means $M_w^g(R)$ for utility function $g(x, y) = xy$ is not straight line. Thus in case when a utility function g is given by (2.14) with $\gamma > 0$, there exist weighted quasi-arithmetic means $M_w^g(R)$ which are not straight lines but curved lines on $R = [1, 9]^2$ (Fig. 1).

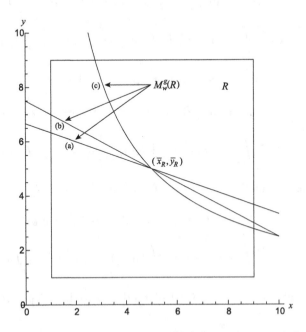

Fig. 1. Weighted quasi-arithmetic means $M_w^g(R)$ for risk neutral functions g ((a) $g(x, y) = x + 3y$, (b) $g(x, y) = x + 2y$, (c) $g(x, y) = xy$).

We can easily check the following equivalent result by [13, Theorem 3.1] and [10, Corollary 5.7].

Theorem 2.2. *Let a utility function $f \in \mathcal{L}$ has an independent form (2.3) with a coefficient $(\alpha, \beta, \gamma, \delta) \in \Gamma$ and utility functions $f_i \in \mathcal{L}_i$ $(i = 1, 2)$. Then the following (a) and (b) are equivalent:*

(a) *f is risk averse (risk loving).*
(b) *f_1 and f_2 are risk averse (risk loving resp.).*

We introduce the following natural ordering on $(0, \infty)^2$.

Definition 2.3. (A partial order $\preceq$ on $(0, \infty)^2$).

(i) For two points $(\underline{x}, \underline{y}), (\overline{x}, \overline{y})(\in (0, \infty)^2)$, an order $(\underline{x}, \underline{y}) \preceq (\overline{x}, \overline{y})$ implies $\underline{x} \leq \overline{x}$ and $\underline{y} \leq \overline{y}$.
(ii) For two sets $A, B(\subset (0, \infty)^2)$, an order $A \preceq B$ implies the following (a) and (b):
    (a) For any $(\underline{x}, \underline{y}) \in A$ there exists $(\overline{x}, \overline{y}) \in B$ satisfying $(\underline{x}, \underline{y}) \preceq (\overline{x}, \overline{y})$.
    (b) For any $(\overline{x}, \overline{y}) \in B$ there exists $(\underline{x}, \underline{y}) \in A$ satisfying $(\underline{x}, \underline{y}) \preceq (\overline{x}, \overline{y})$.

Now we compare weighted quasi-arithmetic means given by two different weighting functions on two-dimensional regions, and we give their characterization. For random vectors $(X_1, Y_1), (X_2, Y_2) \in \mathcal{X}^2$, let w and v be C^1-class density functions for random vectors (X_1, Y_1) and (X_2, Y_2) respectively such that

$$w(x, y) = w_1(x)w_2(y) \quad \text{and} \quad v(x, y) = v_1(x)v_2(y) \qquad (2.16)$$

for $(x, y) \in D = D_1 \times D_2$, where w_1, w_2, v_1 and v_2 are C^1-class density functions of X_1, Y_1, X_2 and Y_2 respectively. Then by [11, Theorem 2.1] and independent form (2.3) we obtain the following characterization for comparison of weighted quasi-arithmetic means.

Theorem 2.3. *Let w and v be C^1-class density functions for random vectors (X_1, Y_1) and (X_2, Y_2) respectively given in (2.16). Then the following (a) and (b) are equivalent:*

(a) *$M_w^f(R) \preceq M_v^f(R)$ for all utility functions $f \in \mathcal{L}$ and all rectangle regions $R \in \mathcal{R}(D)$.*
(b) *$\frac{w_1'}{w_1} \leq \frac{v_1'}{v_1}$ on D_1 and $\frac{w_2'}{w_2} \leq \frac{v_2'}{v_2}$ on D_2.*

3   Weighted Quasi-Arithmetic Means on Two-Dimensional Regions and Stochastic Risks in Economics

In this section, we deal with the relations between weighted quasi-arithmetic means on two-dimensional regions and stochastic risks in micro-economics. Firstly we extend the domain D to $(-\infty, \infty)^2$ as follows. Let w be a C^1-class density function for a random vector $(X_1, Y_1) \in \mathcal{X}^2$. Let $w_1 : D_1 \mapsto (0, \infty)$ be a C^1-class density function of a random variable $X_1 \in \mathcal{X}_1$, and let $w_2 : D_2 \mapsto (0, \infty)$

be a C^1-class density function of a random variable $Y_1 \in \mathcal{X}_2$. Hence we extend their domains to $(-\infty, \infty)$ by

$$w_1(x) = 0 \text{ for } x \in (-\infty, \infty) \setminus D_1 \quad \text{and} \quad w_2(y) = 0 \text{ for } y \in (-\infty, \infty) \setminus D_2. \quad (3.1)$$

Then by (2.2) the domain of w is also extended from D to $(-\infty, \infty)^2$. Let $f \in \mathcal{L}$ be a utility function. Then there exist C^1-class strictly increasing utility functions $f_i \in \mathcal{L}_i$ such that $f_i : D_i \mapsto (0, \infty)$ $(i = 1, 2)$ and coefficients $(\alpha, \beta, \gamma, \delta) \in \Gamma$ satisfying (2.3). Hence we extend their domains to $(-\infty, \infty)$ by

$$f_1(x) = 0 \text{ for } x \in (-\infty, \infty) \setminus D_1 \quad \text{and} \quad f_2(y) = 0 \text{ for } y \in (-\infty, \infty) \setminus D_2. \quad (3.2)$$

Then by (2.3) the domain of f is also extended from D to $(-\infty, \infty)^2$.

For random variables $X \in \mathcal{X}_1$ and $Y \in \mathcal{X}_2$, we denote by $\mathcal{L}_1(X)$ a family of extended functions $f_1 \in \mathcal{L}_1$ such that $f_1(X)$ are integrable and we also denote by $\mathcal{L}_2(Y)$ a family of extended functions $f_2 \in \mathcal{L}_2$ such that $f_2(Y)$ are integrable. For a random vector $(X, Y) \in \mathcal{X}^2$, we denote by $\mathcal{L}(X, Y)$ a family of extended functions $f(\in \mathcal{L})$ such that $f(X, Y)$ are integrable. For $f \in \mathcal{L}(X, Y)$ there exist C^1-class strictly increasing utility functions $f_1 \in \mathcal{L}_1(X)$ and $f_1 \in \mathcal{L}_2(Y)$ and coefficients $(\alpha, \beta, \gamma, \delta) \in \Gamma$ satisfying (2.3). Hence we introduce the following concept.

Definition 3.1

(i) Let $i = 1, 2$, and let real valued random variables $X, Y \in \mathcal{X}_i$. Random variable X is dominated by random variable Y in the sense of *the first-order stochastic dominance* if

$$P(X \le x) \ge P(Y \le x) \text{ for any } x \in (-\infty, \infty). \quad (3.3)$$

Then we write it simply as $X \preceq_{\text{FSD}} Y$.

(ii) Let random vectors $(X_1, Y_1), (X_2, Y_2) \in \mathcal{X}^2$. Random vector (X_1, Y_1) is dominated by random vector (X_2, Y_2) in the sense of *the first-order stochastic dominance* if

$$P((X_1, Y_1) \preceq (x, y)) \ge P((X_2, Y_2) \preceq (x, y)) \text{ for any } (x, y) \in (-\infty, \infty)^2. \quad (3.4)$$

Then we write it simply as $(X_1, Y_1) \preceq_{\text{FSD}} (X_2, Y_2)$.

Hence we obtain the following Lemmas by (2.3) and (2.16).

Lemma 3.1. *For random vectors $(X_1, Y_1), (X_2, Y_2) \in \mathcal{X}^2$, the following (a) and (b) are equivalent:*

(a) $(X_1, Y_1) \preceq_{\text{FSD}} (X_2, Y_2)$.
(b) $X_1 \preceq_{\text{FSD}} X_2$ *and* $Y_1 \preceq_{\text{FSD}} Y_2$.

Lemma 3.2. *For random vectors $(X_1, Y_1), (X_2, Y_2) \in \mathcal{X}^2$, the following (a) and (b) are equivalent:*

(a) $E(f(X_1, Y_1)) \leq E(f(X_2, Y_2))$ for all utility functions $f \in \mathcal{L}(X_1, Y_1) \cap \mathcal{L}(X_2, Y_2)$.

(b) $E(f_1(X_1)) \leq E(f_1(X_2))$ for all utility functions $f_1 \in \mathcal{L}(X_1) \cap \mathcal{L}(X_2)$ and $E(f_2(Y_1)) \leq E(f_2(Y_2))$ for all utility functions $f_2 \in \mathcal{L}(Y_1) \cap \mathcal{L}(Y_2)$.

Then we obtain the following results by Lemmas 3.1 and 3.2, Theorem 2.3 and [11, Proposition 2, Corollary 1].

Theorem 3.1. Let random vectors $(X_1, Y_1), (X_2, Y_2) \in \mathcal{X}^2$ and let w and v be their density functions respectively.

(i) If $M_w^f(R) \preceq M_v^f(R)$ for all utility functions $f \in \mathcal{L}$ and all rectangle regions $R \in \mathcal{R}(D)$, then $(X_1, Y_1) \preceq_{\text{FSD}} (X_2, Y_2)$.

(ii) If $(X_1, Y_1) \preceq_{\text{FSD}} (X_2, Y_2)$, then $E(f(X_1, Y_1)) \leq E(f(X_2, Y_2))$ for all utility functions $f \in \mathcal{L}(X_1, Y_1) \cap \mathcal{L}(X_2, Y_2)$.

Risk premium is one of important concepts in mathematical finance. In the rest of this section we discuss risk premiums on two-dimensional regions. From [13] we introduce the following concept of risk premiums for utility functions in comparison with risk neutral points.

Definition 3.2. Let $f \in \mathcal{L}$ be a utility function on D and let w be a weighting functions given in (2.2). Let a rectangle region $R \in \mathcal{R}(D)$. A nonnegative vector $\pi_w^f(R)(\in [0, \infty)^2)$ is called a *risk premium for utility f* if $(\overline{x}_R, \overline{y}_R) - \pi_w^f(R) \in D$ and it satisfies the following equation:

$$f((\overline{x}_R, \overline{y}_R) - \pi_w^f(R)) \iint_R w(x, y) \, dx \, dy = \iint_R f(x, y) w(x, y) \, dx \, dy. \qquad (3.5)$$

We define the set of risk premiums satisfying (3.5) by

$$\Pi_w^f(R) = \{\pi_w^f(R) \mid (\overline{x}_R, \overline{y}_R) - \pi_w^f(R) \in M_w^f(R), \; \mathbf{0} \preceq \pi_w^f(R)\}, \qquad (3.6)$$

where $\mathbf{0}$ is the zero vector on $(-\infty, \infty)^2$. Hence $\Pi_w^f(R) \neq \emptyset$ if decision making with utility f is risk averse on R with weighting w, i.e. it holds that

$$f(\overline{x}_R, \overline{y}_R) \iint_R w(x, y) \, dx \, dy \geq \iint_R f(x, y) w(x, y) \, dx \, dy. \qquad (3.7)$$

Then $\Pi_w^f(R)$ is also written as

$$\Pi_w^f(R) = \{(\overline{x}_R, \overline{y}_R) - (x, y) \mid (x, y) \in M_w^f(R) \cap R_{-}^{(\overline{x}_R, \overline{y}_R)}\}, \qquad (3.8)$$

where $R_{-}^{(\overline{x}_R, \overline{y}_R)}$ is *a subregion dominated by the invariant risk neutral point* $(\overline{x}_R, \overline{y}_R)$ which is defined by

$$R_{-}^{(\overline{x}_R, \overline{y}_R)} = \{(x, y) \in R \mid (x, y) \preceq (\overline{x}_R, \overline{y}_R)\}. \qquad (3.9)$$

If a utility function f is strictly increasing and concave on D, then utility f is risk averse and $\Pi_w^f(R) \neq \emptyset$ for all weightings w and all rectangle regions $R \in \mathcal{R}(D)$ by [13, Lemma 2.2, Theorem 4.1]. The following lemma, which is checked directly, gives a condition for the concavity of f.

Lemma 3.3. *Let $f \in \mathcal{L}$ be a utility function given in (2.3) with a coefficient $(\alpha, \beta, \gamma, \delta) \in \Gamma$ and C^2-class utility functions $f_i \in \mathcal{L}_i$ $(i = 1, 2)$. If f_1 and f_2 are strictly increasing and concave and they satisfy the following inequality:*

$$(f_1' f_1)' \leq 0 \text{ on } D_1 \quad and \quad (f_2' f_2)' \leq 0 \text{ on } D_2, \tag{3.10}$$

then f is strictly increasing and concave on D.

The following results, which are trivial, give risk premiums for utility functions with independent forms (2.3).

Theorem 3.2. *Let $f \in \mathcal{L}$ be a utility function given in (2.3) with a coefficient $(\alpha, \beta, \gamma, \delta) \in \Gamma$ and utility functions $f_i \in \mathcal{L}_i$ $(i = 1, 2)$. Let w be a density function on D given in (2.2), and let a rectangle region $R(\in \mathcal{R}(D))$. Then it holds that*

$$\Pi_w^f(R) = \left\{ (\overline{x}_R, \overline{y}_R) - (x, y) \,\middle|\, \begin{matrix} g(f_1(x), f_2(y)) = g(F_1, F_2), \\ x \leq \overline{x}_R, y \leq \overline{y}_R, (x, y) \in R \end{matrix} \right\}, \tag{3.11}$$

where F_1 and F_2 are constants given by

$$F_1 = \frac{\int_I f_1(x) w_1(x) \, dx}{\int_I w_1(x) \, dx} \quad and \quad F_2 = \frac{\int_J f_2(y) w_2(y) \, dy}{\int_J w_2(y) \, dy} \tag{3.12}$$

and g is a risk neutral utility function given in (2.14).

Corollary 3.1. *The following (i) and (ii) are important cases in Theorem 3.2.*

(i) *Let a utility $f(x, y) = f_1(x) + f_2(y)$ for $(x, y) \in D$ with $f_i \in \mathcal{L}_i$ $(i = 1, 2)$. Then*

$$\Pi_w^f(R) = \left\{ (\overline{x}_R, \overline{y}_R) - (x, y) \,\middle|\, \begin{matrix} f_1(x) + f_2(y) = F_1 + F_2, \\ x \leq \overline{x}_R, y \leq \overline{y}_R, (x, y) \in R \end{matrix} \right\}. \tag{3.13}$$

(ii) *Let a utility $f(x, y) = f_1(x) f_2(y)$ for $(x, y) \in D$ with $f_i \in \mathcal{L}_i$ $(i = 1, 2)$. Then*

$$\Pi_w^f(R) = \left\{ (\overline{x}_R, \overline{y}_R) - (x, y) \,\middle|\, \begin{matrix} f_1(x) f_2(y) = F_1 F_2, \\ x \leq \overline{x}_R, y \leq \overline{y}_R, (x, y) \in R \end{matrix} \right\}. \tag{3.14}$$

4   Examples

In Table 1 we list up some economic utility functions φ on one-dimensional domains ([10, 12]), and then we can construct utility functions on two-dimensional regions by combining these functions. Examples 4.1 and 4.2 illustrate risk premiums, and in Example 4.3 we deal with the other results in previous sections.

Example 4.1. Take a two-dimensional domain $D = (0, \infty)^2$ and let a region $R = I \times J = [1, 2] \times [1, 2]$. Let a density function

$$w(x,y) = w_1(x)w_2(y) = x^{\kappa_1}y^{\kappa_2} \tag{4.1}$$

for $(x,y) \in D$ with real constants κ_1, κ_2. Then invariant risk neutral point $(\overline{x}_R, \overline{y}_R)$ is given by

$$\overline{x}_R = \frac{\int_I xw_1(x)\,dx}{\int_I w_1(x)\,dx} = \frac{(\kappa_1+1)(2^{\kappa_1+2}-1)}{(\kappa_1+2)(2^{\kappa_1+1}-1)}, \tag{4.2}$$

$$\overline{y}_R = \frac{\int_J yw_2(y)\,dy}{\int_J w_2(y)\,dy} = \frac{(\kappa_2+1)(2^{\kappa_2+2}-1)}{(\kappa_2+2)(2^{\kappa_2+1}-1)}. \tag{4.3}$$

Let $f(x,y) = \sqrt{x} + \sqrt{y}$ be a utility function in case (Γ, a) with $(\alpha, \beta, \gamma, \delta) = (1,1,0,0)$. Then it is trivial that f is concave on $(0,\infty)^2$, and $\Pi_w^f(R) \neq \emptyset$. By Corollary 3.1 risk premiums are

$$\Pi_w^f(R) = \left\{ (\overline{x}_R, \overline{y}_R) - (x,y) \,\middle|\, \begin{array}{l} \sqrt{x} + \sqrt{y} = F_1 + F_2, \\ 1 \leq x \leq \overline{x}_R, 1 \leq y \leq \overline{y}_R \end{array} \right\}, \tag{4.4}$$

where F_1 and F_2 are given by

$$F_1 = \frac{\int_I f_1(x)w_1(x)\,dx}{\int_I w_1(x)\,dx} = \frac{(\kappa_1+1)(2^{\kappa_1+\frac{3}{2}}-1)}{(\kappa_1+\frac{3}{2})(2^{\kappa_1+1}-1)}, \tag{4.5}$$

$$F_2 = \frac{\int_J f_2(y)w_2(y)\,dy}{\int_J w_2(y)\,dy} = \frac{(\kappa_2+1)(2^{\kappa_2+\frac{3}{2}}-1)}{(\kappa_2+\frac{3}{2})(2^{\kappa_2+1}-1)}. \tag{4.6}$$

Example 4.2. Take a two-dimensional domain $D = (0,\infty)^2$ and let a region $R = I \times J = [1,2] \times [1,2]$. Let a density function

$$v(x,y) = v_1(x)v_2(y) = e^{\kappa_1 x}e^{\kappa_2 y} \tag{4.7}$$

for $(x,y) \in D$ with real constants κ_1, κ_2. Then invariant risk neutral point $(\overline{x}_R, \overline{y}_R)$ is given by

$$\overline{x}_R = \frac{\int_I xv_1(x)\,dx}{\int_I v_1(x)\,dx} = \frac{(2\kappa_1-1)e^{\kappa_1} - \kappa_1 + 1}{\kappa_1(e^{\kappa_1}-1)}, \tag{4.8}$$

$$\overline{y}_R = \frac{\int_J yv_2(y)\,dy}{\int_J v_2(y)\,dy} = \frac{(2\kappa_2-1)e^{\kappa_2} - \kappa_2 + 1}{\kappa_2(e^{\kappa_2}-1)}. \tag{4.9}$$

Let $f(x,y) = (1-e^{-x})(1-e^{-y})$ be a utility function in case (Γ, b) with $(\alpha, \beta, \gamma, \delta) = (0,0,1,0)$. By Lemma 3.3, we can easily check f is concave on $[\log 2, \infty)^2$, and $\Pi_v^f(R) \neq \emptyset$. By Corollary 3.1 risk premiums are

$$\Pi_v^f(R) = \left\{ (\overline{x}_R, \overline{y}_R) - (x,y) \,\middle|\, \begin{array}{l} (1-e^{-x})(1-e^{-y}) = G_1G_2, \\ 1 \leq x \leq \overline{x}_R, 1 \leq y \leq \overline{y}_R \end{array} \right\}, \tag{4.10}$$

where G_1 and G_2 are given by

$$G_1 = \frac{\int_I f_1(x)w_1(x)\,dx}{\int_I w_1(x)\,dx} = 1 - \frac{\kappa_1(e^{\kappa_1}-e)}{e^2(\kappa_1-1)(e^{\kappa_1}-1)}, \tag{4.11}$$

$$G_2 = \frac{\int_J f_2(y)w_2(y)\,dy}{\int_J w_2(y)\,dy} = 1 - \frac{\kappa_2(e^{\kappa_2} - e)}{e^2(\kappa_2 - 1)(e^{\kappa_2} - 1)}. \tag{4.12}$$

Example 4.3. Take a two-dimensional domain $D = (0,\infty)^2$. Let w and v be density functions in Examples 3.1 and 3.2:

$$w(x,y) = w_1(x)w_2(y) = x^{\kappa_1}y^{\kappa_2} \tag{4.13}$$

$$v(x,y) = v_1(x)v_2(y) = e^{\kappa_1 x}e^{\kappa_2 y} \tag{4.14}$$

for $(x,y) \in D$ with positive constants κ_1, κ_2. Then we have

$$\frac{w_1'(x)}{w_1(x)} = \frac{\kappa_1}{x} \underset{>}{\lessgtr} \kappa_1 = \frac{v_1'(x)}{v_1(x)} \quad \text{if } x \underset{<}{\gtrless} 1, \tag{4.15}$$

$$\frac{w_2'(y)}{w_2(y)} = \frac{\kappa_2}{y} \underset{>}{\lessgtr} \kappa_2 = \frac{v_2'(y)}{v_2(y)} \quad \text{if } y \underset{<}{\gtrless} 1. \tag{4.16}$$

For a region $R = I \times J = [1,2] \times [1,2]$, by Theorem 2.1 we get $M_w^f(R) \preceq M_v^f(R)$ for all utility functions $f \in \mathcal{L}$. Let $(X_1,Y_1),(X_2,Y_2) \in \mathcal{X}^2$ be random vectors which have the corresponding density functions w and v respectively. Then by Theorem 3.1 it holds that $(X_1,Y_1) \preceq_{\mathrm{FSD}} (X_2,Y_2)$ and $E(f(X_1,Y_1)) \leq E(f(X_2,Y_2))$ for all utility functions $f \in \mathcal{L}(X_1,Y_1) \cap \mathcal{L}(X_2,Y_2)$.

Remark. We can construct other utility functions on two-dimensional domain from utility functions φ on one-dimensional domain given in Table 1.

Table 1. Strictly concave utility functions φ on one-dimensional domains

Utility function, domain and parameters	$\varphi(x)$
Power utility $(0,\infty); 0 < \lambda < 1$	$\frac{x^\lambda}{\lambda}$
Logarithmic utility $(0,\infty); \lambda > 0$	$\lambda \log x$
Exponential utility $(-\infty,\infty); \lambda > 0$	$\frac{1-e^{-\lambda x}}{\lambda}$
Quadratic utility $(0,\lambda); \lambda > 0$	$\lambda x - \frac{1}{2}x^2$
Sigmoid utility $(0,\infty); \lambda > 0$	$\frac{1}{1+e^{-\lambda x}}$

Acknowledgments. This research is supported from JSPS KAKENHI Grant Number JP 16K05282.

References

1. Aczél, J.: On weighted mean values. Bull. Am. Math. Soc. **54**, 392–400 (1948)
2. Bustince, H., Calvo, T., Baets, B., Fodor, J., Mesiar, R., Montero, J., Paternain, D., Pradera, A.: A class of aggregation functions encompassing two-dimensional OWA operators. Inf. Sci. **180**, 1977–1989 (2010)

3. Eeckhoudt, L., Gollier, G., Schkesinger, H.: Economic and Financial Decisions under Risk. Princeton University Press, Princeton (2005)
4. Fishburn, P.C.: Utility Theory for Decision Making. Wiley, New York (1970)
5. Labreuche, C., Grabisch, M.: The Choquet integral for the aggregation of interval scales in multicriteria decision making. Fuzzy Sets Syst. **137**, 11–26 (2003)
6. Kolmogoroff, A.N.: Sur la notion de la moyenne. Acad. Naz. Lincei Mem. Cl. Sci. Fis. Mat. Natur. Sez. **12**, 388–391 (1930)
7. Nagumo, K.: Über eine Klasse der Mittelwerte. Japan. J. Math. **6**, 71–79 (1930)
8. Torra, V., Godo, L.: On defuzzification with continuous WOWA operators. In: Calvo, P.T., Mayor, P.G., Mesiar, P.R. (eds.) Aggregation Operators. STUDFUZZ, vol. 97, pp. 159–176. Springer, Heidelberg (2002)
9. Yoshida, Y.: Aggregated mean ratios of an interval induced from aggregation operations. In: Torra, V., Narukawa, Y. (eds.) MDAI 2008. LNCS (LNAI), vol. 5285, pp. 26–37. Springer, Heidelberg (2008)
10. Yoshida, Y.: Quasi-arithmetic means and ratios of an interval induced from weighted aggregation operations. Soft Comput. **14**, 473–485 (2010)
11. Yoshida, Y.: Weighted quasi-arithmetic means and conditional expectations. In: Torra, V., Narukawa, Y., Daumas, M. (eds.) MDAI 2010. LNCS, vol. 6408, pp. 31–42. Springer, Heidelberg (2010)
12. Yoshida, Y.: Weighted quasi-arithmetic means and a risk index for stochastic environments, International Journal of Uncertainty. Fuzziness Knowl.-Based Syst. (IJUFKS) **16**, 1–16 (2011)
13. Yoshida, Y.: Weighted quasi-arithmetic mean on two-dimensional regions and their applications. In: Torra, V., Narukawa, T. (eds.) MDAI 2015. LNCS, vol. 9321, pp. 42–53. Springer, Heidelberg (2015)

Axiomatisation of Discrete Fuzzy Integrals with Respect to Possibility and Necessity Measures

D. Dubois[1(✉)] and A. Rico[2]

[1] IRIT, Université Paul Sabatier, 31062 Toulouse Cedex 9, France
Didier.Dubois@irit.fr
[2] ERIC, Université Claude Bernard Lyon 1, 69100 Villeurbanne, France

Abstract. Necessity (resp. possibility) measures are very simple representations of epistemic uncertainty due to incomplete knowledge. In the present work, a characterization of discrete Choquet integrals with respect to a possibility or a necessity measure is proposed, understood as a criterion for decision under uncertainty. This kind of criterion has the merit of being very simple to define and compute. To get our characterization, it is shown that it is enough to respectively add an optimism or a pessimism axiom to the axioms of the Choquet integral with respect to a general capacity. This additional axiom enforces the maxitivity or the minitivity of the capacity and essentially assumes that the decision-maker preferences only reflect the plausibility ordering between states of nature. The obtained pessimistic (resp. optimistic) criterion is an average of the maximin (resp. maximax) criterion of Wald across cuts of a possibility distribution on the state space. The additional axiom can be also used in the axiomatic approach to Sugeno integral and generalized forms thereof. The possibility of axiomatising of these criteria for decision under uncertainty in the setting of preference relations among acts is also discussed.

Keywords: Choquet integral · Sugeno integral · Possibility theory

1 Introduction

In multiple-criteria decision making, discrete fuzzy integrals are commonly used as aggregation functions [11]. They calculate a global evaluation for objects or alternatives evaluated according to some criteria. When the evaluation scale is quantitative, Choquet integrals are often used, while in the case of qualitative scale, Sugeno integrals are more naturally considered [9]. The definition of discrete fuzzy integrals is based on a monotonic set function named capacity or fuzzy measure. Capacities are used in many areas such as uncertainty modeling [4], multicriteria aggregation or in game theory [14].

The characterization of Choquet integral on quantitative scales is based on a general capacity, for instance a lower or upper probability defined from a family of probability functions [12,15]. There are no results concerning the characterisation of the Choquet integral with respect to a possibility or a necessity measure.

© Springer International Publishing Switzerland 2016
V. Torra et al. (Eds.): MDAI 2016, LNAI 9880, pp. 94–106, 2016.
DOI: 10.1007/978-3-319-45656-0_8

In contrast, for the qualitative setting, characterizations of Sugeno integrals with respect to possibility measures exist [3,8]. However, Sugeno integrals with respect to necessity (resp. possibility) measures are minitive (resp. maxitive) functionals, while this is not the case for the corresponding Choquet integrals.

This paper proposes a property to be added to axioms characterizing discrete Choquet integrals that may justify the use of a possibility or necessity measure representing a plausibility ordering between states. We then generalize maximin and maximax criteria of Wald. Such specific criteria are currently used in signal processing based on maxitive kernels [10] or in sequential decision [1]. We also show that the same additional property can be added to characterisations of Sugeno integrals and more general functionals, to obtain possibilistic qualitative integrals (weighted min and max). Finally we show that the additional property can be expressed in the Savage setting of preference between acts, and discuss the possibility of act-based characterizations of possibilistic Choquet and Sugeno integrals.

2 Characterization of Possibilistic Choquet Integrals

We adopt the notations used in multi-criteria decision making where some objects or alternatives are evaluated according to a common finite set $C = \{1, \cdots, n\}$ of criteria. In the case of decision under uncertainty (DMU) C is the set of the possible states of the world. A common, totally ordered, evaluation scale V is assumed to provide ratings according to the criteria. Each object is identified with a function $f = (f_1, \cdots, f_n) \in V^n$, called a *profile*, where f_i is the evaluation of f according to the criterion i. The set of all these objects (or acts in the setting of DMU) is denoted by $\mathcal{V}$.

A capacity or fuzzy measure is a non-decreasing set function $\mu : 2^C \to L$, a totally ordered scale with top 1 and bottom 0 such that $\mu(\emptyset) = 0$ and $\mu(C) = 1$, with $L \subseteq V$. When L is equipped with a negation denoted by $1-$, the conjugate of a capacity μ is defined by $\mu^c(A) = 1 - \mu(\overline{A})$. A possibility measure Π is a capacity such that $\Pi(A \cup B) = \max(\Pi(A), \Pi(B))$. If $\pi = (\pi_1, \ldots, \pi_n)$ is the possibility distribution associated with Π, we have $\Pi(A) = \max_{i \in A} \pi_i$, which makes it clear that $\pi_i = 1$ for some i. In multi-criteria decision making, π_i is the importance of the criterion i. In the case of decision under uncertainty, π_i represents the plausibility of the state i. A necessity measure is a capacity N such that $N(A \cap B) = \min(N(A), N(B))$; then we have $N(A) = \min_{i \notin A} 1 - \pi_i$ since functions Π and N are conjugate capacities.

2.1 Possibilistic Choquet Integrals

In this part, L is supposed to be the unit interval. The Moebius transform associated with a capacity μ is the set function $m_\mu(T) = \sum_{K \subseteq T} (-1)^{|T \setminus K|} \mu(K)$, where $\sum_{T \subseteq C} m_\mu(T) = 1$. The sets T such that $m_\mu(T) \neq 0$ are called the focal sets of μ. Using m_μ, the discrete Choquet integral of a function $f : C \to \mathbb{R}$ with respect to a capacity μ can be simply expressed as a generalized weighted mean:

$$C_\mu(f) = \sum_{T \subseteq C} m_\mu(T) \min_{i \in T} f_i. \tag{1}$$

Suppose μ is a necessity measure N and let σ be the permutation on the criteria such that $1 = \pi_{\sigma(1)} \geq \cdots \geq \pi_{\sigma(n)} \geq \pi_{\sigma(n+1)} = 0$. The Choquet integral of f with respect to N boils down to:

$$C_N(f) = \sum_{i=1}^{n} (\pi_{\sigma(i)} - \pi_{\sigma(i+1)}) \min_{j:\pi_j \geq \pi_{\sigma(i)}} f_j = \sum_{i=1}^{n} (\pi_{\sigma(i)} - \pi_{\sigma(i+1)}) \min_{j=1}^{i} f_{\sigma(j)} \quad (2)$$

since the focal sets of N are the sets $\{\sigma(1), \cdots, \sigma(i)\}_{i=1,\cdots,n}$ and their value for the Moebius transform is $\pi_{\sigma(i)} - \pi_{\sigma(i+1)}$ respectively. Using the identity $C_\Pi(f) = 1 - C_N(1-f)$ one obtains the Choquet integral of f with respect to the conjugate possibility measure:

$$C_\Pi(f) = \sum_{i=1}^{n} (\pi_{\sigma(i)} - \pi_{\sigma(i+1)}) \max_{j:\pi_j \geq \pi_{\sigma(i)}} f_j = \sum_{i=1}^{n} (\pi_{\sigma(i)} - \pi_{\sigma(i+1)}) \max_{j=1}^{i} f_{\sigma(j)} \quad (3)$$

Note that if $\pi_1 = \cdots \pi_n = 1$ then $C_N(f) = \min_{i=1}^{n} f_i$ and $C_\Pi(f) = \max_{i=1}^{n} f_i$ are Wald maximin and maximax criteria, respectively. Moreover if many criteria have the same importance π_i, then the expression of C_N (resp. C_Π) proves that we take into account the worst (resp. best) value of f_j according to these criteria.

It is worth noticing that the functional C_N is not minitive and C_Π is not maxitive [5] as shown by the following example.

Example 1. We consider $\mathcal{C} = \{1,2\}$, the possibility distribution π and the following profiles f and g: $\pi_1 = 1, \pi_2 = 0.5$; $f_1 = 0.2, f_2 = 0.3$; and $g_1 = 0.4, g_2 = 0.1$. We have $C_N(f) = 0.5 \cdot 0.2 + 0.5 \cdot 0.2 = 0.2$ and $C_N(g) = 0.5 \cdot 0.4 + 0.5 \cdot 0.1 = 0.25$, but $C_N(\min(f,g)) = 0.5 \cdot 0.2 + 0.5 \cdot 0.1 = 0.15 \neq \min(C_N(f), C_N(g))$. By duality, it also proves the non-maxitivity of C_Π using acts $1 - f$ and $1 - g$.

2.2 Pessimistic and Optimistic Substitute Profiles

Using the permutation σ on the criteria associated with π, a pessimistic profile $f^{\sigma,-}$ and an optimistic profile $f^{\sigma,+}$ can be associated with each profile f:

$$f_i^{\sigma,-} = \min_{j:\pi_j \geq \pi_{\sigma(i)}} f_j = \min_{j=1}^{i} f_{\sigma(j)}; \quad f_i^{\sigma,+} = \max_{j:\pi_j \geq \pi_{\sigma(i)}} f_j = \max_{j=1}^{i} f_{\sigma(j)}. \quad (4)$$

Observe that only the ordering of elements i induced by π on $\mathcal{C}$ is useful in the definition of the pessimistic and optimistic profiles associated with f. These profiles correspond to the values of f appearing in the weighted mean expressions (2) and (3). Substituting pessimistic and optimistic profiles associated with f in these expressions, possibilistic Choquet integrals take the form of usual discrete expectations wrt a probability distribution m with $m_{\sigma(i)} = \pi_{\sigma(i)} - \pi_{\sigma(i+1)}$:

$$C_N(f) = \sum_{i=1}^{n} m_{\sigma(i)} f_i^{\sigma,-} = C_N(f^{\sigma,-}), \quad C_\Pi(f) = \sum_{i=1}^{n} m_{\sigma(i)} f_i^{\sigma,+} = C_\Pi(f^{\sigma,+}).$$

Two profiles f and g are said to be comonotone if and only if for all $i, j \in \mathcal{C}$, $f_i > f_j$ implies $g_i \geq g_j$. So f and g are comonotone if and only if there exists the permutation τ on $\mathcal{C}$ such that $f_{\tau(1)} \leq \cdots \leq f_{\tau(n)}$ and $g_{\tau(1)} \leq \cdots \leq g_{\tau(n)}$.

For any profile f, we have $f_1^{\sigma,-} \geq \cdots \geq f_n^{\sigma,-}$ and $f_1^{\sigma,+} \leq \cdots \leq f_n^{\sigma,+}$. So for any pair of profiles f and g, $f^{\sigma,-}$ and $g^{\sigma,-}$ (resp. $f^{\sigma,+}$ and $g^{\sigma,+}$) are comonotone.

We can define a sequence of progressively changing profiles $(\phi_k)_{1 \leq k \leq n}$ that are equivalently evaluated by C_N. Namely, $\phi_1 = f$, $\phi_{k+1} \leq \phi_k$, $\phi_n = f^{\sigma,-}$ where $\phi_{k+1} = \phi_k$ except for one coordinate. The profiles ϕ_k are defined by

$$\phi_k(i) = \begin{cases} \min_{l:\pi_l \geq \pi_{\sigma(i)}} f_l & \text{if } i \leq k \\ f_i & \text{otherwise.} \end{cases}$$

Similarly we can define a sequence of profiles that are equivalently evaluated by C_Π. Namely, $(\phi^k)_{1 \leq k \leq n}$ such that $\phi^1 = f$, $\phi^{k+1} \geq \phi^k$, $\phi^n = f^{\sigma,+}$ where $\phi^{k+1} = \phi^k$ except for one coordinate. The profiles ϕ^k are defined by

$$\phi^k(i) = \begin{cases} \max_{l:\pi_l \geq \pi_{\sigma(i)}} f_l & \text{if } i \leq k \\ f_i & \text{otherwise.} \end{cases}$$

We observe that $C_N(f) = C_N(\phi_k)$, $C_\Pi(f) = C_\Pi(\phi^k)$, for all $1 \leq k \leq n$.

2.3   Representation Theorem

Consider the case of Boolean functions, corresponding to subsets A, B of $\mathcal{C}$. Their profiles are just characteristic functions $1_A, 1_B$. Given a permutation σ induced by π, let us find the corresponding optimistic and pessimistic Boolean profiles.

Lemma 1. *For all $A \subseteq \mathcal{C}$ non empty, $1_A^{\sigma,-} = 1_B$ for a subset $B = A^{\sigma,-} \subseteq A$ and $1_A^{\sigma,+} = 1_B$ for a superset $B = A^{\sigma,+} \supseteq A$.*

Proof. $1_A^{\sigma,-}(i) = \min_{k=1}^i 1_A(\sigma(k)) = 1$ if $\forall k \leq i : \sigma(k) \in A$ and 0 otherwise. So $1_A^{\sigma,-} = 1_B$ with $B \subseteq A$.

$1_A^{\sigma,+}(i) = \max_{k=1}^i 1_A(\sigma(k)) = 1$ if $\exists k \leq i$ and $\sigma(k) \in A$, and 0 otherwise. So $1_A^{\sigma,+} = 1_B$ with $A \subseteq B$. $\qquad\square$

It is easy to realize that the set $A^{\sigma,-}$ exactly contains the largest sequence of consecutive criteria $(\sigma(1), \ldots, \sigma(k^-))$ in A, while the set $A^{\sigma,+}$ exactly contains the smallest sequence of consecutive criteria $(\sigma(1), \ldots, \sigma(k^+))$ that includes A.

Lemma 2. *A capacity μ is a necessity measure if and only if there exists a permutation σ on $\mathcal{C}$ such that for all A we have $\mu(A) = \mu(A^{\sigma,-})$.*

A capacity μ is a possibility measure if and only if there exists a permutation σ on $\mathcal{C}$ such that for all A we have $\mu(A) = \mu(A^{\sigma,+})$.

Proof. Let σ be such that $\mu(A) = \mu(A^{\sigma,-})$. Let us prove that for all $A, B \subseteq \mathcal{C}$, we have $\mu(A \cap B) = \min(\mu(A), \mu(B))$.

From Lemma 1, $(A \cap B)^{\sigma,-} \subseteq A \cap B$. So $\mu(A \cap B) = \mu((A \cap B)^{\sigma,-}) = \mu(\{\sigma(1), \cdots, \sigma(k^-)\})$. As $\sigma(k^-+1) \notin A \cap B$, then $\sigma(k^-+1) \notin A$ or $\sigma(k^-+1) \notin B$.

Suppose without loss of generality that $\sigma(k^- + 1) \notin A$. Then $A^{\sigma,-} = (A \cap B)^{\sigma,-}$ hence $\mu(A \cap B) = \mu(A) \leq \mu(B)$ so $\mu(A \cap B) = \min(\mu(A), \mu(B))$. Consequently μ is a necessity measure.

Conversely we consider a necessity measure N and the permutation such that $\pi_1 \geq \cdots \geq \pi_n$, $N(A) = 1 - \pi_{i_0}$ with $i_0 = \min\{j : j \notin A\}$. So the set $A^{\sigma,-}$ is $\{1, \cdots, i_0 - 1\}$ so $N(A^{\sigma,-}) = 1 - \pi_{i_0}$.

A similar proof can be developed for the case of possibility measures. $\square$

Now we add suitable axioms to a known representation theorem of Choquet integral [15], and obtain a characterisation theorem for the case when the capacity is a possibility or a necessity measure.

Theorem 1. *A function $I : V \to \mathbb{R}$ satisfies the following properties:*

C1 $I(1, \cdots, 1) = 1$,
C2 *Comonotonic additivity:f and g comonotone implies $I(f + g) = I(f) + I(g)$,*
C3 *Pareto-domination:$f \geq g$ implies $I(f) \geq I(g)$,*
Π4 *There exists a permutation σ on C such that $\forall A$, $I(\mathbf{1}_A) = I(\mathbf{1}_{A^{\sigma,+}})$*

if and only if $I = C_\Pi$, where $\Pi(A) = I(\mathbf{1}_A)$ is a possibility measure.

Proof. It is easy to check that the Choquet integral with respect to Π satisfies the properties C1-C3 and Π4 according to the permutation associated with π.

If I satisfies the properties C1-C3, then according to the results presented in [15] I is a Choquet integral with respect to the fuzzy measure μ defined by $\mu(A) = I(\mathbf{1}_A)$. The property Π4 implies $\mu(A) = I(\mathbf{1}_A^{\sigma,+}) = \mu(A^{\sigma,+})$, and using Lemma 2 this equality is equivalent to have a possibility measure. $\square$

Note that Axiom Π4 can be replaced by: There exists a permutation σ on C such that $\forall f$, $I(f) = I(f^{\sigma,+})$. We have a similar result for necessity measures:

Theorem 2. *A function $I : V \to \mathbb{R}$ satisfies the following properties:*

C1 $I(1, \cdots, 1) = 1$,
C2 *Comonotonic additivity: f and g comonotone implies $I(f + g) = I(f) + I(g)$,*
C3 *Pareto-domination: $f \geq g$ implies $I(f) \geq I(g)$,*
N4 *There exists a permutation σ on C such that $\forall A$, $I(\mathbf{1}_A) = I(\mathbf{1}_{A^{\sigma,-}})$*

if and only if $I = C_N$, where $N(A) = I(\mathbf{1}_A)$ is a necessity measure.

Axiom N4 can be replaced by: There exists a permutation σ on C such that $\forall f$, $I(f) = I(f^{\sigma,-})$. These results indicate that Choquet integrals w.r.t possibility and necessity measures are additive for a larger class of pairs of functions than usual, for instance $C_N(f + g) = C_N(f) + C_N(g)$ as soon as $(f + g)^{\sigma,-} = f^{\sigma,-} + g^{\sigma,-}$, which does not imply that f and g are comonotone.

Example 2. We consider $C = \{1, 2, 3\}$, the permutation associated with Π such that $\pi_1 \geq \pi_2 \geq \pi_3$ and the profiles $f = (1, 2, 3)$, $g = (1, 3, 2)$ which are not comonotone. It is easy to check that $(f + g)^- = f^{\sigma,-} + g^{\sigma,-}$.

The above result should be analyzed in the light of a claim by Mesiar and Šipoš [13] stating that if the capacity μ is modular on the set of cuts $\{\{i : f_i \geq \alpha\} : \alpha > 0\} \cup \{\{i : g_i \geq \alpha\} : \alpha > 0\}$ of f and g, then $C_N(f + g) = C_N(f) + C_N(g)$. For a general capacity, it holds if f and g are comonotonic. For more particular capacities, the set of pairs of acts for which modularity holds on cuts can be larger. This is what seems to happen with possibility and necessity measures.

3   A New Characterisation of Possibilistic Sugeno Integrals

In this part we suppose that $L = V$ is a finite, totally ordered set with 1 and 0 as respective top and bottom. So, $V = L^C$. Again, we assume that L is equipped with a unary order reversing involutive operation $t \to 1 - t$ called a negation. To distinguish from the numerical case, we denote by $\wedge$ and $\vee$ the minimum and the maximum on L. As we are on a qualitative scale we speak of q-integral in this section. The Sugeno q-integral [16,17], of an alternative f can be defined by means of several expressions, among which the two following normal forms [12]:

$$\int_\mu f = \bigvee_{A \subseteq C} \mu(A) \bigwedge \wedge_{i \in A} f_i = \bigvee_{A \subseteq C} (1 - \mu^c(A)) \bigvee \vee_{i \in A} f_i \qquad (5)$$

Sugeno q-integral can be characterized as follows:

Theorem 3 [3]. *Let $I : V \to L$. There is a capacity μ such that $I(f) = \int_\mu f$ for every $f \in V$ if and only if the following properties are satisfied*

1. $I(f \vee g) = I(f) \vee I(g)$, for any comonotone $f, g \in V$.
2. $I(a \wedge f) = a \wedge I(f)$, for every $a \in L$ and $f \in V$.
3. $I(1_C) = 1$.

Equivalently, conditions (1–3) can be replaced by conditions (1'–3') below.

1'. $I(f \wedge g) = I(f) \wedge I(g)$, for any comonotone $f, g \in V$.
2'. $I(a \vee f) = a \vee I(f)$, for every $a \in L$ and $f \in V$.
3'. $I(0_C) = 0$.

The existence of these two equivalent characterisations is due to the possibility of writing Sugeno q-integral in conjunctive and disjunctive forms (5) equivalently.

Moreover, for a necessity measure N, $\int_N f = \wedge_{i=1}^n (1 - \pi_i) \vee f_i$; and for a possibility measure Π, $\int_\Pi f = \vee_{i=1}^n \pi_i \wedge f_i$. The Sugeno q-integral with respect to a possibility (resp. necessity) measure is maxitive (resp. minitive), hence the following known characterization results for them:

Theorem 4. *Let $I : V \to L$. There is a possibility measure Π such that $I(f) = \int_\Pi f$ for every $f \in V$ if and only if the following properties are satisfied*

1. $I(f \vee g) = I(f) \vee I(g)$, for any $f, g \in V$.
2. $I(a \wedge f) = a \wedge I(f)$, for every $a \in L$ and $f \in V$.
3. $I(1_C) = 1$.

Theorem 5. *There is a necessity measure Π such that $I(f) = \int_N f$ for every $f \in V$ if and only if the following properties are satisfied*

1'. $I(f \wedge g) = I(f) \wedge I(g)$, for any $f, g \in V$.
2'. $I(a \vee f) = a \vee I(f)$, for every $a \in L$ and $f \in V$.
3'. $I(0_C) = 0$.

However, we can alternatively characterise those simplified Sugeno q-integrals in the same style as we did for possibilistic Choquet integrals due to the following

Lemma 3.

$$\int_N f = \int_N f^{\sigma,-}, \quad \int_\Pi f = \int_\Pi f^{\sigma,+}.$$

Proof. Assume $\pi_1 \geq \cdots \geq \pi_n$ for simplicity, i.e. $\sigma(i) = i$. By definition $f^{\sigma,-} \leq f$ so $\int_N f^{\sigma,-} \leq \int_N f$ since the Sugeno q-integral is an increasing function. Let i_0 and i_1 be the indices such that $\int_N f^{\sigma,-} = \max(1 - \pi_{i_0}, \min_{j \leq i_0} f_{i_0}) = \max(1 - \pi_{i_0}, f_{i_1})$ where $i_1 \leq i_0$. Hence $\pi_{i_1} \geq \pi_{i_0}$ i.e. $1 - \pi_{i_1} \leq 1 - \pi_{i_0}$ and $\int_N f^{\sigma,-} \geq \max(1 - \pi_{i_1}, f_{i_1}) \geq \int_N f$.

By definition $f \leq f^{\sigma,+}$ so $\int_\Pi f \leq \int_\Pi f^{\sigma,+}$. Let i_0 and i_1 be the indices such that $\int_\Pi f^{\sigma,+} = \min(\pi_{i_0}, \max_{j \leq i_0} f_j) = \min(\pi_{i_0}, f_{i_1})$ where $i_1 \leq i_0$. Hence $\pi_{i_0} \leq \pi_{i_1}$ and $\int_\Pi f^{\sigma,+} \leq \min(\pi_{i_1}, f_{i_1}) \leq \int_\Pi f$.   □

In particular, $\int_N f = \int_N \phi_k$, $\int_\Pi f = \int_\Pi \phi^k$, for all $1 \leq k \leq n$, as for Choquet integral. Now we can state qualitative counterparts of Theorems 1 and 2:

Theorem 6. *There is a possibility measure Π such that $I(f) = \int_\Pi f$ for every $f \in V$ if and only if the following properties are satisfied*

1. $I(f \vee g) = I(f) \vee I(g)$, for any comonotone $f, g \in V$.
2. $I(a \wedge f) = a \wedge I(f)$, for every $a \in L$ and $f \in V$.
3. $I(1_C) = 1$.

$\Pi 4$ *There exists a permutation σ on C such that $\forall A$, $I(1_A) = I(1_{A^{\sigma,+}})$*

Theorem 7. *There is a necessity measure N such that $I(f) = \int_N f$ for every $f \in V$ if and only if the following properties are satisfied*

$I(f \wedge g) = I(f) \wedge I(g)$, *for any comonotone $f, g \in V$.*
$I(a \vee f) = a \vee I(f)$, *for every $a \in L$ and $f \in V$.*
$I(0_C) = 0$.

$N4$ *There exists a permutation σ on C such that $\forall A$, $I(1_A) = I(1_{A^{\sigma,-}})$*

Axiom $\Pi 4$ (resp., $N4$) can be replaced by: there exists a permutation σ on C such that for all f, $I(f) = I(f^{\sigma,+})$ (resp. $I(f) = I(f^{\sigma,-})$).

We can generalize Sugeno q-integrals as follows: consider a bounded complete totally ordered value scale $(L, 0, 1, \leq)$, equipped with a binary operation $\otimes$ called right-conjunction, which has the following properties:

- the top element 1 is a right-identity: $x \otimes 1 = x$,
- the bottom element 0 is a right-anihilator $x \otimes 0 = 0$,
- the maps $x \mapsto a \otimes x$, $x \mapsto x \otimes a$ are order-preserving for every $a \in L$.

Note that we have $0 \otimes x = 0$ since $0 \otimes x \leq 0 \otimes 1 = 0$. An example of such a conjunction is a semi-copula (such that $a \otimes b \leq \min(a, b)$). We can define an implication $\rightarrow$ from $\otimes$ by semi-duality: $a \rightarrow b = 1 - a \otimes (1 - b)$. Note that this implication coincides with a Boolean implication on $\{0, 1\}$, is decreasing according to its first argument and it is increasing according to the second one.

A non trivial example of semi-dual pair (implication, right-conjunction) is the contrapositive Gödel implication $a \rightarrow_{GC} b = \begin{cases} 1 - a \text{ if } a > b \\ 1 \text{ otherwise} \end{cases}$ and the associated right-conjunction $a \otimes_{GC} b = \begin{cases} a \text{ if } b > 1 - a \\ 0 \text{ otherwise} \end{cases}$ (it is not a semi-copula).

The associated q-integral is $\int_\mu^\otimes f = \vee_{A \subseteq C}((\wedge_{i \in A} f_i) \otimes \mu(A))$. This kind of q-integral is studied in [2] for semi-copulas. The associated q-cointegral obtained via semi-duality is of the form $\int_\mu^\rightarrow f = \wedge_{A \subseteq C}(\mu^c(A) \rightarrow (\vee_{i \in A} f_i))$. We can see that $\int_\mu^\rightarrow f = 1 - \int_{\mu^c}^\otimes (1 - f)$. But in general, $\int_\mu^\otimes f \neq \int_\mu^\rightarrow f$ even when $a \rightarrow b = 1 - a \otimes (1 - b)$ [6], contrary to the case of Sugeno q-integral, for which $\otimes$ is the minimum, and $a \rightarrow b = \max(1 - a, b)$. We have $\int_\Pi^\otimes (f) = \max_{i=1}^n f_i \otimes \pi_i$ and $\int_N^\rightarrow f = \min_{i=1}^n (1 - f_i) \rightarrow (1 - \pi_i)$ since $N^c = \Pi$.

With a proof similar as the one for Sugeno q-integral, it is easy to check that a generalized form of Lemma 3 holds: $\int_\Pi^\otimes f = \int_\Pi^\otimes f^{\sigma,+}$, $\int_N^\rightarrow f = \int_N^\otimes f^{\sigma,-}$.

The characterisation results for Sugeno q-integral (Theorem 3) and their possibilistic specialisations (Theorems 4 and 5) can be generalised for right-conjunction-based q-integrals and q-cointegrals albeit separately for each:

Theorem 8. *A function $I : \mathcal{V} \rightarrow L$ satisfies the following properties:*

RC1 *f and g comonotone implies $I(f \vee g) = I(f) \vee I(g)$,*
RC2 *$I(1_A \otimes a) = I(1_A) \otimes a$*
RC3 *$I(1_C) = 1$.*

if and only if I is a q-integral $\int_\mu^\otimes f$ with respect to a capacity $\mu(A) = I(1_A)$. Adding axiom $\Pi 4$ yields an optimistic possibilistic q-integral $\int_\Pi^\otimes f$.

Theorem 9. *A function $I : \mathcal{V} \rightarrow L$ satisfies the following properties:*

IRC1 *f and g comonotone implies $I(f \wedge g) = I(f) \wedge I(g)$,*
IRC2 *$I(a \rightarrow 1_A) = a \rightarrow I(1_A)$*
IRC3 *$I(1_\emptyset) = 0$.*

if and only if I is an implicative q-integral $\int_\mu^\rightarrow f$ with respect to a capacity $\mu(A) = I(1_A)$. Adding axiom $N4$ yields a pessimistic possibilistic q- cointegral $\int_N^\otimes f$.

4   Axiomatisation Based on Preference Relations

In the context of the decision under uncertainty we consider a preference relation on the profiles and we want to represent it with a Choquet integral, a Sugeno q-integral or a q-integral with respect to a possibility or a necessity.

4.1   Preference Relations Induced by Fuzzy Integrals

With the previous integrals with respect to a possibility we can define a preference relation: $f \succeq_\oint^+ g$ if and only if $\oint_\Pi (f^{\sigma,+}) \geq \oint_\Pi (g^{\sigma,+})$ where $\oint$ is one of the integrals presented above. For all f, we have the indifference relation $f \sim_\oint^+ f^{\sigma,+}$. An optimistic decision maker is represented using a possibility measure since the attractiveness $(f_i^{\sigma,+})$ is never less that the greatest utility f_j among the states more plausible than i. Particularly, if the state 1 is the most plausible with $f_1 = 1$ then we have $f = (1, 0, \cdots, 0) \sim (1, \cdots, 1)$. The expected profit in a very plausible state is not affected with the expected losses in less plausible states. The Choquet integral calculates the average of the best consequences for each plausibility level.

Similarly we can define preference relations $\succeq_\oint^-$ using a necessity measure. In such a context for all f, $f \sim_\oint^- f^{\sigma,-}$. In this case, the decision maker is pessimistic since the attractiveness $(f_i^{\sigma,-})$ is never greater that the smallest utility f_j among the states more plausible than i. Particularly, if the state 1 is the most plausible with $f_1 = 0$ then we have $f = (0, 1, \cdots, 1) \sim (0, \cdots, 0)$. The expected profits in the least plausible states cannot compensate the expected losses in more plausible states. In this case the Choquet integral calculates the average of the worst consequences for each plausibility level.

4.2   The case of Choquet integral

Let $\succeq$ be a preference relation on profiles given by the decision maker. In [4] the following axioms are proposed, in the infinite setting, where the set of criteria is replaced by a continuous set of states $\mathcal{S}$:

A1 $\succeq$ is non trivial complete preorder.
A2 Continuity according to uniform monotone convergence
    A2.1 $[f_n, f, g \in \mathcal{V}, f_n \succeq g, f_n \downarrow^u f] \Rightarrow f \succeq g$;
    A2.2 $[f_n, f, g \in \mathcal{V}, g \succeq f_n, f_n \uparrow^u f] \Rightarrow g \succeq f$;
A3 If $f \geq g + \epsilon$ where ϵ is a positive constant function then $f \succ g$
A4 Comonotonic independence: If f, g, h are profiles such that f and h, and g and h are comonotone, then: $f \succeq g \Leftrightarrow f + h \succeq g + h$

And we have the following result [4]:

Theorem 10. *A preference relation $\succeq$ satisfies axioms $A1 - A4$ if and only if there exists a capacity μ such that C_μ represents the preference relation. This capacity is unique.*

The notion of pessimistic and optimistic profile can be extended to the continuous case. A possibility distribution on $\mathcal{S}$ defines a complete plausibility preordering $\leq_\pi$ on $\mathcal{S}$, and given an act f, we can define its pessimistic counterpart as $f^{\leq_\pi, -}(s) = \inf_{s \leq_\pi s'} f(s')$. Let us add the pessimistic axiom:

N4 There is a complete plausibility preordering $\leq_\pi$ on $\mathcal{S}$ such that $f \sim f^{\leq_\pi, -}$.

Similarly an optimistic axiom $\Pi 4$ can be written, using optimistic counterparts of profiles $f^{\leq_\pi, +}(s) = \sup_{s \leq_\pi s'} f(s')$.

$\Pi 4$ There is a complete plausibility preordering $\leq_\pi$ on $\mathcal{S}$ such that $f \sim f^{\leq_\pi, +}$.

We can conjecture the following result for necessity measures:

Theorem 11. *A preference relation $\succeq$ satisfies axioms $A1 - A4$ and $N4$ if and only if there exists a necessity measure N such that C_N represents the preference relation. This necessity measure is unique.*

The proof comes down to showing that the unique capacity obtained from axioms $A1 - A4$ is a necessity measure. However, it is not so easy to prove in the infinite setting. Indeed, a necessity measure then must satisfy the infinite minitivity axiom, $N(\cap_{i \in I}) = \inf_{i \in I} N(A_i)$, for any index set I, which ensures the existence of a possibility distribution underlying the capacity. But it is not clear how to extend Lemma 2 to infinite families of sets. As its stands, Lemma 2 only justifies finite minitivity. The same difficulty arises for the optimistic counterpart of the above tentative result. In a finite setting, the permutation σ that indicates the relative plausibility of states can be extracted from the preference relation on profiles, by observing special ones. More precisely, $C_N(1, \ldots, 1, 0, 1, \ldots 1) = 1 - \pi_i$ (the 0 in the case i) in the pessimistic case, and $C_\Pi(0, \ldots, 0, 1, 0, \ldots 0) = \pi_i$ in the optimistic case. This fact would still hold in the form $C_N(\mathcal{C} \backslash \{s\}) = 1 - \pi(s)$ and $C_\Pi(\{s\}) = \pi(s)$, respectively, with infinite minitivity (resp. maxitivity).

4.3   A New Characterisation for Qualitative Possibilistic Integrals

The axiomatization of Sugeno q-integrals in the style of Savage was carried out in [7]. Here, acts are just functions f from $\mathcal{C}$ to a set of consequences X. The axioms proposed are as follows, where xAf is the act such that $(xAf)_i = x$ if $i \in A$ and f_i otherwise, $x \in X$ being viewed as a constant act:

A1 $\succeq$ is a non trivial complete preorder.
WP3 $\forall A \subseteq \mathcal{C}, \forall x, y \in X, \forall f, x \geq y$ implies $xAf \succeq yAf$,
RCD: if f is constant, $f \succ h$ and $g \succ h$ imply $f \wedge g \succ h$
RDD: if f is a constant act, $h \succ f$ and $h \succ g$ imply $h \succ f \vee g$.

Act $f \vee g$ makes the best of f and g, such that $\forall s \in \mathcal{S}, f \vee g(s) = f(s)$ if $f(s) \succeq g(s)$ and $g(s)$ otherwise; and act $f \wedge g$, makes the worst of f and g, such that $\forall s \in \mathcal{S}, f \wedge g(s) = f(s)$ if $g(s) \succeq f(s)$ and $g(s)$ otherwise. We recall here the main results about this axiomatization for decision under uncertainty.

Theorem 12. *[7]: The following propositions are equivalent:*

– $(X^{\mathcal{C}}, \succeq)$ *satisfies A1, WP3, RCD, RDD.*
– *there exists a finite chain of preference levels L, an L-valued monotonic set-function μ, and an L-valued utility function u on X, such that $f \succeq g$ if and only if $\int_\mu (u \circ f) \geq \int_\mu (u \circ g)$.*

In the case of a Sugeno q-integral with respect to a possibility measure, RDD is replaced by the stronger axiom of disjunctive dominance **DD**:

Axiom DD: $\forall f, g, h, h \succ f$ and $h \succ g$ imply $h \succ f \vee g$

and we get a similar result as the above theorem, whereby $f \succeq g$ if and only if $\int_\Pi (u \circ f) \geq \int_\Pi (u \circ g)$ for a possibility measure Π [8].

In the case of a Sugeno q-integral with respect to a necessity measure, RCD is replaced by the stronger axiom of conjunctive dominance **CD**:

Axiom CD: $\forall f, g, h, f \succ h$ and $g \succ h$ imply $f \wedge g \succ h$

and we get a similar result as the above Theorem 12, whereby $f \succeq g$ if and only if $\int_N (u \circ f) \geq \int_N (u \circ g)$ for a necessity measure Π [8].

We can then replace the above representation results by adding to the characteristic axioms for Sugeno q-integrals on a preference relation between acts the same axioms based on pessimistic and optimistic profiles as the ones that, added to characteristic axioms of Choquet integrals lead to a characterisation of preference structures driven by possibilistic Choquet integrals.

Theorem 13. *The following propositions are equivalent:*

– $(X^{\mathcal{C}}, \succeq)$ *satisfies A1, WP3, RCD, RDD and $\Pi 4$*
– *there exists a finite chain of preference levels L, an L-valued possibility measure Π, and an L-valued utility function u on X, such that $f \succeq g$ if and only if $\int_\Pi (u \circ f) \geq \int_\Pi (u \circ g)$.*

Theorem 14. *The following propositions are equivalent:*

– $(X^{\mathcal{C}}, \succeq)$ *satisfies A1, WP3, RCD, RDD and N4*
– *there exists a finite chain of preference levels L, an L-valued necessity measure N, and an L-valued utility function u on X, such that $f \succeq g$ if and only if $\int_N (u \circ f) \geq \int_N (u \circ g)$.*

The reason for the validity of those theorems in the case of Sugeno q-integral is exactly the same as the reason for the validity of Theorems 1, 2, 6 and 7, adding $\Pi 4$ (resp. N4) to the representation theorem of Sugeno q-integral forces the capacity to be a possibility (resp. necessity) measure. However, this method seems to be unavoidable to axiomatize Choquet integrals for possibility and necessity measures as they are not maxitive nor minitive. In the case of possibilistics q-integrals, the maxitivity or minitivity property of the preference functional makes it possible to propose more choices of axioms. However, it is interesting to notice that the same axioms are instrumental to specialize Sugeno and Choquet integrals to possibility and necessity measures.

5 Conclusion

This paper proposes an original axiomatization of discrete Choquet integrals with respect to possibility and necessity measures, and shows that it is enough to add, to existing axiomatisations of general instances of Choquet integrals, a property of equivalence between profiles, that singles out possibility or necessity measures. Remarkably, this property, which also says that the decision-maker only considers relevant the relative importance of single criteria, is qualitative in nature and can thus be added as well to axiom systems for Sugeno integrals, to yield qualitative weighted min and max aggregation operations, as well as for their ordinal preference setting à la Savage. We suggest these results go beyond Sugeno integrals and apply to more general qualitative functionals. One may wonder if this can be done for the ordinal preference setting of the last section, by changing axioms RCD or RDD using right-conjunctions and their semi-duals.

References

1. Ben Amor, N., Fargier, H., Guezguez, W.: Possibilistic sequential decision making. Int. J. Approximate Reasoning **55**(5), 1269–1300 (2014)
2. Borzová-Molnárová, J., Halčinová, L., Hutník, O.: The smallest semicopula-based universal integrals, part I. Fuzzy Sets Syst. **271**, 1–17 (2015)
3. Chateauneuf, A., Grabisch, M., Rico, A.: Modeling attitudes toward uncertainty through the use of the Sugeno integral. J. Math. Econ. **44**(11), 1084–1099 (2008)
4. Chateauneuf, A.: Modeling attitudes towards uncertainty and risk through the use of Choquet integral. Ann. Oper. Res. **52**, 3–20 (1994)
5. Cooman, G.: Integration and conditioning in numerical possibility theory. Ann. Math. Artif. Intell. **32**(1–4), 87–123 (2001)
6. Dubois, D., Prade, H., Rico, A.: Residuated variants of Sugeno integrals. Inf. Sci. **329**, 765–781 (2016)
7. Dubois, D., Prade, H., Sabbadin, R.: Qualitative decision theory with Sugeno integrals. In: Proceedings of 14th UAI Conference, pp. 121–128 (1998)
8. Dubois, D., Prade, H., Sabbadin, R.: Decision-theoretic foundations of qualitative possibility theory. Eur. J. Oper. Res. **128**, 459–478 (2001)
9. Dubois, D., Marichal, J.-L., Prade, H., Roubens, M., Sabbadin, R.: The use of the discrete Sugeno integral in decision making: a survey. Int. J. Uncertainty Fuzziness Knowl.-Based Syst. **9**, 539–561 (2001)
10. Graba, F., Strauss, O.: An interval-valued inversion of the non-additive interval-valued F-transform: use for upsampling a signal. Fuzzy Sets Syst. **288**, 26–45 (2016)
11. Grabisch, M., Labreuche, C.: A decade of application of the Choquet and Sugeno intégrals in multi-criteria décision aid. Ann. Oper. Res. **175**, 247–286 (2010)
12. Marichal, J.-L.: An axiomatic approach of the discrete Choquet Integral as a tool to aggregate interacting criteria. IEEE Trans. Fuzzy Syst. **8**, 800–807 (2000)
13. Mesiar, R., Šipoš, J.: A theory of fuzzy measures: integration and its additivity. IJGS **23**(1), 49–57 (1993)

14. Schmeidler, D.: Cores of exact games. J. Math. Anal. Appl. **40**(1), 214–225 (1972)
15. Schmeidler, D.: Integral représentations without additivity. Proc. Am. Math. Soc. **97**, 255–261 (1986)
16. Sugeno, M.: Theory of fuzzy integrals and its applications. Ph.D. thesis, Tokyo Institute of Technology (1974)
17. Sugeno, M.: Fuzzy measures and fuzzy integrals: a survey. In: Gupta, M.M., et al. (eds.) Fuzzy Automata and Decision Processes, pp. 89–102. North-Holland (1977)

An Equivalent Definition of Pan-Integral

Yao Ouyang[1(✉)] and Jun Li[2]

[1] Faculty of Science, Huzhou University, Huzhou 313000, Zhejiang, China
oyy@zjhu.edu.cn
[2] School of Sciences, Communication University of China, Beijing 100024, China
lijun@cuc.edu.cn

Abstract. In this note, we introduce the concepts of support disjoint-ness super-$\oplus$-additivity and positively super-$\otimes$-homogeneity of a functional (with respect to pan-addition $\oplus$ and pan-multiplication $\otimes$, respectively). By means of these two properties of functionals, we discuss the characteristics of pan-integrals and present an equivalent definition of the pan-integral. As special cases, we obtain the equivalent definitions of the Shilkret integral, the $+, \cdot$-based pan-integral, and the Sugeno integral.

Keywords: Pan-integral · Sugeno integral · Shilkret integral · Support disjointness super-$\oplus$-additivity · Positively super-$\otimes$-homogeneity

1 Introduction

In non-additive measure theory, several prominent nonlinear integrals, for example, the Choquet integral [3] and the Sugeno [12] integral, have been defined and discussed in detail [4,10,16].

As a generalization of the Legesgue integral and Sugeno integral, Yang [17] introduced the pan-integral with respect to a monotone measure and a com-mutative isotonic semiring $(\overline{R}_+, \oplus, \otimes)$, where $\oplus$ is a pan-addition and $\otimes$ a pan-multiplication [16,17]. The researches on this topic can be also found in [1,5,8–10,13,18].

On the other hand, Lehrer introduced a new kind of nonlinear integral — the concave integral with respect to a capacity, see [6,7,14]. Let $(X, \mathcal{A})$ be a measurable space and $\mathcal{F}_+$ denote the class of all finite nonnegative real-valued measurable functions on $(X, \mathcal{A})$. For fixed capacity ν, the concave integral with respect to ν is a concave and positively homogeneous nonnegative functional on $\mathcal{F}_+$. Observe that such integral was defined as the infimum taken over all concave and positively homogeneous nonnegative functionals H defined on $\mathcal{F}_+$ with the condition: $\forall A \in \mathcal{A}, H(\chi_A) \geq \mu(A)$.

Inspiration received from the definition of concave integral, we try to characterize the pan-integrals via functionals over $\mathcal{F}_+$ (with some additional restricts). We introduce the concepts of *support disjointness super-$\oplus$-additivity* and *positively super-$\otimes$-homogeneity* of a functional on $\mathcal{F}_+$ (with respect to pan-addition $\oplus$ and pan-multiplication $\otimes$, respectively). We will show the pan-integral, as

V. Torra et al. (Eds.): MDAI 2016, LNAI 9880, pp. 107–113, 2016.
DOI: 10.1007/978-3-319-45656-0_9

a functional defined on $\mathcal{F}_+$, is support disjointness super-$\oplus$-additive and positively super-$\otimes$-homogeneous. We shall present an equivalent definition of the pan-integral by using monotone, support disjointness super-$\oplus$-additive and positively super-$\otimes$-homogeneous functionals on $\mathcal{F}_+$.

2 Preliminaries

Let X be a nonempty set and $\mathcal{A}$ a σ-algebra of subsets of X, $R_+ = [0, +\infty)$, $\overline{R}_+ = [0, +\infty]$. Recall that a set function $\mu : \mathcal{A} \to \overline{R}_+$ is a monotone measure, if it satisfies the following conditions:

(1) $\mu(\emptyset) = 0$ and $\mu(X) > 0$;
(2) $\mu(A) \le \mu(B)$ whenever $A \subset B$ and $A, B \in \mathcal{A}$.

In this paper we restrict our discussion on a fixed measurable space $(X, \mathcal{A})$. Unless stated otherwise all the subsets mentioned are supposed to belong to $\mathcal{A}$. Let $\mathcal{M}$ be the set of all monotone measures defined on $(X, \mathcal{A})$. When μ is a monotone measure, the triple $(X, \mathcal{A}, \mu)$ is called a monotone measure space [10, 16].

The concept of a pan-integral involves two binary operations, the pan-addition $\oplus$ and pan-multiplication $\otimes$ of real numbers [16, 17].

Definition 1. *An binary operation $\oplus$ on $\overline{R}_+$ is called a pan-addition if it satisfies the following requirements:*

(PA1) $a \oplus b = b \oplus a = a$ (commutativity);
(PA2) $(a \oplus b) \oplus c = a \oplus (b \oplus c)$ (associativity);
(PA3) $a \le c$ and $b \le d$ imply that $a \oplus b \le c \oplus d$ (monotonicity);
(PA4) $a \oplus 0 = a$ (neutral element);
(PA5) $a_n \to a$ and $b_n \to b$ imply that $a_n \oplus b_n \to a \oplus b$ (continuity).

Definition 2. *Let $\oplus$ be a given pan-addition on $\overline{R}_+$. A binary operation $\otimes$ on $\overline{R}_+$ is said to be a pan-multiplication corresponding to $\oplus$ if it satisfies the following properties:*

(PM1) $a \otimes b = b \otimes a$ (commutativity);
(PM2) $(a \otimes b) \otimes c = a \otimes (b \otimes c)$ (associativity);
(PM3) $a \otimes (b \oplus c) = (a \otimes b) \oplus (a \otimes c)$ (distributive law);
(PM4) $a \le b$ implies $(a \otimes c) \le (b \otimes c)$ for any c (monotonicity);
(PM5) $a \otimes b = 0 \Leftrightarrow a = 0$ or $b = 0$ (annihilator);
(PM6) there exists $e \in [0, \infty]$ such that $e \otimes a = a$ for any $a \in [0, \infty]$ (neutral element);
(PM7) $a_n \to a \in [0, \infty)$ and $b_n \to b \in [0, \infty)$ imply $(a_n \otimes b_n) \to (a \otimes b)$ (continuity).

When $\oplus$ is a pseudo-addition on $\overline{R}_+$ and $\otimes$ is a pseudo-multiplication (with respect to $\oplus$) on $\overline{R}_+$, the triple $(\overline{R}_+, \oplus, \otimes)$ is called a *commutative isotonic semiring* (with respect to $\oplus$ and $\otimes$) [16].

Notice that similar operations called pseudo-addition and pseudo-multiplication can be found in the literature [1, 2, 5, 8–10, 13, 15, 18].

In the following, we recall the concept of *pan-integral* [16, 17].

Definition 3. Consider a commutative isotonic semiring $(\overline{R}_+, \oplus, \otimes)$. Let $\mu \in \mathcal{M}$ and $f \in \mathcal{F}_+$. *The pan-integral of f on X with respect to μ is defined via*

$$\mathbf{I}_{pan}^{(\oplus,\otimes)}(\mu, f) = \sup\left\{\bigoplus_{i=1}^{n}\left(\lambda_i \otimes \mu(A_i)\right) : \bigoplus_{i=1}^{n}\left(\lambda_i \otimes \chi_{A_i}\right) \leq f, \{A_i\}_{i=1}^{n} \in \mathcal{P}\right\},$$

where χ_A is the characteristic function of A which takes value e on A and 0 elsewhere, and $\mathcal{P}$ is the set of all finite partitions of X.

For $A \in \mathcal{A}$, the pan-integral of f on A is defined by $\mathbf{I}_{pan}^{(\oplus,\otimes)}(\mu, f \otimes \chi_A)$.

Note: A finite partition of X is a finite disjoint system of sets $\{A_i\}_{i=1}^{n} \subset \mathcal{A}$ such that $A_i \cap A_j = \emptyset$ for $i \neq j$ and $\cup_{i=1}^{n} A_i = X$.

Note that in the case of commutative isotonic semiring $(\overline{R}_+, \vee, \wedge)$, Sugeno integral [12] is recovered, while for $(\overline{R}_+, \vee, \cdot)$, Shilkret integral [11] is covered by the pan-integral in Definition 3.

Proposition 1. *Consider a commutative isotonic semiring $(\overline{R}_+, \oplus, \otimes)$ and fixed $\mu \in \mathcal{M}$. Then $\mathbf{I}_{pan}^{(\oplus,\otimes)}(\mu, \cdot)$, as a functional on $\mathcal{F}_+$, is monotone, i.e., for any $f, g \in \mathcal{F}_+$,*

$$f \leq g \implies \mathbf{I}_{pan}^{(\oplus,\otimes)}(\mu, f) \leq \mathbf{I}_{pan}^{(\oplus,\otimes)}(\mu, g).$$

Proposition 2. *For any $A \in \mathcal{F}$, $\mathbf{I}_{pan}^{(\oplus,\otimes)}(\mu, \chi_A) \geq \mu(A)$.*

3 Main Results

In this section we present an equivalent definition of pan-integral. In order to do it, we first introduce two new concepts and show two lemmas.

Definition 4. Consider a commutative isotonic semiring $(\overline{R}_+, \oplus, \otimes)$. A functional $F : \mathcal{F}_+ \to \overline{R}_+$ is said to be

(i) *positively super-$\otimes$-homogeneous*, if for any $f \in \mathcal{F}_+$ and any $a > 0$, we have

$$F(a \otimes f) \geq a \otimes F(f). \tag{1}$$

(ii) *support disjointness super-$\oplus$-additive*, if for any $f, g \in \mathcal{F}_+$, $supp(f) \cap supp(g) = \emptyset$, we have

$$F(f \oplus g) \geq F(f) \oplus F(g), \tag{2}$$

here $supp(f) = \{x \in X : f(x) > 0\}$ since we do not concern the topology.

Lemma 1. *Consider a commutative isotonic semiring $(\overline{R}_+, \oplus, \otimes)$ and fixed $\mu \in \mathcal{M}$. Then $\mathbf{I}_{pan}^{(\oplus,\otimes)}(\mu, \cdot)$, as a functional on $\mathcal{F}_+$, is positively super-$\otimes$-homogeneous, i.e., for any $f \in \mathcal{F}_+$ and any $a > 0$, we have*

$$\mathbf{I}_{pan}^{(\oplus,\otimes)}(\mu, a \otimes f) \geq a \otimes \mathbf{I}_{pan}^{(\oplus,\otimes)}(\mu, f). \tag{3}$$

Proof. For any finite partition $\{A_1, \ldots, A_n\}$ of X and $\{\lambda_1, \ldots, \lambda_n\} \subset R_+$ with $\bigoplus_{i=1}^n (\lambda_i \otimes \chi_{A_i}) \leq f$, we have that $\bigoplus_{i=1}^n \left((a \otimes \lambda_i) \otimes \chi_{A_i}\right) \leq a \otimes f$. Thus,

$$\mathbf{I}_{pan}^{(\oplus,\otimes)}(\mu, a \otimes f)$$

$$= \sup\left\{\bigoplus_{j=1}^m \left(\beta_j \otimes \mu(B_j)\right) : \bigoplus_{j=1}^m \left(\beta_j \otimes \chi_{B_j}\right) \leq a \otimes f, \{B_j\}_{j=1}^m \in \mathcal{P}\right\}$$

$$\geq \sup\left\{\bigoplus_{i=1}^n \left((a \otimes \lambda_i) \otimes \mu(A_i)\right) : \bigoplus_{i=1}^n \left((a \otimes \lambda_i) \otimes \chi_{A_i}\right) \leq a \otimes f, \{A_i\}_{i=1}^n \in \mathcal{P}\right\}$$

$$= \sup\left\{a \otimes \bigoplus_{i=1}^n \left(\lambda_i \otimes \mu(A_i)\right) : a \otimes \bigoplus_{i=1}^n \left(\lambda_i \otimes \chi_{A_i}\right) \leq a \otimes f, \{A_i\}_{i=1}^n \in \mathcal{P}\right\}$$

$$\geq a \otimes \sup\left\{\bigoplus_{i=1}^n \left(\lambda_i \otimes \mu(A_i)\right) : \bigoplus_{i=1}^n \left(\lambda_i \otimes \chi_{A_i}\right) \leq f, \{A_i\}_{i=1}^n \in \mathcal{P}\right\}$$

$$= a \otimes \mathbf{I}_{pan}^{(\oplus,\otimes)}(\mu, f). \qquad \square$$

Remark 1. Notice that for the commutative isotonic semiring $(\overline{R}_+, \oplus, \cdot)$, i.e., $\otimes$ is the usual multiplication, then the related pan-integral is *positively homogeneous*, i.e.,

$$\mathbf{I}_{pan}^{(\oplus,\cdot)}(\mu, af) = a \cdot \mathbf{I}_{pan}^{(\oplus,\cdot)}(\mu, f).$$

In fact, by Lemma 1, $\mathbf{I}_{pan}^{(\oplus,\cdot)}(\mu, af) \geq a\, \mathbf{I}_{pan}^{(\oplus,\cdot)}(\mu, f)$. On the other hand, $\mathbf{I}_{pan}^{(\oplus,\cdot)}(\mu, f) = \mathbf{I}_{pan}^{(\oplus,\cdot)}(\mu, \frac{1}{a}(af)) \geq \frac{1}{a}\mathbf{I}_{pan}^{(\oplus,\cdot)}(\mu, af)$, which implies the reverse inequality and hence the equality holds.

Lemma 2. *Consider a commutative isotonic semiring $(\overline{R}_+, \oplus, \otimes)$ and fixed $\mu \in \mathcal{M}$. Then $\mathbf{I}_{pan}^{(\oplus,\otimes)}(\mu, \cdot)$, as a functional on $\mathcal{F}_+$, is support disjointness super-$\oplus$-additive, i.e., for any $f, g \in \mathcal{F}_+$ such that $supp(f) \cap supp(g) = \emptyset$, we have*

$$\mathbf{I}_{pan}^{(\oplus,\otimes)}(\mu, f \oplus g) \geq \mathbf{I}_{pan}^{(\oplus,\otimes)}(\mu, f) \oplus \mathbf{I}_{pan}^{(\oplus,\otimes)}(\mu, g). \tag{4}$$

Proof. If one of the two integrals on the right-hand side of Ineq. (4) is infinite then, by the monotonicity of the pan-integral, $\mathbf{I}_{pan}^{(\oplus,\otimes)}(\mu, f \oplus g)$ also equals to infinity, which implies that (4) holds.

So, without loss of generality, we can suppose that both $\mathbf{I}_{pan}^{(\oplus,\otimes)}(\mu, f)$ and $\mathbf{I}_{pan}^{(\oplus,\otimes)}(\mu, g)$ are finite. Let $l_n \nearrow \mathbf{I}_{pan}^{(\oplus,\otimes)}(\mu, f)$ and $r_n \nearrow \mathbf{I}_{pan}^{(\oplus,\otimes)}(\mu, f)$ be two sequences of real number. Then, for each n, there is a partition $\{A_i^{(n)}\}_{i=1}^{k_n}$ of

$supp(f)$, a partition $\{B_j^{(n)}\}_{j=1}^{m_n}$ of $supp(g)$, and two sequences of positive number $\{\alpha_i^{(n)}\}_{i=1}^{k_n}$ and $\{\beta_j^{(n)}\}_{j=1}^{m_n}$ such that $\bigoplus_{i=1}^{k_n}(\alpha_i^{(n)}\otimes\chi_{A_i^{(n)}}) \leq f$, $\bigoplus_{j=1}^{m_n}(\beta_j^{(n)}\otimes\chi_{B_j^{(n)}}) \leq g$ and both the following two inequalities hold

$$\bigoplus_{i=1}^{k_n}\left(\alpha_i^{(n)}\otimes\mu(A_i^{(n)})\right) \geq l_n, \qquad \bigoplus_{j=1}^{m_n}\left(\beta_j^{(n)}\otimes\mu(B_j^{(n)})\right) \geq r_n.$$

By the fact of $supp(f) \cap supp(g) = \emptyset$, we know that $\{A_i^{(n)}\}_{i=1}^{k_n} \cup \{B_j^{(n)}\}_{j=1}^{m_n}$ is a partition of $supp(f \oplus g)$. Moreover, we have that

$$\left(\bigoplus_{i=1}^{k_n}(\alpha_i^{(n)}\otimes\chi_{A_i^{(n)}})\right) \oplus \left(\bigoplus_{j=1}^{m_n}(\beta_j^{(n)}\otimes\chi_{B_j^{(n)}})\right) \leq f \oplus g,$$

and

$$\mathbf{I}_{pan}^{(\oplus,\otimes)}(\mu, f \oplus g) \geq \left(\bigoplus_{i=1}^{k_n}(\alpha_i^{(n)}\otimes\mu(A_i^{(n)}))\right) \oplus \left(\bigoplus_{j=1}^{m_n}(\beta_j^{(n)}\otimes\mu(B_j^{(n)}))\right)$$
$$\geq l_n \oplus r_n.$$

Letting $n \to \infty$, by the continuity of pan-addition, we get that

$$\mathbf{I}_{pan}^{(\oplus,\otimes)}(\mu, f \oplus g) \geq \mathbf{I}_{pan}^{(\oplus,\otimes)}(\mu, f) \oplus \mathbf{I}_{pan}^{(\oplus,\otimes)}(\mu, g).$$

The proof is complete. $\square$

Consider a commutative isotonic semiring $(\overline{R}_+, \oplus, \otimes)$. Let $\mathcal{C}_{\oplus,\otimes}$ be the set of all nonnegative, monotone, positively super-$\otimes$-homogeneous and support disjointness super-$\oplus$-additive functionals on $\mathcal{F}_+$.

The following is our main result which provides an equivalent definition of the pan-integral.

Theorem 1. *Consider a commutative isotonic semiring* $(\overline{R}_+, \oplus, \otimes)$ *and fixed* $\mu \in \mathcal{M}$. *Then for any* $f \in \mathcal{F}_+$,

$$\mathbf{I}_{pan}^{(\oplus,\otimes)}(\mu, f) = \inf\left\{F(f) : F \in \mathcal{C}_{\oplus,\otimes}, \forall A \in \mathcal{A}, F(\chi_A) \geq \mu(A)\right\}.$$

Proof. By Propositions 1 and 2, Lemmas 1 and 2, we know that $\mathbf{I}_{pan}^{(\oplus,\otimes)}(\mu, \cdot) : \mathcal{F}^+ \to [0, \infty]$ is monotone, positively super-$\otimes$-homogeneous, support disjointness super-$\oplus$-additive, i.e., $\mathbf{I}_{pan}^{(\oplus,\otimes)}(\mu, \cdot) \in \mathcal{C}_{\oplus,\otimes}$ and $\mathbf{I}_{pan}^{(\oplus,\otimes)}(\mu, \chi_A) \geq \mu(A)$ for any $A \in \mathcal{A}$. Therefore,

$$\mathbf{I}_{pan}^{(\oplus,\otimes)}(\mu, f) \geq \inf\left\{F(f) : F \in \mathcal{C}_{\oplus,\otimes}, \forall A \in \mathcal{A}, F(\chi_A) \geq \mu(A)\right\}.$$

On the other hand, for any $f \in \mathcal{F}^+$, any $\bigoplus_{i=1}^n (\lambda_i \otimes \chi_{A_i}) \leq f$ and any $F \in \mathcal{C}_{\oplus,\otimes}$ with $F(\chi_A) \geq \mu(A), \forall A \in \mathcal{A}$, we have

$$F(f) \geq F\Big(\bigoplus_{i=1}^n (\lambda_i \otimes \chi_{A_i}) \Big) \geq \bigoplus_{i=1}^n F\Big(\lambda_i \otimes \chi_{A_i} \Big)$$

$$\geq \bigoplus_{i=1}^n \Big(\lambda_i \otimes F(\chi_{A_i}) \Big) \geq \bigoplus_{i=1}^n \Big(\lambda_i \otimes \mu(A_i) \Big).$$

Thus,

$$F(f) \geq \sup \Big\{ \bigoplus_{i=1}^n (\lambda_i \otimes \mu(A_i)) : \bigoplus_{i=1}^n (\lambda_i \otimes \chi_{A_i}) \leq f \Big\} = \mathbf{I}_{pan}^{(\oplus,\otimes)}(\mu, f).$$

By the arbitrariness of F, we infer that

$$\inf \Big\{ F(f) : F \in \mathcal{C}_{\oplus,\otimes}, \forall A \in \mathcal{A}, F(\chi_A) \geq \mu(A) \Big\} \geq \mathbf{I}_{pan}^{(\oplus,\otimes)}(\mu, f),$$

which proves the conclusion.   □

Let $\mathcal{C}_{\oplus,\cdot}^{(1)}$ be the set of nonnegative, monotone, positively homogeneous and support disjointness super-$\oplus$-additive functionals on $\mathcal{F}^+$. Then $\mathcal{C}_{\oplus,\cdot}^{(1)} \subset \mathcal{C}_{\oplus,\cdot}$. Noting that $\mathbf{I}_{pan}^{(\oplus,\cdot)}(\mu, \cdot): \mathcal{F}^+ \to [0, \infty]$ is positively homogeneous (Remark 1), then $\mathbf{I}_{pan}^{(\oplus,\cdot)}(\mu, \cdot) \in \mathcal{C}_{\oplus,\cdot}^{(1)}$. Thus we have the following result.

Theorem 2. *Let $(X, \mathcal{F}, \mu)$ be a monotone measure space. Then for any $f \in \mathcal{F}_+$,*

$$\mathbf{I}_{pan}^{(\oplus,\cdot)}(\mu, f) = \inf \Big\{ F(f) : F \in \mathcal{C}_{\oplus,\cdot}^{(1)}, \forall A \in \mathcal{A}, F(\chi_A) \geq \mu(A) \Big\}.$$

Let the commutative isotonic semiring be $(\overline{R}_+, \vee, \cdot)$. Noticing that $\mathbf{I}_{pan}^{(\vee,\cdot)}(\mu, \chi_A) = \mu(A), \forall A \in \mathcal{A}$, by Theorem 2, we get an equivalent definition for the Shilkret integral.

Corollary 1. *Let $(X, \mathcal{F}, \mu)$ be a monotone measure space. Then for any $f \in \mathcal{F}_+$,*

$$\mathbf{I}_{pan}^{(\vee,\cdot)}(\mu, f) = \inf \Big\{ F(f) : F \in \mathcal{C}_{\vee,\cdot}^{(1)}, \forall A \in \mathcal{A}, F(\chi_A) = \mu(A) \Big\}.$$

If we let $\oplus = +$, then we get an equivalent definition for the usual addition and multiplication based pan-integral.

Corollary 2. *Let $(X, \mathcal{F}, \mu)$ be a monotone measure space. Then for any $f \in \mathcal{F}_+$,*

$$\mathbf{I}_{pan}^{(+,\cdot)}(\mu, f) = \inf \Big\{ F(f) : F \in \mathcal{C}_{+,\cdot}^{(1)}, \forall A \in \mathcal{A}, F(\chi_A) \geq \mu(A) \Big\}.$$

Noting that the Sugeno integral is positively $\wedge$-homogeneous [16] and satisfies $\mathbf{I}_{pan}^{(\vee,\wedge)}(\mu, \chi_A) = \mu(A), \forall A \in \mathcal{A}$, we also have the following result.

Corollary 3. *Let* $(X, \mathcal{F}, \mu)$ *be a monotone measure space. Then for any* $f \in \mathcal{F}_+$,

$$\mathbf{I}_{pan}^{(\vee, \wedge)}(\mu, f) = \inf \left\{ F(f) : F \in \mathcal{C}_{\vee, \wedge}^{(1)}, \forall A \in \mathcal{A}, F(\chi_A) = \mu(A) \right\}.$$

Acknowledgements. This research was partially supported by the National Natural Science Foundation of China (Grant No. 11371332 and No. 11571106) and the NSF of Zhejiang Province (No. LY15A010013).

References

1. Benvenuti, P., Mesiar, R., Vivona, D.: Monotone set functions-based integrals. In: Pap, E. (ed.) Handbook of Measure Theory, vol. II, pp. 1329–1379. Elsevier, Amsterdam (2002)
2. Benvenuti, P., Mesiar, R.: Pseudo-arithmetical operations as a basis for the general measure and integration theory. Inf. Sci. **160**, 1–11 (2004)
3. Choquet, G.: Theory of capacities. Ann. Inst. Fourier **5**, 131–295 (1954)
4. Grabisch, M., Murofushi, T., Sugeno, M. (eds.): Fuzzy Measures and Integrals: Theory and Applications. Studies in Fuzziness Soft Computing, vol. 40. Physica, Heidelberg (2000)
5. Ichihashi, H., Tanaka, M., Asai, K.: Fuzzy integrals based on pseudo-additions and multiplications. J. Math. Anal. Appl. **130**, 354–364 (1988)
6. Lehrer, E.: A new integral for capacities. Econ. Theor. **39**, 157–176 (2009)
7. Lehrer, E., Teper, R.: The concave integral over large spaces. Fuzzy Sets Syst. **159**, 2130–2144 (2008)
8. Mesiar, R.: Choquet-like integrals. J. Math. Anal. Appl. **194**, 477–488 (1995)
9. Mesiar, R., Rybárik, J.: Pan-operaions structure. Fuzzy Sets Syst. **74**, 365–369 (1995)
10. Pap, E.: Null-Additive Set Functions. Kluwer, Dordrecht (1995)
11. Shilkret, N.: Maxitive measure and integration. Indag. Math. **33**, 109–116 (1971)
12. Sugeno, M.: Theory of fuzzy integrals and its applications. Ph.D. dissertation, Takyo Institute of Technology (1974)
13. Sugeno, M., Murofushi, T.: Pseudo-additive measures and integrals. J. Math. Anal. Appl. **122**, 197–222 (1987)
14. Teper, R.: On the continuity of the concave integral. Fuzzy Sets Syst. **160**, 1318–1326 (2009)
15. Tong, X., Chen, M., Li, H.X.: Pan-operations structure with non-idempotent pan-addition. Fuzzy Sets Syst. **145**, 463–470 (2004)
16. Wang, Z., Klir, G.J.: Generalized Measure Theory. Springer, Berlin (2009)
17. Yang, Q.: The pan-integral on fuzzy measure space. Fuzzy Math. **3**, 107–114 (1985). (in Chinese)
18. Zhang, Q., Mesiar, R., Li, J., Struk, P.: Generalized Lebesgue integral. Int. J. Approx. Reason. **52**, 427–443 (2011)

Monotonicity and Symmetry of IFPD Bayesian Confirmation Measures

Emilio Celotto[1], Andrea Ellero[1], and Paola Ferretti[2(✉)]

[1] Department of Management, Ca' Foscari University of Venice,
Fondamenta San Giobbe, Cannaregio 873, 30121 Venezia, Italy
{celotto,ellero}@unive.it
[2] Department of Economics, Ca' Foscari University of Venice,
Fondamenta San Giobbe, Cannaregio 873, 30121 Venezia, Italy
ferretti@unive.it

Abstract. IFPD confirmation measures are used in ranking inductive rules in Data Mining. Many measures of this kind have been defined in literature. We show how some of them are related to each other via weighted means. The special structure of IFPD measures allows to define also new monotonicity and symmetry properties which appear quite natural in such context. We also suggest a way to measure the degree of symmetry of IFPD confirmation measures.

Keywords: Confirmation measures · IFPD · Monotonicity · Symmetry · Degree of symmetry

1 Introduction

The effects of a piece of knowledge E on a conclusion H can be conveyed as an inductive rule in the form $E \to H$. When expressing the content of a dataset via inductive rules, the strengths of the rules need to be compared and ranked. The most natural way to scoring them is combining quantities which measure the probability change of conclusion H: it is possible to measure the degree to which E supports or contradicts H using prior probability $P(H)$ and posterior probability $P(H|E)$ and the probability $P(E)$ of evidence E. What is needed is a *confirmation measure* $C(H, E)$ which evaluates the degree to which a piece of evidence E provides *evidence for or against* or *support for or against* conclusion H.

As soon as the evidence E occurs, the knowledge changes and conclusion H may be confirmed, when $P(H|E) > P(H)$, or disconfirmed, when $P(H|E) < P(H)$. A Bayesian Confirmation measure $C(H, E)$ is required to possess the following properties:

- $C(H, E) > 0$ if $P(H|E) > P(H)$ *(confirmation case)*
- $C(H, E) = 0$ if $P(H|E) = P(H)$ *(neutrality case)*
- $C(H, E) < 0$ if $P(H|E) < P(H)$ *(disconfirmation case)*

© Springer International Publishing Switzerland 2016
V. Torra et al. (Eds.): MDAI 2016, LNAI 9880, pp. 114–125, 2016.
DOI: 10.1007/978-3-319-45656-0_10

Confirmation measures have been deeply explored in literature from different perspectives (see e.g. [4,6,8]). We focus on special Bayesian confirmation measures called Initial Final Probability Dependence (IFPD) (for more details see [4]). IFPD property identifies confirmation measures that are in some sense *essential* since they use only prior and posterior probabilities to evaluate the rules. The analysis of the special class of IFPD confirmation measures results to be interesting with reference to the recent use of visualization techniques that are particularly meaningful for the comprehension and selection of different interestingness measures (see [20,21]). In the same way, the study of analytical properties of confirmation measures provides some useful insights on the discrimination between different measures ([7,13]).

Actually, some recent confirmation measures were developed by Rough Sets Theory specialists ([13]). Throughout the paper we will consider as benchmark examples four well known confirmation measures, which turn out to be used in *jMAF* (see [2,15]), a well-established software for Rough Set based Decision Support Systems:

- $G(H, E) = \log \left[\dfrac{P(E|H)}{P(E|\neg H)} \right]$ defined by Good in [10]

- $K(H, E) = \dfrac{P(E|H) - P(E|\neg H)}{P(E|H) + P(E|\neg H)}$ defined by Kemeny and Oppenheim in [14]

- $Z(H, E) = \begin{cases} Z_1(H, E) = \dfrac{P(H|E) - P(H)}{1 - P(H)} & \text{in case of confirmation} \\ Z_2(H, E) = \dfrac{P(H|E) - P(H)}{P(H)} & \text{in case of disconfirmation} \end{cases}$

  first defined by Rescher in [17] and reproposed, e.g., in [4,13,22]

- $A(H, E) = \begin{cases} A_1(H, E) = \dfrac{P(H) - P(H|\neg E)}{P(H)} & \text{in case of confirmation} \\ A_2(H, E) = \dfrac{P(H) - P(H|\neg E)}{1 - P(H)} & \text{in case of disconfirmation} \end{cases}$

  proposed by Greco, Słowiński and Szczęch in [13].

The above formulas allow to observe that measure G, even though originally expressed as a function of likelihoods of the hypotheses $P(E|H)$ and $P(E|\neg H)$, can be expressed as a function of only $P(H)$ and $P(H|E)$, the so-called prevalence and confidence (or precision) of the rule:

$$G(H, E) = \log \left[\frac{P(H|E)}{P(H)} \frac{[1 - P(H)]}{[1 - P(H|E)]} \right].$$

The same happens for K

$$K(H, E) = \frac{P(H|E) - P(H)}{P(H|E) - 2P(H|E)P(H) + P(H)}$$

and Z

$$Z(H, E) = \begin{cases} Z_1(H, E) = \dfrac{P(H|E) - P(H)}{1 - P(H)} & \text{in case of confirmation} \\ Z_2(H, E) = \dfrac{P(H|E) - P(H)}{P(H)} & \text{in case of disconfirmation.} \end{cases}$$

Many Bayesian confirmation measures are, indeed, defined as functions of $P(H|E)$ and $P(H)$ only, e.g., justification measures (see [18]), although some of them take explicitly into account the value of $P(E)$ (see e.g. [1]). To express A as a function of prevalence and confidence, in fact, we cannot avoid to include in the definition the probability $P(E)$ of evidence E, the so called coverage of the rule. It is possible to rewrite it in terms of $P(H)$, $P(H|E)$ and $P(E)$:

$$A(H, E) = \begin{cases} A_1(H, E) = \dfrac{P(E)}{1 - P(E)} \dfrac{P(H|E) - P(H)}{P(H)} & \text{in case of confirmation} \\ A_2(H, E) = \dfrac{P(E)}{1 - P(E)} \dfrac{P(H|E) - P(H)}{1 - P(H)} & \text{in case of disconfirmation.} \end{cases}$$

Therefore G, K and Z are examples of IFPD confirmation measures, while this is not true for A.

An IFPD confirmation measure thus, is a real-valued function of two variables $x = P(H|E)$ and $y = P(H)$, which coherently can be analysed with usual calculus techniques: we will use this observation to outline some of the properties of confirmation measures.

2   Relationships Between Confirmation Measures

Several confirmation measures are defined and used in many fields (see, e.g., [8]), notwithstanding their different names some of them conceal the same measure. Moreover some truly different measures provide the same ranking among rules, like Kemeny's K and Good's G: in this case the measures are said to be ordinally equivalent [4].

Nevertheless, even not equivalent measures, like K and Z, may exhibit rather similar outcomes. Remarkably, an algebraic investigation of the analytical expressions of those measures reveals that it is possible to rewrite K in terms of Z. In particular, K can be rewritten in a rather natural way, as a weighted harmonic mean of the expressions for Z in case of confirmation (which was indicated above by Z_1) and disconfirmation (Z_2). Moreover, in a similar way K can also be expressed using the two expressions that define measure A.

Let us go into some more details.

As reported in [4], the idea beyond the definition of Z was to calculate the relative reduction of the distance from certainty, i.e., to determine to what extent the probability distance from certainty concerning the truth of H is reduced by a confirming piece of evidence E.

The way in which we will link the expressions of K and Z seems to allow insights into the meaning of those measures. Let us first formally extend the use of function Z_1 also to the case of disconfirmation, and that of Z_2 to the case of confirmation, so that we can readily write K as a weighted harmonic mean of those functions, with weights $w_1 = P(H|E)$ and $w_2 = 1 - P(H|E)$, respectively,[1]

$$K(H, E) = \left(\frac{P(H|E)}{Z_1(H, E)} + \frac{1 - P(H|E)}{Z_2(H, E)} \right)^{-1} \qquad (1)$$

or

$$K(H, E) = \frac{P(H|E) - P(H)}{P(H|E)(1 - P(H)) + (1 - P(H|E))P(H)}.$$

This way K can be interpreted as a synthesis of the two expressions for Z putting higher weight to the confirmation formula when the confidence of the rule $P(H|E)$ is high, and to the disconfirmation formula when confidence is low. We can observe that K, as an harmonic mean, will be more stable with respect to possible extreme values of Z_1 or Z_2 (when $P(H|E)$ is either rather close to 1 or to 0).

Similar considerations hold for confirmation measure A which also requires the use of two different expressions in the cases of confirmation and disconfirmation. By extending the domains of functions A_1 and A_2 to both the cases of confirmation and disconfirmation, we obtain

$$K(H, E) = \frac{1 - P(E)}{P(E)} \left(\frac{1 - P(H|E)}{A_1(H, E)} + \frac{P(H|E)}{A_2(H, E)} \right)^{-1} \qquad (2)$$

Note that a factor depending on the probability of evidence E is now required. K can be viewed in terms of the two formulas for A_1 and A_2, but the role played by confidence is now upset: a higher weight is assigned to the confirmation formula when the confidence of the rule is low and to the disconfirmation formula when confidence is high.

Once realized that K can be expressed as a weighted harmonic mean of Z_1 and Z_2, or of A_1 and A_2, it seems interesting to observe other, similar, relationships. For example, the simplest confirmation measure, the difference confirmation measure $d(H, E) = P(H|E) - P(H)$ (see Carnap [3]), can be expressed as the harmonic mean of Z_1 and Z_2. Note that weighted (harmonic) means of Bayesian confirmation measures clearly provide new Bayesian confirmation measures. In Table 1 we propose some examples of weighted harmonic means of Z_1 and Z_2 by considering the first weight w_1 set equal to $P(H|E)$, $1 - P(H|E)$, $P(H)$, $1 - P(H)$, $P(E)$, $1 - P(E)$, respectively.

The last two measures, hwm_4 and hwm_5, are the only not-IFPD confirmation measures of the set. Some of those measures correspond, or are ordinally equivalent, to confirmation measures which have been already defined in the literature; it is obvious that this way we can also easily think up completely new

---

[1] As a matter of fact, when the evidence E disconfirms conclusion H, i.e. $P(H|E) < P(H)$, both Z_1 and Z_2 assume a negative value: strictly speaking their harmonic mean is not defined, but the proposed link (1) among measures holds, with the same meaning. In the neutrality case we have the boundary values $K = Z = 0$ and their link cannot be defined by a harmonic mean like (1).

Table 1. Confirmation measures as means of Z_1 and Z_2

Name	w_1	w_2	Formula	Kind of mean
d Carnap	$1/2$	$1/2$	$2[P(H\|E) - P(H)]$	Harmonic
K - Kemeny	$P(H\|E)$	$1 - P(H\|E)$	$\frac{P(H\|E)-P(H)}{P(H\|E)-2P(H\|E)P(H)+P(H)}$	w. harmonic
hwm_1	$1 - P(H\|E)$	$P(H\|E)$	$\frac{P(H\|E)-P(H)}{2P(H\|E)P(H)-P(H\|E)-P(H)+1}$	w. harmonic
hwm_2	$P(H)$	$1 - P(H)$	$1/2\frac{P(H\|E)-P(H)}{P(H)(1-P(H))}$	w. harmonic
hwm_3	$1 - P(H)$	$P(H)$	$\frac{P(H\|E)-P(H)}{2P(H)^2-2P(H)+1}$	w. harmonic
hwm_4	$P(E)$	$1 - P(E)$	$\frac{P(H\|E)-P(H)}{-2P(H)P(E)+P(H)+P(E)}$	w. harmonic
hwm_5	$1 - P(E)$	$P(E)$	$\frac{P(H\|E)-P(H)}{2P(H)P(E)-P(H)-P(E)+1}$	w. harmonic

measures, using different means which can be found in literature, or to choose weights in order to calibrate a specific confirmation measure with specific properties. More in general one could also exploit the large variety of aggregation functions in order to obtain measures which satisfy desired properties (see, e.g., [11]).

Clearly, the new generated rules may or may not have some additional properties which are often requested for Confirmation measures. In the following we restrict our attention to monotonicity and symmetry properties in terms of $P(H|E)$, $P(H)$ and $P(E)$, primarily focusing on IFPD measures.

3   Monotonicity

Since monotonicity and symmetry properties of confirmation measures are usually expressed in contingency table notation, let us recall how the rules $E \to H$ induced from a dataset on a universe U, can be represented involving the contingency table notation, as in Table 2

Table 2. Contingency table

	H	$\neg H$	Σ
E	a	c	$a + c$
$\neg E$	b	d	$b + d$
Σ	$a + b$	$c + d$	a+b+c+d

where $a = \sup(H, E)$, $b = \sup(H, \neg E)$, $c = \sup(\neg H, E)$, $d = \sup(\neg H, \neg E)$ and $\sup(B, A)$ denotes the support of the rule $A \to B$, i.e., the number of elements in the dataset for which both the premise A and the conclusion B of the rule are true (see Greco et al. in [13]). In this way the cardinality of the universe U is $|U| = a + b + c + d$.

By estimating probabilities in terms of frequencies, e.g. $P(H|E) = \frac{a}{a+c}$, $P(H) = \frac{a+b}{|U|}$, $P(E) = \frac{a+c}{|U|}$, the four confirmation measures considered above as our benchmark examples admit the following representation

- $G(a, b, c, d) = \log \left[\frac{a(c+d)}{c(a+b)} \right]$
- $K(a, b, c, d) = \frac{ad-bc}{ad+bc+2ac}$
- $Z(a, b, c, d) = \begin{cases} Z_1(a, b, c, d) = \frac{ad-bc}{(c+d)(a+c)} & \text{in case of confirmation} \\ Z_2(a, b, c, d) = \frac{ad-bc}{(a+b)(a+c)} & \text{in case of disconfirmation} \end{cases}$
- $A(a, b, c, d) = \begin{cases} A_1(a, b, c, d) = \frac{ad-bc}{(a+b)(b+d)} & \text{in case of confirmation} \\ A_2(a, b, c, d) = \frac{ad-bc}{(c+d)(b+d)} & \text{in case of disconfirmation.} \end{cases}$

Several definitions of monotonicity for interestingness measures with different meanings have been proposed in literature (see e.g. [8] in particular with reference to interestingness measures in the data mining framework). Greco et al. in [12] suggest that a confirmation measure should be not decreasing with respect to both a and d and not increasing with respect to both b and c in the contingency table. In other words the proposed property of monotonicity requires that

Monotonicity. $C(H, E)$ is monotonic (M) if it is a function not decreasing with respect to both $P(E \cap H)$ and $P(\neg E \cap \neg H)$, non increasing with respect to both $P(E \cap H)$ and $P(\neg E \cap \neg H)$.

Since IFPD confirmation measures can directly be defined in terms of confidence and prevalence of a rule, we propose here a new monotonicity property for confirmation measures which is set in terms of confidence $P(H|E)$ and prevalence $P(H)$ only:

Confidence Prevalence Monotonicity. An IFPD confirmation measure C satisfies Confidence Prevalence Monotonicity (CPM) if it is non decreasing with respect to confidence $P(H|E)$ and non increasing with respect to prevalence $P(H)$.

Monotonicity with respect to confidence means that any higher value in $P(H|E)$ increases or at lest does not decrease the credibility of the decision rule $E \to H$; again, monotonicity with respect to prevalence reflects the idea that any higher value for $P(H)$ decreases or at least does not increase the credibility of the rule $E \to H$. Observe that in [16] Piatetsky-Shapiro proposed three principles that should be obeyed by any interestingness measure which only partially overlap with (CPM) in the case of IFPD confirmation measures.

Property (CPM) for an IFPD confirmation measure appears to be quite a natural assumption. Moreover property (M) implies (CPM), more precisely:

Proposition 1. *If an IFPD confirmation rule C satisfies Monotonicity property (M) then C satisfies Confidence Prevalence Monotonicity (CPM).*

To prove the proposition, note that a confirmation measure $C(H, E)$, expressed in terms of $x = P(H|E) = \frac{a}{a+c}$ and $y = P(H) = \frac{a+b}{|U|}$, can be considered as a compound function $F(a, b, c, d)$ of the entries a, b, c and d of the contingency table. By considering the associated real function in the differentiable case, the chain rule formula allows to set a link between the different monotonicity properties, that is

$$F'_a = C'_x \left[\frac{c}{(a+c)^2} \right] + C'_y \left[\frac{c+d}{|U|^2} \right]$$
$$F'_b = C'_y \left[\frac{c+d}{|U|^2} \right]$$
$$F'_c = C'_x \left[\frac{-a}{(a+c)^2} \right] + C'_y \left[\frac{-(a+b)}{|U|^2} \right]$$
$$F'_d = C'_y \left[\frac{-(a+b)}{|U|^2} \right].$$

The monotonicity properties required in Greco et al. [12] imply in particular that $C'_x \geq 0$ and $C'_y \leq 0$ as it can be deduced by referring to the monotonic dependence of x and y on a, b, c and d: $x'_a \geq 0$, $x'_c \leq 0$; $y'_a \geq 0$, $y'_b \geq 0$, $y'_c \leq 0$, $y'_d \leq 0$. Note that, conversely, (CPM) monotonicity property doesn't imply the validity of monotonicity properties required in Greco et al. [12] in (M): in fact by considering $a = 100$, $b = c = d = 10$ and Carnap's confirmation measure, if the value of a increases to 101 then $d(x, y) = x - y$ decreases, while it is clearly increasing with respect to x and decreasing as a function of y.

4   Symmetries

Also symmetry properties of confirmation measures have been discussed in the literature (see e.g. [9,13]) observing that some of them should be required while some other ones should be avoided. We propose three quite natural symmetry definitions in the framework of IFPD confirmation measures. The proposed definitions are suggested by recalling simple geometric symmetry properties and only one of them turns out to coincide with the classical *hypothesis symmetry* (see [3]) which by the way is considered a desirable property by the literature (see e.g. [5]). The first definition we propose is inspired by skew-symmetry, where we again use the notation $x = P(H|E)$ and $y = P(H)$.

Prior Posterior Symmetry. An IFPD confirmation measure C satisfies Prior Posterior Symmetry (PPS) if

$$C(x, y) = -C(y, x).$$

The skew-symmetric condition characterizing *PPS* is due to a sign constraint: when an order relation between x and y is encountered, necessarily the symmetric point with respect to the main diagonal requires an evaluation which is opposite in sign.

Example. Consider a situation in which $P(H) = 0.2$ and that, given that evidence E is true, the probability of hypothesis H increases to $P(H|E) = 0.5$, with a confirmation measure $C(0.5, 0.2)$; the Prior Posterior Symmetry requires the same disconfirmation evaluation if, given $P(H) = 0.5$, the evidence E lowers the probability of H down to $P(H|E) = 0.2$, i.e. $C(0.2, 0.5) = -C(0.5, 0.2)$.

In the next two definitions the considered points (x, y) and $(1 - y, 1 - x)$ are symmetric with respect to the line $y = 1 - x$.

Complementary Probability Symmetry. An IFPD confirmation measure C satisfies Complementary Probability Symmetry (CPS) if

$$C(x, y) = C(1 - y, 1 - x).$$

Example. Consider $P(H) = 0.8$ and assume that, given evidence E, the probability of hypothesis H becomes $P(H|E) = 0.3$, with confirmation $C(0.3, 0.8)$; the Complementary Probability Symmetry requires the same (dis)confirmation level if, given $P(\neg H) = 1 - P(H) = 0.2$, evidence E increases the probability of $\neg H$ to $P(\neg H|E) = 1 - P(H|E) = 0.7$, i.e. $C(0.7, 0.2) = -C(0.3, 0.8)$.

The next definition (see [3] where it is named *hypothesis symmetry*) is clearly related to both the previous ones

Probability Centre Symmetry. An IFPD confirmation measure C satisfies Probability Center Symmetry (PCS) if

$$C(x, y) = -C(1 - x, 1 - y).$$

Note that in this case each point (x, y) is compared to its symmetric counterpart with respect to the point $(0.5, 0.5)$, namely the point $(1 - x, 1 - y)$. We can observe, for example, that Carnap's d, Good's G, K by Kemeny and Oppenheim all satisfy each of the above defined symmetries. Instead Rescher's Z satisfies only PCS symmetry.

Kemeny's measure K, which is also a normalized confirmation measure, appears to be a particular smooth and balanced average of Z_1 and Z_2: its monotonicity and symmetry properties strengthen the idea of an equilibrated IFPD confirmation measure.

4.1 Degree of Symmetry

Various confirmation measures that are proposed in literature have not all the properties satisfied by Kemeny's K, like those proposed by Shogenji [18] or Crupi et al. [4]. This is why we are interested in evaluating also the degree of symmetry of an IFPD confirmation measure. At this aim, let us first remark that for any given confirmation measure C it is possible to consider a transposed confirmation function C^T (see [19] where an analogous approach has been considered for studying the degree of exchangeability of continuous identically distributed random variables) which is defined by

$$C^T(x, y) = C(y, x).$$

In this way, it is possible to restate the PPS condition as

$$C \text{ satisfies } PPS \iff C = -C^T.$$

Observe that any confirmation measure admits the following decomposition

$$C = C_{PPS} + \overline{C}_{PPS}$$

with

$$C_{PPS}(x,y) = \frac{C(x,y) - C^T(x,y)}{2} \qquad \overline{C}_{PPS}(x,y) = \frac{C(x,y) + C^T(x,y)}{2}$$

where C_{PPS} and $\overline{C}_{PPS}$ satisfy the relations

$$C_{PPS} = -C_{PPS}^T \qquad \overline{C}_{PPS} = \overline{C}_{PPS}^T.$$

In this way any function C admits a decomposition into a PPS confirmation function C_{PPS} and a complementary function $\overline{C}_{PPS}$ and, if C satisfies PPS, then necessarily $\overline{C}_{PPS} \equiv 0$ and $C = C_{PPS}$.

By referring now to the involution function i for which $i(x,y) = (1-x, 1-y)$ the CPS condition can be expressed as

$$C \text{ satisfies } CPS \iff C = C^T \circ i.$$

It is possible to rewrite any confirmation measure as a sum of a CPS confirmation measure C_{CPS} and of a complementary function $\overline{C}_{CPS}$

$$C = C_{CPS} + \overline{C}_{CPS}$$

where

$$C_{CPS}(x,y) = \frac{C(x,y) + C^T(i(x,y))}{2} \qquad \overline{C}_{CPS}(x,y) = \frac{C(x,y) - C^T(i(x,y))}{2}.$$

Clearly

$$C_{CPS} = C_{CPS}^T \circ i \qquad \overline{C}_{CPS} = -\overline{C}_{CPS}^T \circ i.$$

Finally,

$$C \text{ satisfies } PCS \iff C = -C \circ i.$$

and by setting

$$C_{PCS}(x,y) = \frac{C(x,y) - C(i(x,y))}{2} \qquad \overline{C}_{PCS}(x,y) = \frac{C(x,y) + C(i(x,y))}{2}$$

where

$$C_{PCS} = -C_{PCS} \circ i \qquad \overline{C}_{PCS} = \overline{C}_{PCS} \circ i$$

we can again express C as a sum of functions, one of which is now a PCS confirmation measure C_{PCS}

$$C = C_{PCS} + \overline{C}_{PCS}.$$

Table 3 presents the symmetric components of some well known IFPD confirmation measures and of some of the new measures presented in Table 1. Clearly, the symmetric component coincides with the corresponding confirmation measures when it satisfies the corresponding symmetry property. In Table 3 we consider, in particular, Shogenji's justification measures J [18] defined as

$$J(H,E) = 1 - \ln[P(H|E)]/\ln[P(H)]$$

which is an example of an IFPD confirmation measure that does not satisfy any of the proposed symmetries.

Observe that, calling σ any of the above defined symmetries, it is now possible to evaluate the *degree of symmetry of an IFPD measure* C with respect to σ by defining the norm $\|\cdot\|_\sigma$ (see [19])

$$\|C\|_\sigma = \frac{\|C_\sigma\|^2 - \|\overline{C}_\sigma\|^2}{\|C\|^2}$$

where $\|\cdot\|$ is a norm on a function space $\mathcal{F}$ of real valued functions. Note that when C itself satisfies PPS, CPS or PCS, then necessarily $\overline{C}_{PPS} = 0$, $\overline{C}_{CPS} = 0$, $\overline{C}_{PCS} = 0$ so that $\|C\|_{PPS}=1$, $\|C\|_{CPS}=1$, $\|C\|_{PCS}=1$. In general it is $\|C\|_\sigma \leq 1$.

Table 3. Symmetry properties of IFPD confirmation measures: a repetition of measure's name indicates that the measure itself is symmetric.

Measure C	C_{PPS}	C_{CPS}	C_{PCS}
d	d	d	d
K	K	K	K
hwm_1	hwm_1	hwm_1	hwm_1
hwm_2	$\frac{1}{4}\left[\frac{(x-y)}{y(1-y)} + \frac{(x-y)}{x(1-x)}\right]$	$\frac{1}{4}\left[\frac{(x-y)}{y(1-y)} + \frac{(x-y)}{x(1-x)}\right]$	hwm_2
hwm_3	$\frac{1}{2}\left[\frac{(x-y)}{2y^2-2y+1} + \frac{x-y}{2x^2-2x+1}\right]$	$\frac{1}{2}\left[\frac{x-y}{2y^2-2y+1} + \frac{(x-y)}{2x^2-2x+1}\right]$	hwm_3
Z	$\begin{cases}\frac{1}{2}\frac{(x-y)(x-y+1)}{x(1-y)}, & x>y;\\[4pt] \frac{1}{2}\frac{(x-y)(y-x+1)}{y(1-x)}, & x<y.\end{cases}$	$\begin{cases}\frac{1}{2}\frac{(x-y)(x-y+1)}{x(1-y)}, & x>y;\\[4pt] \frac{1}{2}\frac{(x-y)(y-x+1)}{y(1-x)}, & x<y.\end{cases}$	Z
G	G	G	G
J	$\frac{1}{2}\left[\frac{\log y}{\log x} - \frac{\log x}{\log y}\right]$	$1 - \frac{1}{2}\left[\frac{\log x}{\log y} + \frac{\log(1-y)}{\log(1-x)}\right]$	$\frac{1}{2}\left[\frac{\log(1-x)}{\log(1-y)} - \frac{\log x}{\log y}\right]$

5 Conclusions

In this paper we investigated IFPD confirmation measures, some of their relationships, which have also a practical application, for example when inductive rules need to be compared (see, e.g., [13]). Those measures are sometimes ordinally equivalent and also when this does not happen it turns out that the links among them are rather strong. It is in fact possible to observe that some measures can be obtained using some means of other ones: an interesting extension would be to make use of the huge amount of more general aggregation functions (see, e.g., [11]), which could be the subject of a further promising research, as far as we can see.

The new measures, as told, can be defined in order to have desired properties, as their particular use may require. But, in practice, which properties do/should the new defined measures possess? The debate on that is not new, of course. Here we tried to exploit the special structure of IFPD confirmation measures, to study some of those properties. We focused in particular on a new monotonicity definition, which appears to be a quite natural request for IFPD confirmation measures. The particular structure of IFPD confirmation measures also allows to define symmetry properties which are inspired directly by classical geometric symmetries on the plane $(P(H|E), P(E))$, in this way providing a way to identify their symmetric component and also to suggest a way to measure their degree of symmetry. Some examples concerning both old and newly defined IFPD measures illustrate the possibilities of future development, but further investigations are needed to study the defined properties on an extended set of IFPD confirmation measures.

References

1. Atkinson, D.: Confirmation and justification. A commentary on Shogenji's measure. Synthese **184**, 49–61 (2012)
2. Błaszczyński, J., Greco, S., Matarazzo, B., Słowiński, R.: jMAF - dominance-based rough set data analysis framework. In: Skowron, A., Suraj, Z. (eds.) Rough Sets and Intelligent Systems - Professor Zdzisław Pawlak in Memoriam. ISRL, vol. 42, pp. 185–209. Springer, Heidelberg (2013)
3. Carnap, R.: Logical Foundations of Probability, 2nd edn. University of Chicago Press, Chicago (1962)
4. Crupi, V., Festa, R., Buttasi, C.: Towards a grammar of Bayesian confirmation. In: Suárez, M., Dorato, M., Rédei, M. (eds.) Epistemology and Methodology of Science, pp. 73–93. Springer, Dordrecht (2010)
5. Eells, E., Fitelson, B.: Symmetries and asymmetries in evidential support. Philos. Stud. **107**, 129–142 (2002)
6. Fitelson, B.: The plurality of Bayesian measures of confirmation and the problem of measure sensitivity. Philos. Sci. **66**, 362–378 (1999)
7. Fitelson, B.: Likelihoodism, Bayesianism, and relational confirmation. Synthese **156**, 473–489 (2007)
8. Geng, L., Hamilton, H.J.: Interestingness measures for data mining: a survey. ACM Comput. Surv. **38**, 1–32 (2006)
9. Glass, D.H.: Entailment and symmetry in confirmation measures of interestingness. Inf. Sci. **279**, 552–559 (2014)
10. Good, I.J.: Probability and the Weighing of Evidence. Charles Griffin, London (1950)
11. Grabisch, M., Marichal, J.-L., Mesiar, R., Pap, E.: Aggregation Functions. Cambridge University Press, Cambridge (2009)
12. Greco, S., Pawlak, Z., Słowiński, R.: Can Bayesian confirmation measures be useful for rough set decision rules? Eng. Appl. Artif. Intell. **17**, 345–361 (2004)

13. Greco, S., Słowiński, R., Szczęch, I.: Properties of rule interestingness measures and alternative approaches to normalization of measures. Inf. Sci. **216**, 1–16 (2012)
14. Kemeny, J., Oppenheim, P.: Degrees of factual support. Philos. Sci. **19**, 307–324 (1952)
15. Laboratory of Intelligent Decision Support Systems of the Poznan University of Technology. http://idss.cs.put.poznan.pl/site/software.html
16. Piatetsky-Shapiro, G.: Discovery, Analysis, and Presentation of Strong Rules, pp. 229–248. AAAI/MIT Press, Cambridge (1991)
17. Rescher, N.: A theory of evidence. Philos. Sci. **25**, 83–94 (1958)
18. Shogenji, T.: The degree of epistemic justification and the conjunction fallacy. Synthese **184**, 29–48 (2012)
19. Siburg, K.F., Stoimenov, P.A.: Symmetry of functions and exchangeability of random variables. Stat. Pap. **52**, 1–15 (2011)
20. Susmaga, R., Szczech, I.: The property of χ^2_{01}-concordance for Bayesian confirmation measures. In: Torra, V., Narukawa, Y., Navarro-Arribas, G., Megías, D. (eds.) MDAI 2013. LNCS, vol. 8234, pp. 226–236. Springer, Heidelberg (2013)
21. Susmaga, R., Szczech, I.: Can interestingness measures be usefully visualized? Int. J. Appl. Math. Comput. Sci. **25**, 323–336 (2015)
22. Wu, X., Zhang, C., Zhang, S.: Efficient mining of both positive and negative association rules. ACM Trans. Inf. Syst. **22**, 381–405 (2004)

About the Use of Admissible Order for Defining Implication Operators

M. Asiain[1], Humberto Bustince[2,3]([✉]), B. Bedregal[4], Z. Takáč[5], M. Baczyński[6], D. Paternain[2], and G.P. Dimuro[7]

[1] Dept. de Matemáticas,Universidad Pública de Navarra,
Campus Arrosadia, s/n, 31.006 Pamplona, Spain
asiain@unavarra.es
[2] Dept. de Automática y Computación, Universidad Pública de Navarra,
Campus Arrosadia, s/n, 31.006 Pamplona, Spain
{bustince,daniel.paternain}@unavarra.es
[3] Institute of Smart Cities, Universidad Pública de Navarra,
Campus Arrosadia, s/n, 31.006 Pamplona, Spain
[4] Departamento de Informática e Matemática Aplicada,
Universidade Federal do Rio Grande do Norte,
Campus Universitario, s/n, Lagoa Nova, Natal CEP 59078-970, Brazil
bedregal@dimap.ufm.br
[5] Institute of Information Engineering, Automation and Mathematics,
Slovak University of Technology in Bratislava, Radlinskeho 9, Bratislava, Slovakia
zdenko.takac@stuba.sk
[6] Institute of Mathematics, University of Silesia,
ul. Bankowa 14, 40-007 Katowice, Poland
michal.baczynski@us.edu.pl
[7] Centro de Ciências Computacionais, Universidade Federal do Rio Grande,
Av. Itália, km 08, Campus Carreiros, 96201-900 Rio Grande, Brazil
gracaliz@furg.br

Abstract. Implication functions are crucial operators for many fuzzy logic applications. In this work, we consider the definition of implication functions in the interval-valued setting using admissible orders and we use this interval-valued implications for building comparison measures.

Keywords: Interval-valued implication operator · Admissible order · Similarity measure

1 Introduction

Implication operators are crucial for many applications of fuzzy logic, including approximate reasoning or image processing. Many works have been devoted to

H. Bustince was supported by Project TIN2013-40765-P of the Spanish Government. Z. Takáč was supported by Project VEGA 1/0420/15. B. Bedregal and G. Dimuro were supported by Brazilian funding agency CNPQ under Processes 481283/2013-7, 306970/2013-9, 232827/2014-1 and 307681/2012-2.

V. Torra et al. (Eds.): MDAI 2016, LNAI 9880, pp. 126–134, 2016.
DOI: 10.1007/978-3-319-45656-0_11

the analysis of these operators, both in the case of fuzzy sets [1,2,14,15] and in the case of extensions [3–5,13,16]. A key problem in order to define these operators is that of monotonicity. When implication operators are extended to fuzzy extensions, this problem is not trivial, since for most of the fuzzy extensions do not exist a linear order, whereas for some applications, as it is the case of fuzzy rules-based classification systems, it is necessary to have the possibility of comparing any two elements [12].

In this work, we propose the definition of implication operators in the interval-valued setting defining its monotonicity in terms of the so-called admissible orders [11]. This is a class of linear orders which extends the usual order between intervals and which include the most widely used examples of linear orders between intervals, as lexicographical and Xu and Yager ones.

As a first step in a deeper study of these interval-valued implications with admissible orders, we show how implications which are defined in terms of admissible orders can be used to build comparison measures which are of interest from the point of view of applications.

The structure of the present work is as follows. In Sect. 2 we present some preliminary definitions and results. In Sect. 3 we present the definition of interval-valued implication function with respect to an admissible order. Section 4 is devoted to obtaining equivalence and restricted equivalence functions with respect to linear orders. In Sect. 5 we use our previous results to build comparison measures. We finish with some conclusions and references.

2 Preliminaries

In this section we introduce several well known notions and results which will be useful for our subsequent developments.

We are going to work with closed subintervals of the unit interval. For this reason, we define:

$$L([0,1]) = \{[\underline{X}, \overline{X}] \mid 0 \leq \underline{X} \leq \overline{X} \leq 1\}.$$

By $\leq_L$ we denote an arbitrary order relation on $L([0,1])$ with $0_L = [0,0]$ as its minimal element and $1_L = [1,1]$ as maximal element. This order relation can be partial or total. If we must consider an arbitrary total order, we will denote it by $\leq_{TL}$.

Example 1. The partial order relation on $L([0,1])$ induced by the usual partial order in $\mathbb{R}^2$ is:

$$[\underline{X}, \overline{X}] \precsim_L [\underline{Y}, \overline{Y}] \text{ if } \underline{X} \leq \underline{Y} \text{ and } \overline{X} \leq \overline{Y}. \tag{1}$$

As an example of total order in $L([0,1])$ we have Xu and Yager's order (see [17]):

$$[\underline{X}, \overline{X}] \leq_{XY} [\underline{Y}, \overline{Y}] \text{ if } \begin{cases} \underline{X} + \overline{X} < \underline{Y} + \overline{Y} \text{ or} \\ \underline{X} + \overline{X} = \underline{Y} + \overline{Y} \text{ and } \overline{X} - \underline{X} \leq \overline{Y} - \underline{Y}. \end{cases} \tag{2}$$

Definition 1. *An admissible order in $L([0,1])$ is a total order $\leq_{TL}$ which extends the partial order $\precsim_L$.*

In the following, whenever we speak of a total order we assume it is an admissible order.

Definition 2. *Let $\leq_L$ be an order relation in $L([0,1])$. A function $N \colon L([0,1]) \to L([0,1])$ is an interval-valued negation function (IV negation) if it is a decreasing function with respect to the order $\leq_L$ such that $N(0_L) = 1_L$ and $N(1_L) = 0_L$. A negation N is called strong negation if $N(N(X)) = X$ for every $X \in L([0,1])$. A negation N is called non-filling if $N(X) = 1_L$ iff $X = 0_L$, while N is called non-vanishing if $N(X) = 0_L$ iff $X = 1_L$.*

We recall now the definition of interval-valued aggregation function.

Definition 3. *Let $n \geq 2$. An (n-dimensional) interval-valued (IV) aggregation function in $(L([0,1]), \leq_L, 0_L, 1_L)$ is a mapping $M \colon (L([0,1]))^n \to L([0,1])$ which verifies:*

(i) $M(0_L, \cdots, 0_L) = 0_L$.
(ii) $M(1_L, \cdots, 1_L) = 1_L$.
(iii) M is an increasing function with respect to $\leq_L$.

Example 2. Fix $\alpha \in [0,1]$. With the order $\leq_{XY}$, the function

$$M_\alpha \colon L([0,1])^2 \to L([0,1])$$

defined by

$$M_\alpha([\underline{X}, \overline{X}], [\underline{Y}, \overline{Y}]) = [\alpha\underline{X} + (1-\alpha)\underline{Y}, \alpha\overline{X} + (1-\alpha)\overline{Y}]$$

is an IV aggregation function.

3   Interval-Valued Implication Functions

Definition 4 (cf. [5] and [2])**.** *An interval-valued (IV) implication function in $(L([0,1]), \leq_L, 0_L, 1_L)$ is a function $I \colon (L([0,1]))^2 \to L([0,1])$ which verifies the following properties:*

(i) I is a decreasing function in the first component and an increasing function in the second component with respect to the order $\leq_L$.
(ii) $I(0_L, 0_L) = I(0_L, 1_L) = I(1_L, 1_L) = 1_L$.
(iii) $I(1_L, 0_L) = 0_L$.

Some properties that can be demanded to an IV implication function are the following [10]:

I4: $I(X,Y) = 0_L \Leftrightarrow X = 1_L$ and $Y = 0_L$.
I5: $I(X,Y) = 1_L \Leftrightarrow X = 0_L$ or $Y = 1_L$.

NP: $I(1_L, Y) = Y$ for all $Y \in L([0,1])$.

EP: $I(X, I(Y, Z)) = I(Y, I(X, Z))$ for all $X, Y, Z \in L([0,1])$.

OP: $I(X, Y) = 1_L \Leftrightarrow X \leq_L Y$.

SN: $N(X) = I(X, 0_L)$ is a strong IV negation.

$I10$: $I(X, Y) \geq_L Y$ for all $X, Y \in L([0,1])$.

IP: $I(X, X) = 1_L$ for all $X \in L([0,1])$.

CP: $I(X, Y) = I(N(Y), N(X))$ for all $X, Y \in L([0,1])$, where N is an IV negation.

$I14$: $I(X, N(X)) = N(X)$ for all $X \in L([0,1])$, where N is an IV negation.

We can obtain IV implication functions from IV aggregation functions as follows.

Proposition 1. *Let M be an IV aggregation function such that*

$$M(1_L, 0_L) = M(0_L, 1_L) = 0_L$$

and let N be an IV negation in $L([0,1])$, both with respect to the same order $\leq_L$. Then the function $I_M : L([0,1])^2 \to L([0,1])$ given by

$$I_M(X, Y) = N(M(X, N(Y)))$$

is an IV implication function.

Proof. It follows from a straight calculation. □

However, in this work we are going to focus on a different construction method for IV implication functions.

Proposition 2. *Let $\leq_{TL}$ be a total order in $L([0,1])$, and let N be an IV negation function with respect to that order. The function $I : L([0,1])^2 \to L([0,1])$ defined by*

$$I(X, Y) = \begin{cases} 1_L, & \text{if } X \leq_{TL} Y, \\ \vee(N(X), Y), & \text{if } X >_{TL} Y. \end{cases}$$

is an IV implication function.

Proof. It is clear that the function I is an increasing function in the second component and a decreasing function in the first component. Moreover

$$I(0_L, 0_L) = I(0_L, 1_L) = I(1_L, 1_L) = 1_L$$

and $I(1_L, 0_L) = 0_L$. □

This result can be further generalized as follows [15]:

Proposition 3. *Let $\leq_{TL}$ be a total order in $L([0,1])$, and let N be an IV negation function with respect to that order. If $M : L([0,1])^2 \to L([0,1])$ is an IV aggregation function, then the function $I : L([0,1])^2 \to L([0,1])$ defined by*

$$I(X, Y) = \begin{cases} 1_L, & \text{if } X \leq_{TL} Y, \\ M(N(X), Y), & \text{if } X >_{TL} Y, \end{cases}$$

is an IV implication function.

4   Equivalence and Restricted Equivalence Functions in $L([0, 1])$ with Respect to a Total Order

Along this section only total orders are considered.

The equivalence functions [6–8] are a fundamental tool in order to build measures of similarity between fuzzy sets. In this section we construct interval-valued equivalence functions from IV aggregation and negation functions.

Definition 5. *A map $F \colon L([0,1])^2 \to L([0,1])$ is called an interval-valued (IV) equivalence function in $(L([0,1]), \leq_{TL})$ if F verifies:*

(1) $F(X,Y) = F(Y,X)$ for every $X,Y \in L([0,1])$.
(2) $F(0_L, 1_L) = F(1_L, 0_L) = 0_L$.
(3) $F(X,X) = 1_L$ for all $X \in L([0,1])$.
(4) If $X \leq_{TL} X' \leq_{TL} Y' \leq_{TL} Y$, then $F(X,Y) \leq_{TL} F(X',Y')$.

Theorem 1. *Let $M_1 \colon L([0,1])^2 \to L([0,1])$ be an IV aggregation function such that $M_1(X,Y) = M_1(Y,X)$ for every $X,Y \in L([0,1])$, $M_1(X,Y) = 1_L$ if and only if $X = Y = 1_L$ and $M_1(X,Y) = 0_L$ if and only if $X = 0_L$ or $Y = 0_L$. Let $M_2 \colon L([0,1])^2 \to L([0,1])$ be an IV aggregation function such that $M_2(X,Y) = 1_L$ if and only if $X = 1_L$ or $Y = 1_L$ and $M_2(X,Y) = 0_L$ if and only if $X = Y = 0_L$. Then the function $F \colon L([0,1])^2 \to L([0,1])$ defined by*

$$F(X,Y) = M_1(I(X,Y), I(Y,X)),$$

with I the IV implication function defined in the Proposition 3 taking $M = M_2$, is an IV equivalence function.

Proof. Since

$$F(X,Y) = \begin{cases} 1_L, & \text{if } X = Y, \\ M_1(M_2(N(Y), X), 1_L), & \text{if } X <_{TL} Y, \\ M_1(M_2(N(X), Y), 1_L), & \text{if } Y <_{TL} X, \end{cases}$$

then F verifies the four properties in Definition 5.   □

In [8] the definition of equivalence function (in the real case) was modified in order to define the so-called restricted equivalence function. Now we develop a similar study for the case of IV equivalence functions.

Definition 6. *Let N be an IV negation. A map $F \colon L([0,1])^2 \to L([0,1])$ is called an interval valued (IV) restricted equivalence function (in $(L([0,1]), \leq_{TL})$) if F verifies the following properties:*

1. $F(X,Y) = F(Y,X)$ for all $X,Y \in L([0,1])$.
2. $F(X,Y) = 1_L$ if and only if $X = Y$.
3. $F(X,Y) = 0_L$ if and only if $X = 0_L$ and $Y = 1_L$, or, $X = 1_L$ and $Y = 0_L$.
4. $F(X,Y) = F(N(X), N(Y))$ for all $X,Y \in L([0,1])$.
5. If $X \leq_{TL} Y \leq_{TL} Z$, then $F(X,Z) \leq_{TL} F(X,Y)$ and $F(X,Z) \leq_{TL} F(Y,Z)$.

Theorem 2. *Let N be an IV negation function. Let $M_1 \colon L([0,1])^2 \to L([0,1])$ be an IV aggregation function such that $M_1(X,Y) = M_1(Y,X)$ for every $X, Y \in L([0,1])$, $M_1(X,Y) = 1_L$ if and only if $X = Y = 1_L$ and $M_1(X,Y) = 0_L$ if and only if $X = 0_L$ or $Y = 0_L$. Let $M_2 \colon L([0,1])^2 \to L([0,1])$ be an IV aggregation function such that $M_2(X,Y) = 1_L$ if and only if $X = 1_L$ or $Y = 1_L$ and $M_2(X,Y) = 0_L$ if and only if $X = Y = 0_L$. Then the function $F \colon L([0,1])^2 \to L([0,1])$ defined by*

$$F(X,Y) = M_1(I(X,Y), I(Y,X))$$

with I an IV implication function defined by

$$I(X,Y) = \begin{cases} 1_L & \text{if } X \leq_{TL} Y \\ M_2(N(X),Y) & \text{otherwise,} \end{cases}$$

verifies the properties (1) and (5) of Definition 6. Moreover, it satisfies property (2) if N is non-filling and property (3) if N is non-vanishing.

Proof. Since

$$F(X,Y) = \begin{cases} 1_L, & \text{if } X = Y \\ M_1(M_2(N(Y),X), 1_L), & \text{if } X <_{TL} Y \\ M_1(M_2(N(X),Y), 1_L), & \text{if } Y <_{TL} X \end{cases}$$

then F verifies:

(1) $F(X,Y) = F(Y,X)$ trivially.
(5) If $X \leq_{TL} Y \leq_{TL} Z$, then $N(Z) \leq_{TL} N(Y) \leq_{TL} N(X)$. Since M_1 is an increasing function then $F(X,Z) \leq_{TL} F(X,Y)$ and $F(X,Z) \leq_{TL} F(Y,Z)$.

Since $M_1(X,Y) = 1_L$ if and only if $X = Y = 1_L$, then, if N is non-filling, $F(X,Y) = 1_L$ if and only if $X = Y$ because

$$\begin{cases} M_2(N(Y),X) \neq 1_L, & \text{if } X <_{TL} Y \\ M_2(N(X),Y) \neq 1_L, & \text{if } X >_{TL} Y. \end{cases}$$

Moreover, $F(X,Y) = 0_L$ if and only if $X >_{TL} Y$ and $M_2(N(X),Y) = 0_L$ or $X <_{TL} Y$ and $M_2(N(Y),X) = 0_L$. Therefore, as N is non-vanishing, $F(X,Y) = 0_L$ if and only if

$$\begin{cases} X = 0_L \text{ or } Y = 1_L \text{ or} \\ Y = 0_L \text{ or } X = 1_L. \end{cases}$$

with $X \neq Y$. $\square$

5   Similarity Measures, Distances and Entropy Measures in $L([0,1])$ with Respect to a Total Order

Our constructions in the previous section can be used to build comparison measures between interval-valued fuzzy sets, and, more specifically, to obtain similarity measures, distances in the sense of Fang and entropy measures. Along this section, we only deal with a total order $\leq_{TL}$.

To start, let us consider a finite referential set of n elements, $U = \{u_1, \ldots, u_n\}$. We denote by $IVFS(U)$ the set of all interval-valued fuzzy sets over U. Recall that an interval-valued fuzzy set A over U is a mapping $A : U \to L([0,1])$ [9]. Note that the order $\leq_{TL}$ induces a partial order $\leq_{TL}$ in $IVFS(U)$ given, for $A, B \in IVFS(U)$, by

$$A \leq_{TL} B \text{ if } A(u_i) \leq_{TL} B(u_i) \text{ for every } u_i \in U.$$

First of all, we show how we can build a similarity between interval-valued fuzzy sets defined over the same referential U. We start recalling the definition.

Definition 7 [8]. *An interval-valued (IV) similarity measure on $IVFS(U)$ is a mapping $SM : IVFS(U) \times IVFS(U) \to L([0,1])$ such that, for every $A, B, A', B' \in IVFS(U)$,*

(SM1) SM is symmetric.
(SM2) $SM(A, B) = 1_L$ if and only if $A = B$.
(SM3) $SM(A, B) = 0_L$ if and only if $\{A(u_i), B(u_i)\} = \{0_L, 1_L\}$ for every $u_i \in U$.
(SM4) If $A \leq_{TL} A' \leq_{TL} B' \leq_{TL} B$, then $SM(A, B) \leq_{TL} SM(A', B')$.

Then we have the following result.

Theorem 3. *Let $M : L([0,1])^n \to L([0,1])$ be an IV aggregation function with respect to the total order $\leq_{TL}$ and such that $M(X_1, \ldots, X_n) = 1_L$ if and only if $X_1 = \cdots = X_n = 1_L$ and $M(X_1, \ldots, X_n) = 0_L$ if and only if $X_1 = \cdots = X_n = 0_L$. Then, the function $SM : IVFS(U) \times IVFS(U) \to L([0,1])$ given by*

$$SM(A, B) = M(F(A(u_1), B(u_1)), \ldots, F(A(u_n), B(u_n)))$$

where F is defined as in Theorem 2 with non-filling and non-vanishing negation, is an IV similarity measure.

Proof. It follows from a straightforward calculation. □

We can make use of this construction method to recover both distances and entropy measures. First of all, let's recall the definition of both concepts.

Definition 8 [6]. *A function $D : IVFS(U) \times IVFS(U) \to L([0,1])$ is called an IV distance measure on $IVFS(U)$ if, for every $A, B, A', B' \in IVFS(U)$, D satisfies the following properties:*

(D1) $D(A, B) = D(B, A)$;
(D2) $D(A, B) = 0_L$ if and only if $A = B$;
(D3) $D(A, B) = 1_L$ if and only if A and B are complementary crisp sets;
(D4) If $A \leq_{TL} A' \leq_{TL} B' \leq_{TL} B$, then $D(A, B) \geq_{TL} D(A', B')$.

Definition 9 [6]. *A function $E : IVFS(U) \to L([0,1])$ is called an entropy on $IVFS(U)$ with respect to a strong IV negation N (with respect to $\leq_{TL}$ such that there exists $\varepsilon \in L([0,1])$ with $N(\varepsilon) = \varepsilon$ if E has the following properties:*

(E1) $E(A) = 0_L$ *if and only if A is crisp;*

(E2) $E(A) = 1_L$ *if and only if* $A = \{(u_i, A(u_i) = \varepsilon) | u_i \in U\};$

(E3) $E(A) \leq_{TL} E(B)$ *if A refines B; that is,* $A(u_i) \leq_{TL} B(u_i) \leq_{TL} \varepsilon$ *or* $A(u_i) \geq_{TL} B(u_i) \geq_{TL} \varepsilon;$

(E4) $E(A) = E(N(A)).$

Then the following two results are straight from Theorem 3.

Corollary 1. *Let* $M : L([0,1])^n \to L([0,1])$ *be an IV aggregation function with respect to the total order* $\leq_{TL}$ *such that* $M(X_1, \ldots, X_n) = 1_L$ *if and only if* $X_1 = \cdots = X_n = 1_L$ *and* $M(X_1, \ldots, X_n) = 0_L$ *if and only if* $X_1 = \cdots = X_n = 0_L$ *and let N be an IV negation with respect to the order* $\leq_{TL}$ *which is non filling and non-vanishing. Then, the function* $D : IVFS(U) \times IVFS(U) \to L([0,1])$ *given by*

$$D(A, B) = N(M(F(A(u_1), B(u_1)), \ldots, F(A(u_n), B(u_n))))$$

where F is defined as in Theorem 2, is an IV distance measure.

Proof. It is straight from Theorem 3, since a similarity measure defines a distance in a straightforward way. $\square$

Theorem 4. *Let N be a strong IV negation (with respect to* $\leq_{TL}$*) and such that there exists* $\varepsilon \in L([0,1])$ *with* $N(\varepsilon) = \varepsilon$*. Let* $M : L([0,1])^n \to L([0,1])$ *be an IV aggregation function with respect to the total order* $\leq_{TL}$ *and such that* $M(X_1, \ldots, X_n) = 1_L$ *if and only if* $X_1 = \cdots = X_n = 1_L$ *and* $M(X_1, \ldots, X_n) = 0_L$ *if and only if* $X_1 = \cdots = X_n = 0_L$*. Then, the function* $E : IVFS(U) \to L([0,1])$ *given by*

$$E(A) = M(F(A(u_1), N(A(u_1))), \ldots, F(A(u_n), N(A(u_n))))$$

where F is defined as in Theorem 2 with non-filling and non-vanishing negation, is an IV entropy measure.

Proof. It follows from the well known fact that, for a given IV similarity SM, the function $E(A) = SM(A, N(A))$ is an IV entropy measure [6]. $\square$

6   Conclusions

In this paper we have considered the problem of defining interval-valued implications when the order relation is a total order. In particular, we have considered the case of admissible orders. We have also studied the construction of interval-valued equivalence and similarity functions constructed with appropriate interval-valued implication functions. Finally we have shown how our constructions can be used to get IV similarity measures, distances and entropy measures with respect to total orders. In future works we will consider the use of these functions in different image processing, classification or decision making problems.

References

1. Baczyński, M., Beliakov, G., Bustince, H., Pradera, A.: Advances in Fuzzy Implication Functions. Studies in Fuzziness and Soft Computing, vol. 300. Springer, Berlin (2013)
2. Baczyński, M., Jayaram, B.: Fuzzy Implications. Studies in Fuzziness and Soft Computing, vol. 231. Springer, Berlin (2008)
3. Bedregal, B., Dimuro, G., Santiago, R., Reiser, R.: On interval fuzzy S-implications. Inf. Sci. **180**(8), 1373–1389 (2010)
4. Burillo, P., Bustince, H.: Construction theorems for intuitionistic fuzzy sets. Fuzzy Sets Syst. **84**, 271–281 (1996)
5. Bustince, H., Barrenechea, E., Mohedano, V.: Intuitionistic fuzzy implication operators. An expression and main properties. Int. J. Uncertain. Fuzziness Knowl.-Based Syst. **12**(3), 387–406 (2004)
6. Bustince, H., Barrenechea, E., Pagola, M.: Relationship between restricted dissimilarity functions, restricted equivalence functions and normal E_N-functions: image thresholding invariant. Pattern Recogn. Lett. **29**(4), 525–536 (2008)
7. Bustince, H., Barrenechea, E., Pagola, M.: Image thresholding using restricted equivalence functions and maximizing the measure of similarity. Fuzzy Sets Syst. **128**(5), 496–516 (2007)
8. Bustince, H., Barrenechea, E., Pagola, M.: Restricted equivalence functions. Fuzzy Sets Syst. **157**(17), 2333–2346 (2006)
9. Bustince, H., Barrenechea, E., Pagola, M., Fernandez, J., Xu, Z., Bedregal, B., Montero, J., Hagras, H., Herrera, F., De Baets, B.: A historical account of types of fuzzy sets and their relationship. IEEE Trans. Fuzzy Syst. **24**(1), 179–194 (2016)
10. Bustince, H., Burillo, P., Soria, F.: Automorphisms, negations and implication operators. Fuzzy Sets Syst. **134**, 209–229 (2003)
11. Bustince, H., Fernández, J., Kolesárová, A., Mesiar, R.: Generation of linear orders for intervals by means of aggregation functions. Fuzzy Sets Syst **220**, 69–77 (2013)
12. Bustince, H., Galar, M., Bedregal, B., Kolesárová, A., Mesiar, R.: A new approach to interval-valued Choquet integrals and the problem of ordering in interval-valued fuzzy set applications. IEEE Trans. Fuzzy Syst. **21**(6), 1150–1162 (2013)
13. Cornelis, C., Deschrijver, G., Kerre, E.E.: Implication in intuitionistic fuzzy and interval-valued fuzzy set theory: construction, classification, application. Int. J. Approx. Reason. **35**(1), 55–95 (2004)
14. Massanet, S., Mayor, G., Mesiar, R., Torrens, J.: On fuzzy implication: an axiomatic approach. Int. J. Approx. Reason. **54**, 1471–1482 (2013)
15. Pradera, A., Beliakov, G., Bustince, H., De Baets, B.: A review of the relationship between implication, negation and aggregation functions from the point of view of material implication. Inf. Sci. **329**, 357–380 (2016)
16. Riera, J.V., Torrens, J.: Residual implications on the set of discrete fuzzy numbers. Inf. Sci. **247**, 131–143 (2013)
17. Xu, Z., Yager, R.R.: Some geometric aggregation operators based on intuitionistic fuzzy sets. Int. J. Gen. Syst. **35**, 417–433 (2006)

Completing Preferences by Means of Analogical Proportions

Marc Pirlot[1], Henri Prade[2,3], and Gilles Richard[2(✉)]

[1] Faculté Polytechnique, Université de Mons, Mons, Belgium
Marc.Pirlot@umons.ac.be
[2] IRIT, Université Paul Sabatier, 31062 Toulouse Cedex 9, France
{prade,richard}@irit.fr
[3] QCIS, University of Technology, Sydney, Australia

Abstract. We suppose that all we know about the preferences of an agent, is given by a (small) collection of relative preferences between choices represented by their evaluations on a set of criteria. Taking lesson from the success of the use of analogical proportions for predicting the class of a new item from a set of classified examples, we explore the possibility of using analogical proportions for completing a set of relative preferences. Such an approach is also motivated by a striking similarity between the formal structure of the axiomatic characterization of weighted averages and the logical definition of an analogical proportion. This paper discusses how to apply an analogical proportion-based approach to the learning of relative preferences, assuming that the preferences are representable by a weighted average, and how to validate experimental results. The approach is illustrated by examples.

1 Introduction

Guessing the preferences of a user, starting from a set of known examples of his/her preferences between choices described by multiple criteria evaluations, is now recognized as a problem of interest. This may be viewed as the elicitation of a particular type of aggregation function that fits with the examples (see, e.g., [8,18]), or more generally with a preference learning problem [5]. Known preferences can be expressed by the values of global evaluations on an absolute scale, or by relative preferences between pairs of choices. In the following we assume the latter.

More generally, one may be interested in mechanisms which, from a set of qualitative preferences expressed in a relative manner, are able to complete the original set of preferences, by applying some general information principle. Early examples of that can be found in [3]. Such an approach may be consistent with the hypothesis of some implicit family of aggregation functions (e.g., Choquet integrals in [6]).

It has been recently noticed [16] that the characteristic axiom of weighted averages, which forbids contradictory tradeoffs, and which also characterizes Choquet integral when its application is restricted to a smaller class of co-monotone patterns, has some striking consonance with the logical modeling

© Springer International Publishing Switzerland 2016
V. Torra et al. (Eds.): MDAI 2016, LNAI 9880, pp. 135–147, 2016.
DOI: 10.1007/978-3-319-45656-0_12

[12,14] of analogical proportions (i.e., statements of the form "a is to b as c is to d"). This is still true for aggregation operators based on differences.

Analogical proportions have be proved to be successful for predicting classes in machine learning [2]. The idea is to find triples of examples a, b, c with known class, which together with a new item d to classify make analogical proportions for each 4-tuple of feature values. Then, by solving an analogical proportion equation, one may predict the class of d from those of a, b, c. The idea is to follow a similar process for relative preference, but with the constraint of agreeing with the underlying hypothesis of a special class of aggregation functions.

The paper is structured as follows. In Sect. 2, we recall the axiomatic characterizations of aggregation operators based on differences, weighted averages, and Choquet integrals. In Sect. 3 we provide the reader with the necessary background on analogical proportions and analogical inference. In Sect. 4, we first propose an "horizontal" reading of a family of characteristic axioms underlying aggregation operators based on differences and related operators, which make these axioms more intuitive, and which is consonant with the idea of arithmetic proportion. Then, a "vertical" reading indicates another relation with analogical proportions. In Sect. 5, based on the "vertical" reading, a prediction algorithm is proposed and illustrated by an example. In Sect. 6, we discuss the problems raised by the validation of such a procedure. The paper ends with directions for further research.

2   Axiomatics of Some Aggregation Models

A choice x is assumed to be represented by the vector of its evaluations $x = (x_1, \ldots, x_n)$ wrt n criteria. The criteria scale S is common for all criteria and then a choice is an element of Cartesian product S^n. For a choice x, x_{-i} denotes the n-1-dimensional vector comprising the evaluations of x on all criteria but the ith one. $\succeq$ denotes a preference relation. Many aggregation functions have the property that they *don't reveal con-tradictory tradeoffs*. A positive reformulation of this property is as follows.

$$\forall i, j, \forall x, y, v, w \in S^n, \forall \alpha, \beta, \gamma, \delta \in S,$$
$$\text{if } x_{-i}\gamma \succeq y_{-i}\delta \text{ and } v_{-j}\alpha \succeq w_{-j}\beta,$$
$$\text{then, at least one of the following holds:}$$
$$v_{-j}\gamma \succeq w_{-j}\delta \text{ or } x_{-i}\alpha \succeq y_{-i}\beta.$$

This property tells us the differences of preference[1] between α and β, on the one side, and between γ and δ, on the other side, can consistently be compared. In other words, this property is tantamount to impose that the relation $\succeq^*$ on the pairs of levels in S defined by:

$$\alpha\beta \succeq^* \gamma\delta \quad \text{if} \quad \forall i, \forall x_{-i}, y_{-i}, \ x_{-i}\gamma \succeq y_{-i}\delta \Rightarrow x_{-i}\alpha \succeq y_{-i}\beta$$

is a complete preorder.

---

[1] Note that "preference difference" is not to be confused with "arithmetic difference". For example, a person who wants to buy a car at a maximal price of 20 k€ will in general consider that an arithmetic difference of 1 k€ between 19 and 20 k€ is a worse preference difference than the same arithmetic difference between 14 and 15 k€.

For preferences $\succeq$ which do not reveal contradictory tradeoffs, it is always possible to substitute a difference of preferences on a criterion by a larger one without reversing the preference direction. This property is verified by preferences that can be represented by a weighted sum of utilities, i.e., in case there exist a real-valued function u defined on S and a set of n weights p_i summing up to 1, such that, $\forall x, y$, $x \succeq y$ if and only if $U(x) = \sum_{i=1}^{n} p_i u(x_i) \geq U(y) = \sum_{i=1}^{n} p_i u(y_i)$.

For more general preferences that can be represented by a Choquet integral, such a substitution does not alter the preference provided that the choices involved in the comparisons are comonotonic (see [15], Chap. 6). Not revealing contradictory tradeoffs is a crucial property in the characterization of both the preferences that can be represented by a weighted sum of utilities ([15], Theorem IV.2.7) and by a Choquet integral ([15], Theorem VI.5.1).

A still more general category of preferences is obtained if the property of non-revelation of contradictory tradeoffs is only imposed for all $i = j$, a property called non-revelation of *coordinate* contradictory tradeoffs. In this case, the preorder $\succeq^*$ generally depends on i and the corresponding model is the *additive value function model*, in which the preference can be represented using n marginal value (or utility) functions u_i; one has $x \succeq y$ if and only if $V(x) = \sum_{i=1}^{n} u_i(x_i) \geq V(y) = \sum_{i=1}^{n} u_i(y_i)$. This model is characterized in [15], Theorem III.6.6.

3 Background on Analogical Proportions

An analogical proportion is a statement of the form "a is to b as c is to d". We assume here that a, b, c, d are Boolean variables pertaining to the values of some binary feature for four items (i.e., $(a, b, c, d) \in \{0, 1\}^4$). Its logical expression [12], denoted $a : b :: c : d$:

$$(a \wedge \neg b \equiv c \wedge \neg d) \wedge (\neg a \wedge b \equiv \neg c \wedge d) \tag{1}$$

A logically equivalent expression provides another view of analogy.

$$(a \wedge d \equiv b \wedge c) \wedge (a \vee d \equiv b \vee c) \tag{2}$$

(1) expresses that "a differs from b as c differs from d, and b differs from a as d differs from c". This expression is only true for 6 valuation patterns (over $2^4 = 16$):

$$(a, b, c, d) = (1,1,1,1), (0,0,0,0), (1,1,0,0), (0,0,1,1), (1,0,1,0), (0,1,0,1).$$

It is worth noticing that this formalization agrees with characteristic properties of the analogical proportion (acknowledged since Aristotle):

- (i) $a : b :: a : b$ (reflexivity);
- (ii) $a : b :: c : d = c : d :: a : b$ (symmetry);
- (iii) $a : b :: c : d = a : c :: b : d$ (central permutation).

However note that $a : b :: c : d = b : a :: c : d$ *does not hold*. This agrees with the fact that "b is to a as c is to d" is a reversed analogical proportion [14], with respect to $a : b :: c : d$, which is true for the two valuations $(0, 1, 1, 0)$ and $(1, 0, 0, 1)$ that makes the analogical proportion false, as well as for the four valuations $(1, 1, 1, 1)$, $(0, 0, 0, 0)$, $(1, 1, 0, 0)$, $(0, 0, 1, 1)$ that are common with the analogical proportion. Thus, only schemas (s, s, s, s), (s, s, t, t), and (s, t, s, t) are valid for the analogical proportion, which excludes the schema (s, t, t, s) on the one hand, et heterogeneous schemas (s, s, s, t), (s, s, t, s), (s, t, s, s), and (t, s, s, s) on the other hand. There are two main graded extensions of the analogical proportion; see [4] for their expressions. The first one, called "conservative", which is a direct extension of (2) is such that, where $a, b, c, d \in [0, 1]$

$$a : b ::_C c : d = 1 \Leftrightarrow \min(a, d) = \min(b, c) \text{ and } \max(a, d) = \max(b, c).$$

Again patterns (s, s, t, t), and (s, t, s, t) (and (s, s, s, s)) are the unique way to have the analogical proportion fully true (equal to 1). The second one, more in the spirit of (1), more "liberal" is such that

$a : b ::_L c : d = 1 \Leftrightarrow a - b = c - d$ (which is an arithmetic proportion).

Clearly, $a : b ::_C c : d = 1 \Rightarrow a : b ::_L c : d = 1$. Both extensions include the Boolean case. Transitivity holds $a : b :: c : d = 1, c : d :: e : f = 1 \Rightarrow a : b :: e : f = 1$. The analogical proportion extends to vectors in a straightforward way:

$\boldsymbol{a} : \boldsymbol{b} :: \boldsymbol{c} : \boldsymbol{d}$ if and only if $\forall i, a_i : b_i :: c_i : d_i, i = 1, n$, and to nominal variables [2].

The analogical inference principle is, logically speaking, an unsound inference principle, but providing plausible conclusions [17,19]. It postulates that, given 4 vectors $\boldsymbol{a}, \boldsymbol{b}, \boldsymbol{c}, \boldsymbol{d}$ such that the proportion holds on some components, then it should also hold on the remaining ones. This can be stated as (where $\boldsymbol{a} = (a_1, a_2, \cdots, a_n)$, and $J \subset \{1, \cdots, n\}$):

$$\frac{\forall j \in J, a_j : b_j :: c_j : d_j}{\forall i \in \{1, \cdots, n\} \setminus J, a_i : b_i :: c_i : d_i} \quad (analogical \ inference)$$

This principle leads to a prediction rule in the following context:

- 4 vectors $\boldsymbol{a}, \boldsymbol{b}, \boldsymbol{c}, \boldsymbol{d}$ are given where $\boldsymbol{d}$ is partially known: only the components of $\boldsymbol{d}$ with indexes in J are known.
- Using analogical inference, we can predict the missing components of $\boldsymbol{d}$ by solving (w.r.t. d_i) the set of equations (if they are solvable):

$$\forall i \in \{1, \cdots, n\} \setminus J, \quad a_i : b_i :: c_i : d_i.$$

In the case where the items are such that their last component is a label, applying this principle to a new element $\boldsymbol{d}$ whose label is unknown leads to predict a candidate label for $\boldsymbol{d}$. This prediction technique has been successfully applied to classification [1,2].

4   Analogy and Contradictory Tradeoffs

We now investigate the relations between analogical proportions and contradictory tradeoffs.

4.1   A Proportion-Like Reading of the Axiom Forbidding Contradictory Tradeoffs

Reading one by one the four preference statements appearing in the axiom ("horizontal" reading) prompts an interpretation in terms of comparison of pairs rather than in terms of equality as in analogical proportions. $x_{-i}\alpha \preceq y_{-i}\beta$ means that the "difference" between α and β on criterion i is smaller than (i.e. does not compensate) the "difference" between the vectors x_{-i} and y_{-i} on the rest of the criteria. In contrast, $x_{-i}\gamma \succeq y_{-i}\delta$ tells us that the difference between γ and δ is larger than the difference between x_{-i} and y_{-i}. The two other preference statements say that, whenever the difference between α and β on any criterion j balances that between the vectors v_{-j} and w_{-j} on the rest of the criteria, it is a fortiori the case that the difference between γ and δ on criterion j balances the difference between v_{-j} and w_{-j}. What is implied is that the proportion or the difference between γ and δ is at least as large as that between α and β, independently of the criterion on which this difference shows up.

Three observations can be made about this interpretation. First, the preference statements are interpreted as comparisons of proportions (or differences). Second, the pairs of objects that are compared are of different natures: pairs of levels on a single criterion vs. pairs of vectors of levels. Finally, the comparison of pairs of levels is independent of the criterion on which they appear.

Since the theory of analogical proportions was not developed to deal with comparisons of proportions but rather with the identity of changes within two pairs, we shall not pursue the "horizontal" reading of the axiom but instead turn to a "vertical" one, in Sect. 4.2. Interestingly enough, the use of quaternary relation as expressed by the horizontal reading, where the difference (or the dissimilarity) between a and b is greater (or smaller) than the difference between c and d, rather than being equal to as in analogical proportion, has been also recently introduced in machine learning [11].

4.2   A Vertical Analogical Reading of the Axiom Forbidding Contradictory Exchanges

Looking at the patterns appearing in the expression of the axiom characterizing the weighted average, we may notice that the 4 pairs of vectors

$$A : x_{-i}\alpha \preceq y_{-i}\beta$$
$$B : x_{-i}\gamma \succeq y_{-i}\delta$$
$$C : v_{-j}\alpha \succeq w_{-j}\beta$$
$$D : v_{-j}\gamma \prec w_{-j}\delta$$

exhibit analogical proportions vertically, *symbol by symbol* (with the exception of the preference relations). Namely (x, x, v, v), (i, i, j, j), $(\alpha, \gamma, \alpha, \gamma)$, (y, y, w, w), $(\beta, \delta, \beta, \delta)$ are analogical proportion patterns. This is quite striking. However, it is no longer exactly so if we look at the vectors *component by component*, as can be seen on the following example.

Example. We consider 3 criteria, graded on an increasing scale with 5 levels 1, 2, 3, 4, 5. The columns 2 and 3 play the role of i and j respectively in the above axiom. Let us consider the 4 pairs of vectors, which obey to the axiom pattern:

$$(1, 1, 3) \preceq (1, 2, 3)$$
$$(1, 5, 3) \succeq (1, 4, 3)$$
$$(2, 4, 1) \succeq (1, 2, 2)$$
$$(2, 4, 5) \succeq (1, 2, 4)$$

As can be seen, we have an analogical proportion pattern only in column 1 (with $(1, 1, 2, 2)$). But no analogical proportion in column 2 or in column 3. Still, one may consider that there are two analogical proportions that are intertwined in positions $i = 2$ and $j = 3$. For instance, $(1, 5, 4, 4)$ and $(3, 3, 1, 5)$ are not analogical proportion patterns, but $(1, 5, 1, 5)$ and $(3, 3, 4, 4)$ are. Let us assume that all the criteria have the *same importance* then criteria values can be permuted in the vectors, and for instance, $(2, 4, 1)$ is the same as $(2, 1, 4)$, $(2, 4, 5)$ the same as $(2, 5, 4)$, and so on in the above example. This clearly enables us to restore analogical proportion patterns in all the columns.

It is worth noticing that in the case the axiom is restricted to the particular case $i = j$, which corresponds to the characterization of the aggregations based on differences, there is no problem, the analogical proportions are preserved without the help of any permutation.

Would it be possible to fictively make the hypothesis of equal importance? Let a, b, c be the *unknown* respective relative weights of the 3 criteria in the example $(a+b+c = 1)$. Let multiply the respective criteria evaluation in order to give the same importance to these rescaled evaluations (note that the evaluations are no longer on the same scale)

$$(1 \times 3a, 1 \times 3b, 3 \times 3c) < (1 \times 3a, 2 \times 3b, 3 \times 3c)$$
$$(1 \times 3a, 5 \times 3b, 3 \times 3c) > (1 \times 3a, 4 \times 3b, 3 \times 3c)$$
$$(2 \times 3a, 4 \times 3b, 1 \times 3c) > (1 \times 3a, 2 \times 3b, 2 \times 3c)$$
$$(2 \times 3a, 4 \times 3b, 5 \times 3c) > (1 \times 3a, 2 \times 3b, 4 \times 3c)$$

After permutation, we get

$$(1 \times 3a, 1 \times 3b, 3 \times 3c) < (1 \times 3a, 2 \times 3b, 3 \times 3c)$$
$$(1 \times 3a, 5 \times 3b, 3 \times 3c) > (1 \times 3a, 4 \times 3b, 3 \times 3c)$$
$$(2 \times 3a, 1 \times 3c, 4 \times 3b) > (1 \times 3a, 2 \times 3c, 2 \times 3b)$$
$$(2 \times 3a, 5 \times 3c, 4 \times 3b) > (1 \times 3a, 4 \times 3c, 2 \times 3b)$$

We can see that analogical proportion patterns now appear again vertically (including for the multiplicative factors).

4.3 Example

Let us consider an extension of the example proposed by [13] in order to illustrate preference completion by means of analogical proportions. We have to compare cars w.r.t. three criteria, namely, *cost, performance* and *comfort*, assessed on a qualitative scale. The criteria scales all have five levels 1, 2, 3, 4, 5, which carry different interpretations. On the *cost* scale, 1, 2, 3, 4, 5 respectively correspond to *very expensive, rather expensive, moderately expensive, not expensive* and *cheap*. On the *performance* scale 1, 2, 3, 4, 5 are respectively interpreted as *very weak, weak, average, high* and *very high*. Finally, regarding the *comfort* criterion, 1, 2, 3, 4, 5 respectively correspond to *very poor, poor, medium, good* and *very good*.

Let us assume that the following preferences are known. A moderately expensive car with weak performance and medium comfort is represented by the vector $(3, 2, 3)$. It is preferred to a very expensive car with average performance and good level of comfort, which is represented by the vector $(1, 3, 4)$. In contrast, we know that a very expensive car with average performance and a medium comfort level $(1, 3, 3)$ is preferred to a moderately expensive car with weak performance and a very poor comfort level $(3, 2, 1)$. Finally, a rather expensive car with average performance and a good comfort level $(2, 3, 4)$ is preferred to a not expensive car with very weak performance and a medium comfort level $(4, 1, 3)$.

We have the following triplet of preferences:

$$A : (1, 3, 4) \preceq (3, 2, 3)$$
$$B : (1, 3, 3) \succeq (3, 2, 1)$$
$$C : (2, 3, 4) \succeq (4, 1, 3).$$

The pattern is that of the axiom forbidding contradictory tradeoffs, with $i = j = 3$. In case the preference can be represented by an *additive value function model*, we should not have the reverse analogical proportion[2]:

$$D : (2, 3, 3) \prec (4, 1, 1),$$

but instead:

$$D : (2, 3, 3) \succeq (4, 1, 1).$$

Let us now consider another configuration (where C is modified). Assume that we have six cars which verify the following preferences:

$$A : (1, 3, 4) \preceq (3, 2, 3)$$
$$B : (1, 3, 3) \succeq (3, 2, 1)$$
$$C : (2, 4, 3) \succeq (4, 3, 1).$$

---

[2] *A* and *B* suggest that the difference of preference $(3, 1)$ on the third criterion (comfort) is "larger" (or more important) than the difference of preference $(4, 3)$ on the same criterion. Given *C*, assuming that $(2, 3, 3)$ is not preferred to $(4, 1, 1)$ would reveal contradictory tradeoffs since it implies that, in this context, the difference of preference $(4, 3)$ on the third criterion is larger than $(3, 1)$.

The pattern is that of the axiom forbidding contradictory tradeoffs, with $i = 3$ and $j = 2$. In case the preference can be represented by a sum of weighted utilities, we should not have the reverse analogical proportion:

$$D : (2,3,3) \prec (4,1,1),$$

but instead:

$$D : (2,3,3) \succeq (4,1,1).$$

For lack of space we do not provide here examples in which the choices involved are comonotonic and the preference model is a Choquet integral of a utility function.

4.4 Compatibility of Analogy with the Difference

Let 4 pairs of vectors such that (x^1, y^1), (x^2, y^2), (v^1, w^1), (v^2, w^2), as in the pattern of the axiom expressing the absence of contradictory tradeoffs. Then $x^1 : x^2 :: v^1 : v^2$ and $y^1 : y^2 :: w^1 : w^2$ means componentwise (for each criterion i) $x_i^1 : x_i^2 :: v_i^1 : v_i^2$ and $y_i^1 : y_i^2 :: w_i^1 : w_i^2$. Since taking the liberal extension, $x_i^1 : x_i^2 :: v_i^1 : v_i^2$ means algebraically $x_i^1 - x_i^2 = v_i^1 - v_i^2$ for each component i, we can easily check that if $x^1 : x^2 :: v^1 : v^2$ and $y^1 : y^2 :: w^1 : w^2$ hold true, it entails that $x^1 - y^1 : x^2 - y^2 :: v^1 - w^1 : v^2 - w^2$ holds true as well. This expresses "vertically" the compatibility of analogical proportion with difference-based comparisons, and further motivates the algorithm proposed in the section for extrapolating preferences from known preferences.

5 Algorithms

Given a set of n criteria, each criterion i being evaluated on a scale $S = \{1, 2, \ldots, k\}$, a choice X is then represented as a vector of value $\{x_1, \ldots, x_n\}$, each x_i being the evaluation of criteria i (i.e. $X \in S^n$). It is assumed that the scale S has the following semantics: the higher x_i, the better the criteria i is satisfied. Let us denote now $\succeq$ a preference relation over the universe S^n: $\succeq$ is assumed to be transitive relation (supposed to be total in our case).

A set of preference examples $e_i = X_i \succeq Y_i$ (abbreviated as a simple pair (X_i, Y_i)), telling us that situation X_i is preferred to situation Y_i, is assumed to be provided. It gives rise to a set of other valid examples by transitivity. So if E is a set of preference examples, we denote $comp(E)$ the transitive completion of E. We will use the following notation:

- $isValid(x, y, z, t)$ is a Boolean function leading to *true* if $x : y :: z : t$ is a valid analogy, and *false* otherwise.
- $solvable(x, y, z)$ is a Boolean function leading to *true* if there is a (unique) t such that $x : y :: z : t$, and *false* otherwise.
- when $solvable(x, y, z) = true$, $sol(x, y, z)$ is the unique vector t such that $x : y :: z : t$.

Two problems, discussed in the next two subsections, can be considered.

5.1   Completion of a Preference Relation

We want to further complete $comp(E)$. Keeping in mind the remarks made in the previous sections, we have two options:

- first avoiding contradictory trade-offs in case we have triples of examples in $comp(E)$ which might lead to it for a fourth pair;
- second, apply the analogical proportion-based inference to the preference relations from suitable triples of examples for guessing the preference relation between the elements of a fourth pair, making sure that we are not creating contradictory trade-offs with respect to other triples.

We then get a bigger file denoted $PrefCompletion(comp(E))$. The algorithm *Preference Completion* can be described as follows:

Algorithm 1. Preference completion

---

input: a sample E of preference examples

step 1) compute $comp(E)^{(a)}$

step 2) init : $PrefCompletion(comp(E)) = comp(E)$

step 3) for each triple $[(X_1, Y_1), (X_2, Y_2), (X_3, Y_3)] \in comp(E)^3$ s.t.
$solvable(Y_1, X_2, X_3)$ and $solvable(X_1, Y_2, Y_3)$
$\qquad PrefCompletion(comp(E)).add\ (sol(Y_1, X_2, X_3) \succeq sol(X_1, Y_2, Y_3))^{(b)}$

step 4) for each triple $(X_1, Y_1), (X_2, Y_2), (X_3, Y_3) \in comp(E)$ s.t.
(3 cases of valid analogy including the preference relation symbol)
$solvable(X_1, X_2, X_3)$ and $solvable(Y_1, Y_2, Y_3)$
$\qquad PrefCompletion(comp(E)).add\ (sol(X_1, X_2, X_3) \succeq sol(Y_1, Y_2, Y_3))$
$solvable(Y_1, Y_2, X_3)$ and $solvable(X_1, X_2, Y_3)$
$\qquad PrefCompletion(comp(E)).add\ (sol(Y_1, Y_2, X_3) \succeq sol(X_1, X_2, Y_3))$
$solvable(Y_1, X_2, Y_3)$ and $solvable(X_1, Y_2, X_3)$
$\qquad PrefCompletion(comp(E)).add\ (sol(Y_1, X_2, Y_3) \succeq sol(X_1, Y_2, X_3))$

step 5) for each triple $(X_1, Y_1), (X_2, Y_2), (X_3, Y_3) \in comp(E)$ s.t.
(2 remaining cases of valid analogy including the preference relation symbol)
$solvable(X_1, X_2, Y_3)$ and $solvable(Y_1, Y_2, X_3)$
$\qquad PrefCompletion(comp(E)).add\ (sol(X_1, X_2, Y_3) \succeq sol(Y_1, Y_2, X_3))$
$solvable(X_1, Y_2, X_3)$ and $solvable(Y_1, X_2, Y_3)$
$\qquad PrefCompletion(comp(E)).add\ (sol(X_1, Y_2, X_3) \succeq sol(Y_1, X_2, Y_3))^{(d)}$

return $PrefCompletion(comp(E))$

(a) The set of examples E is a finite. Standard algorithms exist for computing $comp(E)$.

(b) As explained in the previous sections, as soon as the preference relation is representable by a weighted average, the 3 preferences $X_1 \succeq Y_1$, $X_2 \succeq Y_2$, $X_3 \succeq Y_3$, equivalent to the pattern

$$Y_1 \preceq X_1$$
$$X_2 \succeq Y_2$$
$$X_3 \succeq Y_3$$

entails

$X_4 \succeq Y_4$ (in order to avoid contradictory tradeoffs)

where $X_4 = sol(Y_1, X_2, X_3)$ and $Y_4 = sol(X_1, Y_2, Y_3)$

(c) This step applies analogical proportion to preference symbols for triples for which step 2 does not apply. Three patterns are possible:

(i)	(ii)	(iii)
$X_1 \succeq Y_1$	$Y_1 \preceq X_1$	$Y_1 \preceq X_1$
$X_2 \succeq Y_2$	$X_2 \succeq Y_2$	$Y_2 \preceq X_2$
$X_3 \succeq Y_3$	$Y_3 \preceq X_3$	$X_3 \succeq Y_3$
entails	entails	entails
$sol(X_1, X_2, X_3) \succeq$	$sol(Y_1, X_2, Y_3) \succeq$	$sol(Y_1, Y_2, X_3) \succeq$
$sol(Y_1, Y_2, Y_3)$	$sol(X_1, Y_2, X_3)$	$sol(X_1, X_2, Y_3)$

(d) We have two remaining cases (where $\preceq$ is obtained between solutions)

(iv)	(v)
$X_1 \succeq Y_1$	$X_1 \succeq Y_1$
$Y_2 \preceq X_2$	$X_2 \succeq Y_2$
$X_3 \succeq Y_3$	$Y_3 \preceq X_3$
entails	entails
$sol(X_1, Y_2, X_3) \preceq sol(Y_1, X_2, Y_3)$	$sol(X_1, X_2, Y_3) \preceq sol(Y_1, Y_2, X_3)$

5.2   Checking a Preference Relation

In that case, the problem is to decide if a given preference relation $X \succeq Y$ is a valid consequence of a finite set E of valid preferences. Obviously two strategies are available:

– Either we compute the preference completion $PrefCompletion(comp(E))$ as explained in the previous section. Then the validity of $X \succeq Y$ is just the test: $X \succeq Y \in PrefCompletion(comp(E))$.
– A less expensive option is to check $X \succeq Y$ as follows:

Algorithm 2. Preference validity

---

input: a sample E of preference examples, a relation $X \succeq Y$

step 1) compute $comp(E)$

step 2) if (there exists $(X_1, Y_1), (X_2, Y_2), (X_3, Y_3) \in comp(E)$ s.t. $(isValid(Y_1, X_2, X_3, X)$ and $isValid(X_1, Y_2, Y_3, Y))$:

   return *true*

step 3) else if (there exists $(X_1, Y_1), (X_2, Y_2), (X_3, Y_3) \in comp(E)$ s.t. $(isValid(X_1, X_2, X_3, X)$ and $isValid(Y_1, Y_2, Y_3, Y))$
or $(isValid(Y_1, Y_2, X_3, X)$ and $isValid(X_1, X_2, Y_3, Y))$
or $(isValid(Y_1, X_2, Y_3, X)$ and $isValid(X_1, Y_2, X_3, Y))$

   return *true*

step 4) else if (there exists $(X_1, Y_1), (X_2, Y_2), (X_3, Y_3) \in comp(E)$ s.t. $(isValid(X_1, X_2, Y_3, Y)$ and $isValid(Y_1, Y_2, X_3, X))$
or $(isValid(X_1, Y_2, X_3, Y)$ and $isValid(Y_1, X_2, Y_3, X))$

   return *true*

otherwise *unknown*

Note that steps 3 and 4 above cover the 5 analogical situations of point (c) in the previous algorithm.

5.3 Illustrative Example

Let us consider the following set of examples $E = \{(1,3,4) \preceq (3,2,3), (1,3,3) \succeq (3,2,1), (2,3,4) \succeq (4,1,3), (4,1,3) \preceq (1,4,4), (4,2,2) \succeq (2,3,2)\}$

Then the 3 first examples lead to adopt $(2,3,3) \succeq (4,1,1)$ for avoiding contradictory trade-offs as in the example of Sect. 4.3. Completing known preferences either by monotonicity and transitivity or by forbidding contradictory tradeoffs can be considered as a prudent extension since many preference models satisfy these properties. It is only when the application of transitivity or avoidance of contradictory tradeoffs failed to provide an answer to the comparison of two choices that we may consider the application of the analogical proportion-based inference.

For instance, in the above example, noticing that

$$(4,1,2) \preceq (1,4,4) \text{ (by transitivity from } (4,1,3) \preceq (1,4,4))$$
$$(4,2,2) \succeq (2,3,2)$$
$$(3,1,2) \preceq (1,4,4) \text{ (by transitivity from } (4,1,3) \preceq (1,4,4))$$

we obtain by application of the analogical proportion inference

$$(3,2,2) \succeq (2,3,2).$$

Note that the result of this analogical proportion inference together with the three premises can never form contradictory tradeoffs. This mode of completion deserves further theoretical and experimental investigations.

6 Experimental Setting and Validation

The procedure presented above exploits transitivity, and avoids the introduction of contradictory trade-off. Moreover the analogical proportion-based preference provides a heuristic way of acknowledging the idea that the observation of similar differences should lead to analogous preferences. It seems easy to experiment with such a procedure. First choose a class of aggregation functions that avoids contradictory trade-offs. We may first try weighted averages. For instance, we may

- choose $k = 5$ for the scale and $n = 3$ (i.e., 3 criteria and 5 candidate values per criterion),
- choose a weighted average function f from $S = \{1, \ldots, 5\}^3$ to $\mathbb{R}$ as follows (the weights α, β, γ being positive real numbers such that $\alpha + \beta + \gamma = 1$): $f(a,b,c) = \alpha \times a + \beta \times b + \gamma \times c$;
- the $\succeq$ relation is deduced from f as follows: $X \succeq Y$ iff $f(X) \geq f(Y)$,

- choose a subset E of examples (X, Y),
- apply the previous algorithm and check if the new preferences $X \succeq Y$ we get are "*valid*",
- compute the accuracy rate (cross validation).

What does 'valid' mean here? $f(X) \geq f(Y)$? No!

Obviously, there are generally more than one set of weights (here α, β, γ) that agree with a finite set of comparative rankings between pairs of choices. Indeed let us take, for example, $m = 3$ criteria, and n comparative preferences $v_i = (x_i, y_i, z_i) \succeq v_i' = (x_i', y_i', z_i')$ for $i = 1, n$. Then, this is equivalent to $\exists \alpha, \beta, \gamma \geq 0$ such that $\alpha + \beta + \gamma = 1$ satisfying $\alpha \cdot x_i + \beta \cdot y_i + \gamma \cdot z_i \geq \alpha \cdot x_i' + \beta \cdot y_i' + \gamma \cdot z_i'$ for $i = 1, n$, or if we prefer $\alpha \cdot (x_i - x_i') + \beta \cdot (y_i - y_i') + \gamma \cdot (z_i - z_i') \geq 0$. For finding the vertices of the polytope $\mathcal{T}$ of the (α, β, γ) simplex satisfying the above system of linear inequalities, when it is not empty (if the preferences are consistent with the assumed family of models, this set is not empty), is a matter of linear programming. This problem has been extensively studied in e.g. [7,9,10]. Thus, the validation requires to define a family of considered preference models (e.g. weighted sums, weighted utilities, additive value functions, Choquet integral of utilities, see Sect. 2) and to check if there is at least one model that agrees with each prediction $X \succeq Y$ (for all models alluded to in Sect. 2, this can be done by using linear programming).

7   Concluding Remarks

The paper has reported the first steps of a work in progress. It relies on the idea that rather than learning an aggregation function (or any other general preference representation device, e.g., a CP-net [5]), one may try to take another road, by proceeding in a transductive manner from examples without trying to induce a general representation of the set of examples. Starting from the observation that many methods for handling multiple criteria try to avoid contradictory trade-offs, and noticing the striking similarity of the structure of the axioms preserving from such trade-offs with the notion of analogical proportion, we have proposed a procedure able to predict new preferences directly from the examples that does not create contradictory trade-offs, leading to an analogical completion process of the preferences. The success of this type of approach in classification is another motivation for proposing such a method. The next steps will be to effectively validate the approach along the lines discussed in the previous section, and to investigate how the capability of predicting preferences evolve with the size of the set of examples (taken randomly from a generating function).

References

1. Bayoudh, S., Miclet, L., Delhay, A.: Learning by analogy: a classification rule for binary and nominal data. In: Proceedings of the 20th International Joint Conference on Artificial Intelligence (IJCAI 2007), Hyderabad, India, pp. 678–683 (2007)

2. Bounhas, M., Prade, H., Richard, G.: Ana-logical classification. A new way to deal with examples. In: Proceedings of the ECAI 2014, pp. 135–140 (2014)
3. Dubois, D., Kaci, S., Prade, H.: Expressing Preferences from Generic Rules and Examples – A Possibilistic Approach Without Aggregation Function. In: Godo, L. (ed.) ECSQARU 2005. LNCS (LNAI), vol. 3571, pp. 293–304. Springer, Heidelberg (2005)
4. Dubois, D., Prade, H., Richard, G.: Multiple-valued extensions of analogical proportions. Fuzzy Sets Syst. **292**, 193–202 (2016)
5. Fürnkranz, J., Hüllermeier, E., Rudin, C., Slowinski, R., Sanner, S.: Preference Learning (Dagstuhl Seminar 14101). Dagstuhl Rep. **4**(3), 1–27 (2014)
6. Gérard, R., Kaci, S., Prade, H.: Ranking alternatives on the basis of generic constraints and examples - a possibilistic approach. In: Proceedings of the IJCAI 2007, Hyderabad, 6–12 January, pp. 393–398 (2007). In French, Actes LFA 2006, pp. 103–110. Cépaduès **103–110**, 2006 (2007)
7. Greco, S., Mousseau, V., Słowiński, R.: Ordinal regression revisited: Multiple criteria ranking using a set of additive value functions. Eur. J. Oper. Res. **191**(2), 416–436 (2008)
8. Grabisch, M., Kojadinovic, I., Meyer, P.: A review of methods for capacity identification in Choquet integral based multi-attribute utility theory: Applications of the Kappalab R package. Eur. J. Oper. Res. **186**(2), 766–785 (2008)
9. Jacquet-Lagrèze, E., Siskos, Y.: Assessing a set of additive utility functions for multicriteria decision making: the UTA method. Eur. J. Oper. Res. **10**, 151–164 (1982)
10. Jacquet-Lagrèze, E., Siskos, Y.: Preference disaggregation: 20 years of MCDA experience. Eur. J. Oper. Res. **130**(2), 233–245 (2001)
11. Law, M.T., Thome, N., Cord, M.: Quadruplet-wise image similarity learning. In: Proceedings of the IEEE International Conference on Computer Vision (ICCV) (2013)
12. Miclet, L., Prade, H.: Handling analogical proportions in classical logic and fuzzy logics settings. In: Sossai, C., Chemello, G. (eds.) ECSQARU 2009. LNCS, vol. 5590, pp. 638–650. Springer, Heidelberg (2009)
13. Modave, F., Dubois, D., Grabisch, M., Prade, H.: A Choquet integral representation in multicriteria decision making. In: Medsker, L. (ed.) Working Notes of the Fall AAAI Symposium "Frontiers in Soft Computing and Decision Systems", pp. 30–39. In French, Actes LFA 1997, Lyon, pp. 81–90. Cépaduès (1997)
14. Prade, H., Richard, G.: From analogical proportion to logical proportions. Logica Univers. **7**(4), 441–505 (2013)
15. Wakker, P.: Additive Representations of Preferences: A New Foundation of Decision Analysis. Springer, Heidelberg (1989)
16. Pirlot, M., Prade, H.: Complétion analogique de préférences obéissant à une intégrale de Choquet. In: Actes 24emes Conf. sur la Logique Floue et ses Applications, Poitiers, 5–7 November. Cépaduès (2015)
17. Prade, H., Richard, G.: Reasoning with logical proportions. In: Lin, F., Sattler, U., Truszczynski, M. (eds.) Proceedings of the 12th International Conference on Principles of Knowledge Representation and Reasoning (KR 2010), Toronto, 9–13 May. AAAI Press (2010)
18. Prade, H., Rico, A., Serrurier, M., Raufaste, E.: Elicitating sugeno integrals: methodology and a case study. In: Sossai, C., Chemello, G. (eds.) ECSQARU 2009. LNCS, vol. 5590, pp. 712–723. Springer, Heidelberg (2009)
19. Yvon, F., Stroppa, N.: Formal models of analogical proportions. Technical reportD008, Ecole Nationale Supérieure des Télécommunications, Paris (2006)

Clustering and Classification

Rating Supervised Latent Topic Model for Aspect Discovery and Sentiment Classification in On-Line Review Mining

Wei Ou[(⊠)] and Van-Nam Huynh

Japan Advanced Institute of Science and Technology,
Asahidai 1-1, Nomi, Ishikawa, Japan
ouwei@jaist.ac.jp

Abstract. Topic models have been used for unsupervised joint aspect (or attribute) discovery and sentiment classification in on-line review mining. However in existing methods the straightforward relations between ratings, aspect importance weights and sentiments in reviews are not explicitly exploited. In this paper we propose Rating Supervised Latent Topic Model (RS-LTM) that incorporates these relations into the framework of LDA to fulfill the task. We test the proposed model on a review set crawled from Amazon.com. The preliminary experiment results show that the proposed model outperforms state-of-the-art models by a considerable margin.

Keywords: Aspect discovery · Sentiment classification · Topic models · Review mining

1 Introduction

Joint aspect discovery and sentiment classification in review mining, that is identifying aspects and sentiments simultaneously from review text, has been an active topic in research communities in recent years. An aspect is an attribute or component of a product and the goal of aspect discovery is to find a set of relevant aspects for a target product [14]. The goal of sentiment classification is to find out whether shoppers have positive or negative feedbacks on the aspects of a target product. Joint aspect discovery and sentiment classification allows for generating a high-level summary of the hundreds, even thousands of reviews posted by previous shoppers on a product therefore it has very high research value in product or service recommender systems. Many machine learning techniques have been proposed to fulfill this task. Among those techniques, the one that has recently drawn the most attention from research communities can be probabilistic topic models due to their outstanding performances at relatively low computation costs.

One of the most representative works in topic modeling is Latent Dirichlet Allocation (LDA) [1]. LDA can cluster words that frequently co-occur in the same documents into the same topic classes and have been widely used for topic

© Springer International Publishing Switzerland 2016
V. Torra et al. (Eds.): MDAI 2016, LNAI 9880, pp. 151–164, 2016.
DOI: 10.1007/978-3-319-45656-0_13

classification in text mining. Fletcher et al. [3] apply it to product attribute discovery in review mining by equating the concept of "topic" in text mining with the concept of "aspect" in review mining and assuming words related to the same attribute must frequently co-occur in the same review sentences. Oh et al. extend LDA by introducing sentiment polarities of words in reviews as another latent component besides aspect labels into the topic model to get an aspect-sentiment unification model (ASUM) for the aforementioned task [7].

ASUM or other LDA extensions usually suffer from such a problem: they often cluster words with opposite sentiments into the same sentiment classes. In a dataset with mixed ratings, an aspect word can appear with some particular positive words in many reviews while it can also be described by some particular negative words in other reviews. Therefore the positive words and the negative words are very likely to be assigned with the same sentiment labels because of their common dependency on the same aspect word. The rationale of LDA determines that those problems can be hardly alleviated if without feeding extra prior knowledge about the sentiments to those models.

In a realistic review dataset, the rating of a review always indicates the overall sentiment orientations of product attributes covered in the very review. For example, in a 5-star review most attributes are very likely to be in the positive side; in an 1-star review most attributes are very likely to be in the negative side; in a 3-star review some attributes can be in the positive side while others can be in the negative side. In another word, review ratings can be a very informative prior knowledge for sentiment classification. Besides, aspect weights may also play an important role along with the ratings in the sentiments: the most important aspects are very likely to be positive in reviews with high ratings and to be negative in reviews with low ratings. We believe incorporating the straightforward relations between ratings, aspect weights and sentiments into topic models would result in considerable improvements.

In this paper we propose an extension of LDA, Rating Supervised Latent Topic Model (RS-LTM) that takes into account the aforementioned relations. We assume the sentiment polarity of each word in a review is drawn from a *binomial* distribution that is further sampled from a *Beta* distribution whose hyper parameters are determined by the review's rating and aspect importance weights. Given aspect weights are usually unknown beforehand and have to be initialized with random values, we impose a rating regression problem that captures the relation between ratings and aspect weights as a constraint on the topic model to adjust the weights to be consistent with the training data. They are the main contribution of this paper. With the learnt parameters, we can not only derive aspect and sentiment labels of words, but also estimate aspect weights and sentiment distributions under all rating-aspect pairs.

The rest of the paper is organized as follows. Section 2 presents a brief introduction to existing related work. Section 3 elaborates the proposed approach and Sect. 4 shows the experiment results. Section 5 concludes this paper with future directions.

2   Related Work

Since we focus on the task of joint aspect discovery and sentiment classification for product review mining, there is a need to have a brief introduction to existing work in aspect discovery, sentiment classification, joint aspect discovery and sentiment classification respectively. To our best knowledge, sentiment classification usually involves aspect discovery and the methods that focus solely on sentiment classification in this particular field are very rare. So in this section we will only present related work about aspect discovery, joint aspect discovery and sentiment classification.

Based on the methodologies used, we divide existing works in aspect discovery into two schools: non-topic-model-based approaches and topic-model-based approaches. One early representative work in the former school can be the one proposed by Hu et al. [6]. In this work the authors first identify frequent feature words in a dataset then use association rule mining techniques to allot those words into relevant aspect classes. Zhai et al. first identify leader feature expressions for each product aspect by harnessing the knowledge delivered by co-occurring words and WordNet then assign corresponding aspect labels to remaining unlabeled words through an EM based method [19]. Guo et al. propose a latent semantic association model that clusters words into relative aspects according to their semantic structures and contexts [4].

In the latter school, a representative work can be MaxEnt-LDA proposed by Zhao et al. [21]. In this model the authors differentiate between words in a review sentence by their semantic labels and assume each of those labels has a multinomial distribution over the vocabulary. Deriving those distributions would allow for identifying aspects in each review. Zhai et al. extend LDA by adding must-link and must-not-link constraints that mandate which words must be labeled with the same aspects and which must be not respectively [20]. With the help of the constraints, the words clustered into the same aspects are much more relevant than that obtained by the original LDA model. Titov et al. propose Multi-grain Topic Model in which each review document is associated with two types of aspect distributions [16]: global aspect distribution that captures the overall statistical characteristics of reviews and local aspect distribution that reflects the dependency of locally co-occurring words.

Similarly, works in joint aspect and sentiment classification can be divided into the same two schools as well. There are a few works can be found in the non-topic-model-based school. Fahrni et al. take advantage of the rich knowledge available on Wikipedia to identify aspects in reviews and use a seed set of sentimental words to determine the sentiment polarity of each word through an iterative propagation method [2]. Recently, Li et al. propose a CRF-based model that exploits the conjunction and syntactic relations among words to generate aspect-sentiment summaries of reviews [8].

Joint aspect and sentiment classification based on topic models is a relatively new topic in the academia. Besides the aforementioned ASUM model, the JST model proposed by Lin et al. [10] and the Sentiment-LDA proposed by Li et al. [9] are very representative as well. Those works share similar latent components as ASUM but differentiate themselves from it by assuming different generative processes. Mei et al. propose the TSM model [13] in which the authors divide all the latent labels into two categories: generic backgrounds and sentimental aspects therefore not only aspects and their sentiments but also backgrounds in reviews can be identified. Mukherjee et al. first manually label seed sets of aspect words with different sentiments then use LDA to learn the association between the seed words and unlabeled words [15].

It is worth noting that none of those methods mentioned above takes into account ratings that are usually available beforehand in a review dataset. Generally speaking, the rating score of a review reflects the overall sentiment orientation of the review therefore ignoring it in an algorithm would compromise the algorithm's performance. It has been widely agreed that algorithms which exploit both review text and ratings would generate better performances than the algorithms only consider either textual content or ratings. A number of methods based on this principle have been proposed. Wang et al. propose the LRR model [17] in which the authors assume each aspect of a review has a latent rating and the overall rating of the review is the weighted summation of the latent aspect ratings. In this model, the authors first assign each sentence an aspect label then infer the latent ratings through a regression approach. Yu et al. propose an aspect ranking algorithm [18] that is similar as Wang's work except for it using a much more sophisticated approach to compute aspect weights. Moghaddam et al. propose the ILDA model [14] in which they also use the concept of latent aspect rating and derive those latent labels under the context of topic models. Julie et al. propose a method that combines LDA and matrix factorization algorithm into the same framework that uses LDA to process the textual content and uses the matrix factorization algorithm to process the ratings [11].

Though review ratings are taken into account in those methods, to our best knowledge, none of them explicitly exploits the straightforward relations between ratings, aspect weights and sentiment polarities. Therefore in this paper we propose the RS-LTM model that considers the relations to try to fill in the gap. We will show details in upcoming sections.

3   Rating Supervised Latend Topic Model

In this paper, we use m to denote a review and use d to denote a review sentence. We treat each sentence instead of each review as a document and will use "sentence" and "document" interchangeably in upcoming sections. We assume there are two sentiment polarities: positive (denoted by "+") and negative (denoted by "−") in the proposed model.

3.1   Model Description

By following the approaches of Sentiment-LDA [9] and other related works, we assume the aspect label and sentiment polarity of each word are sampled from a document-specific multinomial distribution over aspects $Multi(\theta)$ and a binomial distribution over sentiment polarities $Bino(\pi)$, respectively. Usually, parameters of the multinomial distribution θ are assumed to be further sampled from a *Dirichlet* distribution; parameters of the binomial distribution π are assumed to be sampled from a *Beta* distribution. Therefore the hyper parameters of the *Dirichlet* distribution α would determine the expectation of θ and the hyper parameters of the *Beta* distribution λ would determine the expectation of π. In existing techniques α and λ are usually symmetric and assigned with empirical values. In this paper, we maintain the same practice on α while making a difference on λ that allows review ratings and aspect weights to interfere in the generation of sentimental labels in the model.

In our approach, we assume the sentiment distribution π_{kr} of an aspect k under a rating r is sampled from a *Beta* distribution whose hyper parameters $\lambda_{kr} = \{\lambda_{kr+}, \lambda_{kr-}\}$ are determined as follows:

$$
\lambda_{kr+} = \begin{cases} \exp(r * \delta_k * \rho) & r > 0.5 \\ \exp(1) & r = 0.5 \\ \exp(r) & r < 0.5 \end{cases} \quad \lambda_{kr-} = \begin{cases} \exp(1-r) & r > 0.5 \\ \exp(1) & r = 0.5 \\ \exp((1-r) * \delta_k * \rho) & r < 0.5 \end{cases} \quad (1)
$$

where r is the normalized rating, δ_k the importance weight of aspect k. ρ is an adjustment coefficient that is defined on an ad-hoc basis to ensure $\delta_k * \rho \geq 1$ for any aspect k. The equation indicates that when a rating is greater than one-half of the full scale rating, the hyper parameter λ_{kr+} corresponding to the positive sentiment weight π_{kr+} would be greater than the hyper parameter λ_{kr-} for the negative sentiment weight π_{kr-}; otherwise λ_{kr-} would be greater than λ_{kr+}. According to the nature of *Beta* distribution, the expectation of a sentiment distribution weight π_{kr} sampled from $Beta(\lambda_{kr})$ can be computed as follows (Table 1):

$$
E(\pi_{krs}) = \frac{\lambda_{krs}}{\lambda_{kr+} + \lambda_{kr-}} \quad (2)
$$

By plugging λ described in Eq. (1) into the expectation equation, we can easily get that: when $r > 0.5$, then $E(\pi_{kr+}) > E(\pi_{kr-})$; when $r < 0.5$, $E(\pi_{kr+}) < E(\pi_{kr-})$. It leads to that a sentiment polarity drawn from $Multi(\pi_{rs})$ is inclined to be positive when the rating is greater than one-half of the full scale rating and to be negative when it is less than one-half of the full scale rating. Furthermore, it also implies that the inclinations on more important aspects would be more clear than that on less important aspects. This can be interpreted in plain English that with the help of λ, aspects are very likely to be positive in reviews with high ratings and to be negative in reviews with low ratings; important aspects are more likely to be positive than less important aspects in reviews with high ratings, and conversely, important aspects are more likely to be negative in reviews with low ratings. Therefore the proposed *Beta* distribution can capture the relations between aspect weights, sentiment polarities and ratings. We replace the symmetric *Beta* distribution in existing work with the proposed asymmetric ones and specify the generative process as follows:

Table 1. Table of notations

M	Total number of reviews
D	Total number of review sentences
K	Total number of aspects
T	Total number of terms in vocabulary
S	Total number of sentiment polarities
R	Full rating scale
z_{di}	Aspect label for the ith word in d
x_{di}	Sentiment polarity for the ith word in d
g_d	The rating of review sentence d
θ_d	K-dimensional aspect distribution for d
ϕ_{ks}	Word distribution in k under sentiment s
π_{kr}	Sentiment distribution in k under rating r
δ_k	The importance weight of aspect k

For the review corpus:

1. For each aspect k under each rating score r, sample a sentiment distribution $\pi_{kr} \sim Beta(\lambda_{kr})$
2. For each aspect k under each sentiment s, sample a word distribution $\phi_{ks} \sim Dir(\beta)$

For each document d with N_d words:

1. Sample an aspect distribution $\theta_d \sim Dir(\alpha)$
2. For each word position i in d
   2.1 sample an aspect $z_{di} \sim Multi(\theta)$
   2.2 sample a sentiment polarity $x_{di} \sim Bino(\pi_{z_{di}g_d})$
   2.3 sample a word $w_{di} \sim Multi(\phi_{z_{di}x_{di}})$

We show the Bayesian network of the proposed model in Fig. 1. Based on the Bayesian network, we can get:

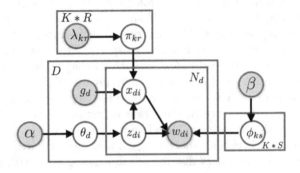

Fig. 1. The Bayesian network of RS-LTM

$$p(\tau|\Theta, \Pi, \Phi, g) = \prod_{d=1}^{D} \prod_{i=1}^{N_m} \sum_{z_{di}=1}^{K} p(z_{di}|\theta_d) \sum_{x_{di}=1}^{S} p(x_{di}|z_{di}, g_d, \Pi) p(w_{di}|z_{di}, x_{di}, \Phi)$$

(3)

$$p(\Theta|\alpha) = \prod_{d=1}^{D} Dir(\theta_d|\alpha)$$

(4)

$$p(\Pi|\lambda) = \prod_{k} \prod_{r} Beta(\pi_{kr}|\lambda_{kr})$$

(5)

$$p(\Phi|\beta) = \prod_{k} \prod_{s} Dir(\phi_{ks}|\beta)$$

(6)

where τ denotes the corpus. Θ, Π, Φ in Eq. (3) can be estimated by Gibbs sampling [5]. However, the aspect weights that serve as a prior knowledge in the topic model are unknown beforehand and have to be initialized with random values. To justify the topic model, extra constraints on the weights need to be imposed to reduce the randomness of the weights and make them consistent with the training dataset. In this paper we adopt the rating regression model proposed by Wang et al. [17] as the constraint. In this model the authors assume each aspect k in a review m has a latent rating g'_{mk} and the overall rating g_m is sampled from a Gaussian distribution with the weighted summation of each aspect's latent rating as the mean.

$$g_m \sim N(\sum_{k} \delta_k * g'_{mk}, \sigma)$$

(7)

where δ_k is aspect k's weight. σ denotes the variance and it is predefined to be 1 in this paper. The latent rating g'_{mk} of an aspect k in a review is simplified as the normalized frequency of positive words labelled with the aspect. It can be written as:

$$g'_{mk} = n_{mk+}/N_m$$

(8)

Optimal aspect weights δ can be computed by maximizing the rating likelihood $\sum_m \log p(g_m)$. Therefore we maximize both the corpus likelihood (3) and the rating likelihood simultaneously. We can combine the topic model and the constraint by solving the following combined optimization problem:

$$\arg\max_{\Psi}, h(\tau|\Psi, g) = p(\tau|\Psi, g) + \sum_{m} \log p(g_m)$$

(9)

$$\Psi = \{\Theta, \Phi, \Pi, \delta\}$$

3.2   Model Inference

Since the inference of the topic model is intractable, direct inference of the optimization problem is out of the question [1]. In this paper we solve the problem by using an alternating algorithm. As aforementioned, Gibbs sampling can be used to estimate Θ, Φ, Π, Z, X in the topic model. Given Z and X, we can easily estimate each aspect's latent rating. We plug the latent ratings into the regression model and use gradient ascent to fit aspect weights δ. Then we feed the resulting aspect weights δ into the Gibbs sampling process and alternate between the two steps until convergent.

To use Gibbs sampling, we first initialize each word w_{di} in each document d with a random aspect label z_{di} and a random sentiment polarity x_{di}. Then, we update the aspect label z_{di} and sentiment polarity x_{di} of each word by conditioning on the aspect labels Z_{-di} and sentiment polarities X_{-di} of the remaining words in the corpus. Therefore we compute $p(z_{di} = k, x_{di} = s | Z_{-di}, X_{-di}, W)$ for the Gibbs sampling process. Assuming $w_{di} = t$, $g_d = r$, by following method presented by Heinrich [5], we can get

$$p(z_{di} = k, x_{di} = s | Z_{-di}, X_{-di}, W) \propto \frac{n_{ks,-di}^{t} + \beta_t}{\sum_{t=1}^{T} n_{ks,-di}^{t} + \beta_t}$$

$$\times \frac{\lambda_{krs} + n_{kr,-di}^{s}}{\sum_s \lambda_{krs} + n_{kr,-di}^{s}} \tag{10}$$

$$\times (n_{d,-i}^{k} + \alpha_k)$$

where n_{ks}^{t} denotes the number of times that term t is assigned with aspect k and sentiment polarity s in the corpus; n_{kr}^{s} denotes the number of times aspect k under rating r is assigned with sentiment polarity s in the corpus; n_d^{k} denotes the number of words is assigned with aspect k in document d.

In the aforementioned gradient ascent algorithm for the constraint, we initialize aspect weights δ with same values used in the topic model. δ can be updated in each iteration of the gradient ascent algorithm as follows:

$$\delta_k = \delta_k + \gamma * \sum_m g'_{mk}(g_m - \sum_k g'_{mk}\delta_k) \tag{11}$$

where γ is the step size. Based on the above description we summarize the inference process in Algorithm 1.

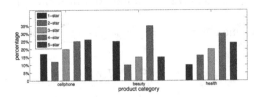

Fig. 2. Rating distribution in the training dataset

Algorithm 1. Inference algorithm

---

procedure INFERENCE(τ, g)
    Initialize δ with random values
    while not convergent **do**
        run phase 1 then feed the results to phase 2
        run phase 2 then feed the results to phase 1
    end while
    return Ψ
end procedure

Phase 1 -Gibbs sampling

Input: τ, g, δ
Initialize Z and X with random values
repeat
    for each document d in the corpus **do**
        for each word i in d **do**
            update z_{di} and x_{di} according to equation (10), then compute Θ, Φ, Π
        end for
    end for
until convergent
return Θ, Φ, Π, Z, X

Phase 2 -Gradient ascent

Input: $\Theta, \Phi, \Pi, Z, X, g$
  repeat
    update δ according to equation (11)
  until convergent
  return δ

---

4 Evaluation

In this section we evaluate the performance of the proposed model and compare it with other state-of-the-art methods on a review data set crawled from amazon.com [12]. In this dataset 3 categories of products are included: health, beauty and cellphone. After removing items whose number of reviews is less then 100, there are 2,687 items, and 253,730 reviews left in the beauty category, 1,581 items and totally 41,9491 reviews included in the health category, 1,200 items and 366,824 reviews included in the cellphone category. The distribution of rating scores in each category is shown in Fig. 2.

We pre-process the data before feeding them into the proposed model as follows. We first use Part-of-Speech (POS) taggers to give each word in each sentence a POS tag then extract the trunks of each sentence. We use 3 sequential patterns to extract sentence trunks: noun+verb+(adv)+(adj)+(noun); verb+(adj/adv)+noun+(adv)+(adj); adj+noun. The components in brackets can appear for 0 or 1 time. In a sentence, we first check if there is any substring matches the first pattern. If no match is found, we try the second one and then the third one. For example, in a sentence "the phone allows you to use many ring tones", two parts: "phone allows", "use many ring tones" will be extracted from the sentence. Finally we manually remove words that are usually neutral from the sentence trunks.

We use 80 % of the reviews in each category as the training set and hold out the remaining ones as the test set. In the test set, we manually cluster all the words into different semantic classes. For example, we allot words like "battery", "charger", "die" into a cluster labelled with "power" and allot words like "affordable", "expensive", "cheap" into another cluster labelled with "price". We first train the proposed model on the training set then use the learnt model parameters to assign aspect and sentiment labels to words in the test set. We evaluate the performance of the proposed model and compare it with 3 state-of-the-art models: ILDA [14], sentiment-LDA [9], ASUM [7]. We use 3 metrics: accuracy of aspect discovery, accuracy of sentiment classification, model perplexity to compare their performances.

Accuracies of Aspect Discovery and Sentiment Classification. We use Rand Index [14] to calculate the accuracy of aspect discovery. It can be calculated by the following equation:

$$Accu_a = \frac{2 \times (n_x + n_y)}{N} \tag{12}$$

where n_x is the number of word pairs that are assigned into the same clusters by both the models and the manual labeling. n_y is the number of word pairs that are assigned into different clusters by both the models and the manual labeling. N is the total number of word pairs in the test set.

We set the parameters for all the 4 models as follows: $K \in [4, 7]$; $\alpha = 2$; $\beta = 0.5$. We calculate the average accuracy of each model over all the parameter settings on each category and report the results in Table 2. The results indicate the proposed model performs better in average precision than the other three models.

Table 2. Aspect discovery accuracies of RS-LTM, ASUM, ILDA, Senti-LDA

	Cellphone				Health				Beauty				Average
	K=4	K=5	K=6	K=7	K=4	K=5	K=6	K=7	K=4	K=5	K=6	K=7	accuracy
RS-LTM	0.71	0.74	0.68	0.75	0.70	0.68	0.71	0.76	0.67	0.66	0.63	0.7	0.7042
ASUM	0.70	0.66	0.69	0.67	0.72	0.70	0.74	0.68	0.60	0.70	0.62	0.64	0.6767
ILDA	0.58	0.68	0.63	0.66	0.58	0.66	0.56	0.69	0.63	0.65	0.62	0.61	0.6342
Senti-LDA	0.74	0.75	0.68	0.69	0.76	0.70	0.60	0.68	0.60	0.58	0.66	0.60	0.6700

We manually label review sentences in the test set with sentiment polarities and calculate the accuracy of sentiment classification as follows:

$$Accu_s = \frac{n_c}{N_t} \tag{13}$$

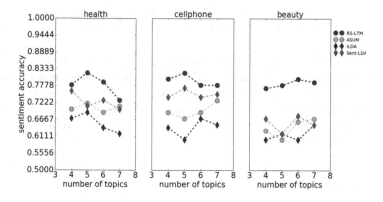

Fig. 3. Sentiment classification accuracies of RS-LTM, ASUM, ILDA, Senti-LDA

where n_c is the number of test review sentences whose sentiments are correctly classified. N_t denotes the total number of test sentences. We compute the average sentiment classification accuracies of each model over all the parameter settings on the three product categories and report the results in Fig. 3.

The results indicate that the proposed model has much better performances than that of other models on all the three categories. We attribute the considerable improvement to that we use review ratings as a prior knowledge to learn the sentiment distributions. It is worth noting that the proposed model improves the most on the category of beauty because this category has a higher portion of 1-star reviews than others therefore it can provide more "confident" evidence for negative sentiments during the training process.

Model Perplexity. We compute the perplexity of each model [1] by the following equation and report the results in Fig. 4.

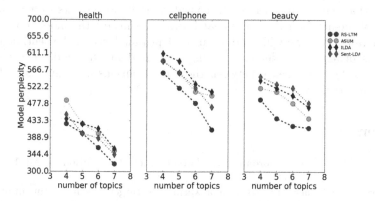

Fig. 4. Model and classification perplexities of RS-LTM, ASUM, ILDA, Senti-LDA

$$perplexity_m(D_{test}) = \exp(-\frac{\sum_d \log p(w_d)}{\sum_d N_d}) \qquad (14)$$

A lower perplexity value means a better generalization performance. The results indicate that the proposed model has a slightly lower model perplexity than other models. This could be due to that the proposed model takes into account relations between sentiment polarities, aspect weights and ratings. Finally we show a snippet of the aspects extracted from the cellphone reviews by the proposed model in Table 3.

Table 3. Examples of cellphone topics extracted by RS-LTM

Aspect 1 (screen)		Aspect 2 (power)		Aspect 3 (price)		Aspect 4 (shipping)	
Positive	Negative	Positive	Negative	Positive	Negative	Positive	Negative
Clear	Scratch	Life	Die	Affordable	Expensive	Bubble	Days
Size	Small	Long	Shortly	Cheap	Waste	Fast	Broken
Big	Borken	Removable	Lost	Worthy	Deal	Delivery	Protection
HD	Disappointed	Charge	Burning	Discount	Refund	Easy	Dirt
Backlight	Low	Work	Quickly	Good	Buck	Proper	Service
Film	Noticeable	Happy	Lose	Recommend	Trick	Arrive	Disappear
Model	Crack	Excellent	Backup	Down	Surprise	Responsible	Lie
Reading	Pixel	Adapter	Junk	Gift	Bother	Free	Photo
Clean	Poor	Strong	Hour	Benefit	Dollar	Express	Replace
Bright	Toy	Day	Plastic	Save	Account	Prime	Missing

5   Conclusion

In this paper we propose Rating Supervised Latent Topic Model for joint aspect discovery and sentiment classification. In this model the relations between ratings, aspect weights and sentiments are taken into account. Experiment results indicate that our approach leads to considerable improvements over a number of state-of-the-art models. In the near future we will test the model on a much larger dataset. Also we will conduct an in-depth analysis on the sentiment distributions of aspects under each rating generated by the model because the Kullback-Leibler divergence between those distributions may give us a clue to identify fake reviews.

References

1. Blei, D.M., Ng, A.Y., Jordan, M.I.: Latent dirichlet allocation. J. Mach. Learn. Res. **3**, 993–1022 (2003)
2. Fahrni, A., Klenner, M.: Old wine or warm beer: target-specific sentiment analysis of adjectives. In: Proceedings of the Symposium on Affective Language in Human and Machine, AISB, pp. 60–63 (2008)

3. Fletcher, J., Patrick, J.: Evaluating the utility of appraisal hierarchies as a method for sentiment classification. In: Proceedings of the Australasian Language Technology Workshop (2005)
4. Guo, H., Zhu, H., Guo, Z., Zhang, X., Su, Z.: Product feature categorization with multilevel latent semantic association. In: Proceedings of the 18th ACM Conference on Information and Knowledge Management, CIKM 2009, pp. 1087–1096. ACM, New York (2009)
5. Heinrich, G.: Parameter estimation for text analysis. Technical report (2005)
6. Hu, M., Liu, B.: Mining and summarizing customer reviews. In: Proceedings of the Tenth ACM SIGKDD International Conference on Knowledge Discovery and Data Mining, pp. 168–177. ACM (2004)
7. Jo, Y., Oh, A.H.: Aspect and sentiment unification model for online review analysis. In: Proceedings of the Fourth ACM International Conference on Web Search and Data Mining, pp. 815–824. ACM (2011)
8. Li, F., Han, C., Huang, M., Zhu, X., Xia, Y.J., Zhang, S., Yu, H.: Structure-aware review mining and summarization. In: Proceedings of the 23rd International Conference on Computational Linguistics, pp. 653–661. Association for Computational Linguistics (2010)
9. Li, F., Huang, M., Zhu, X.: Sentiment analysis with global topics and local dependency. In: AAAI, vol. 10, pp. 1371–1376 (2010)
10. Lin, C., He, Y.: Joint sentiment/topic model for sentiment analysis. In: Proceedings of the 18th ACM Conference on Information and Knowledge Management, pp. 375–384. ACM (2009)
11. McAuley, J., Leskovec, J.: Hidden factors and hidden topics: understanding rating dimensions with review text, pp. 165–172. ACM Press (2013)
12. McAuley, J., Pandey, R., Leskovec, J.: Inferring networks of substitutable and complementary products. In: Proceedings of the 21th ACM SIGKDD International Conference on Knowledge Discovery and Data Mining, pp. 785–794. ACM (2015)
13. Mei, Q., Ling, X., Wondra, M., Su, H., Zhai, C.: Topic sentiment mixture: modeling facets and opinions in weblogs. In: Proceedings of the 16th International Conference on World Wide Web, pp. 171–180. ACM (2007)
14. Moghaddam, S., Ester, M.: ILDA: interdependent LDA model for learning latent aspects and their ratings from online product reviews. In: Proceedings of the 34th International ACM SIGIR Conference on Research and Development in Information Retrieval, SIGIR 2011, pp. 665–674. ACM, New York (2011)
15. Mukherjee, A., Liu, B.: Aspect extraction through semi-supervised modeling. In: Proceedings of the 50th Annual Meeting of the Association for Computational Linguistics: Long Papers, vol. 1, pp. 339–348. Association for Computational Linguistics (2012)
16. Titov, I., McDonald, R.: Modeling online reviews with multi-grain topic models. In: Proceedings of the 17th International Conference on World Wide Web, pp. 111–120. ACM (2008)
17. Wang, H., Lu, Y., Zhai, C.: Latent aspect rating analysis on review text data: a rating regression approach. In: Proceedings of the 16th ACM SIGKDD International Conference on Knowledge Discovery and Data Mining, pp. 783–792. ACM (2010)
18. Yu, J., Zha, Z.J., Wang, M., Chua, T.S.: Aspect ranking: identifying important product aspects from online consumer reviews. In: Proceedings of the 49th Annual Meeting of the Association for Computational Linguistics: Human Language Technologies, vol. 1, pp. 1496–1505. Association for Computational Linguistics (2011)

19. Zhai, Z., Liu, B., Xu, H., Jia, P.: Grouping product features using semi-supervised learning with soft-constraints. In: Proceedings of the 23rd International Conference on Computational Linguistics, pp. 1272–1280. Association for Computational Linguistics (2010)
20. Zhai, Z., Liu, B., Xu, H., Jia, P.: Constrained LDA for grouping product features in opinion mining. In: Huang, J.Z., Cao, L., Srivastava, J. (eds.) PAKDD 2011, Part I. LNCS, vol. 6634, pp. 448–459. Springer, Heidelberg (2011)
21. Zhao, W.X., Jiang, J., Yan, H., Li, X.: Jointly modeling aspects and opinions with a MaxEnt-LDA hybrid. In: Proceedings of the 2010 Conference on Empirical Methods in Natural Language Processing, EMNLP 2010, pp. 56–65. Association for Computational Linguistics, Stroudsburg (2010)

On Various Types of Even-Sized Clustering Based on Optimization

Yasunori Endo[1(✉)], Tsubasa Hirano[2], Naohiko Kinoshita[3],
and Yikihiro Hamasuna[4]

[1] Faculty of Engineering, Information and Systems, University of Tsukuba,
1-1-1, Tennodai, Tsukuba, Ibaraki 305-8573, Japan
`endo@risk.tsukuba.ac.jp`
[2] Canon Inc., 3-30-2, Shimomaruko, Ota-ku, Tokyo 146-0092, Japan
[3] Research Fellowship for Young Scientists of JSPS, University of Tsukuba,
1-1-1, Tennodai, Tsukuba, Ibaraki 305-8573, Japan
`kinoshita@risk.tsukuba.ac.jp`
[4] Department of Informatics, Kindai University,
3-4-1, Kowakae, Higashiosaka, Osaka 577-8502, Japan
`yhama@info.kindai.ac.jp`

Abstract. Clustering is a very useful tool of data mining. A clustering method which is referred to as K-member clustering is to classify a dataset into some clusters of which the size is more than a given constant K. The K-member clustering is useful and it is applied to many applications. Naturally, clustering methods to classify a dataset into some even-sized clusters can be considered and some even-sized clustering methods have been proposed. However, conventional even-sized clustering methods often output inadequate results. One of the reasons is that they are not based on optimization. Therefore, we proposed Even-sized Clustering Based on Optimization (ECBO) in our previous study. The simplex method is used to calculate the belongingness of each object to clusters in ECBO. In this study, ECBO is extended by introducing some ideas which were introduced in k-means or fuzzy c-means to improve problems of initial-value dependence, robustness against outliers, calculation cost, and nonlinear boundaries of clusters. Moreover, we reconsider the relation between the dataset size, the cluster number, and K in ECBO.

1 Introduction

In recent years, collecting and accumulating vast amounts of data has become very easy along with the improvement of computers and the spread of the internet. The data which are collected and accumulated from many and unspecified users is known as the big data. It becomes very difficult for us to deal with the big data directly because of the scale of the big data, and then, data mining technique, which is to obtain useful information and new knowledge beyond our image automatically, has been very important.

Clustering is one of the data mining technique and it classifies a dataset into some clusters automatically. The classification is based on degree of similarity or dissimilarity between objects and do not need any supervised data. Therefore, clustering is an unsupervised classification method. This approach has been

© Springer International Publishing Switzerland 2016
V. Torra et al. (Eds.): MDAI 2016, LNAI 9880, pp. 165–177, 2016.
DOI: 10.1007/978-3-319-45656-0_14

extensively studied from the second half of the 20th century, and many variations of clustering methods have been devised. K-member clustering (KMC) is one of the clustering method and it classifies a dataset into some clusters of which the size is at least K. The following three methods are known as typical KMC methods: greedy k-member clustering (GKC), one-pass k-means algorithm for K-anonymization (OKA), and clustering-based K-anonymity (CBK).

However, those algorithms have some problems. The problem of GCK and OKA is that the clusters have sometimes no sense of unity, and the problem of CBK is that the cluster number is not maximized under the constraint that the size of each cluster is or more than K. To solve the problem of CBK, two-division clustering for K-anonymity of cluster maximization (2DCKM) was proposed and 2DCKM was extended by one of the authors which is referred to .as extended two-division clustering for K-anonymity of cluster maximization [1] (E2DCKM). Both of the methods are based on CBK, then they obtain final cluster division from iteration of classification of one cluster into two clusters and adjustment of each cluster size.

However, the classification accuracy of the above methods is not so high. One of the reason is that those methods is not based on optimization unlike the useful clustering methods such as hard c-means (HCM) and fuzzy c-means.

Typical useful clustering methods, e.g. FCM and HCM, are constructed based on optimization of the given objective function. In addition, spectral clustering is also constructed based on optimization. Therefore, construction of methods based on optimization is one of the solution to the problem about classification accuracy. Moreover, the objective function itself is also an evaluation guideline of results of clustering methods which strongly depend on initial value. Considering together extensibility from the mathematical point of view, there is the great advantage to construct clustering methods in the framework of optimization.

From the above viewpoint, some of the authors was constructed an even-sized clustering method, which is with more strengthened constraints of cluster size than KMC, in the framework of optimization [2]. The constraint is that each cluster size is K or $K+1$. Here we have to notice that the existence of the cluster number c obviously depends on the dataset size n and K. For example, in case that $n = 10$ and $K = 6$, the cluster number c which satisfies the conditions does not exist. Conversely, K exists for any c $(c < n)$. Therefore, a condition of n and K for preventing the case is needed. The even-sized clustering algorithm based on optimization is referred to as ECBO. ECBO is based on HCM and its algorithm is constructed as iterative optimization. The belongingness of each object to clusters are calculated by the simplex method in each iteration. Thus, ECBO has high classification accuracy.

However, ECBO has also the following problems: (1) the results strongly depend on initial values (initial-value dependence), (2) the algorithm is not robust against outliers, (3) it needs a lot of calculation cost, and (4) it is very difficult to classify datasets which consist of clusters with nonlinear boundary. In this study, we describe various types of ECBO to solve the above problems and estimate the methods in some numerical examples. Moreover, we mention one extension of ECBO for dealing with datasets on a sphere.

2   Even-Sized Clustering Based on Optimization (ECBO)

Let $x \in \Re^p$ be an object, and $X = \{x\}$ be a set of objects. $v \in \Re^p$ and V is a cluster center and a set of cluster centers, respectively. C_i is the i-th cluster. Moreover, let $U = (u_{ki})_{k=1,\ldots,n,\ i=1,\ldots,c}$ be a partition matrix of membership grades. $u_{ki} = 1$ iff x_k is in C_i and $u_{ki} = 0$ iff x_k is not in C_i.

Even-sized clustering classifies datasets into some clusters of which the object number are almost even. Depending on the size of dataset, it is not possible to classify the dataset into completely evenly, then each cluster size is defined as K or $K + 1$. Here, K is a given constant. As mentioned above, the existence of c depends on the dataset size n and the given number K for cluster size. We mention the condition of K later.

Even-Sized Clustering Based on Optimization (ECBO) is one of even-sized clustering and it is based on HCM. The difference of ECBO from the conventional even-sized clustering is that ECBO classifies datasets in the framework of optimization, that is, ECBO minimizes a objective function under some constraints.

The objective function and constraints are as follows:

$$\text{minimize} \quad J_{\text{ECBO}}(U, V) = \sum_{k=1}^{n} \sum_{i=1}^{c} u_{ki} \|x_k - v_i\|^2 \tag{1}$$

$$\text{s.t.} \quad \sum_{i=1}^{c} u_{ki} = 1 \quad (k = 1, \ldots, n) \tag{2}$$

$$K \leq \sum_{k=1}^{n} u_{ki} \leq K + 1 \quad (i = 1, \ldots, c) \tag{3}$$

(1) and (2) are the same objective function and constraint as HCM. (3) is the constraints for even cluster size.

These equations are linear with u_{ki}, hence the optimal solution of u_{ki} is obtained by the simplex method. The cluster centers v_i can be calculated in the same way of HCM.

Before starting the ECBO algorithm, we have to give a constant K or a cluster number c. The relation between the dataset size n, c, and K in ECBO was considered in Ref. [2]. Here, we reconsider more precise relation.

The relation between n, c and K is $K = \lfloor \frac{n}{c} \rfloor$. If n and c are given, K exists. On the other hand, even if n and K are given, c does not always exist. Thus, it is necessary to satisfy the following relation between n and K:

$$0 < n \leq (K + 1) \frac{n - (n \bmod K)}{K} \tag{4}$$

If (4) holds true,

$$c = \frac{n - (n \bmod K)}{K}.$$

We present the ECBO algorithm as Algorithm 1.

Algorithm 1. ECBO

Step 0. Give the constants K or c.
Step 1. Give the initial cluster centers V randomly.
Step 2. Update U by the simplex method.
Step 3. Update V by $v_i = \sum_{k=1}^n u_{ki} x_k / \sum_{k=1}^n u_{ki}$.
Step 4. If V changes from previous V, go back to **Step 2**. Otherwise, stop.

---

3 Proposed Methods: Extended ECBO

3.1 ECBO++

We set the initial cluster centers randomly in the ECBO algorithm. However, the ECBO is based on HCM and then, the ECBO has the same problem as HCM, i.e., initial-value dependence. To the problem of HCM, k-means++ was proposed by Arthur et al. [3]. In the method, the initial cluster centers are stochastically selected and the improvement of initial-value dependence was proved theoretically. Therefore, we propose a new method ECBO++ to extend the ECBO by using the same selection method of the initial cluster centers as k-means++.

The first cluster center is selected randomly. Next, we select each cluster center on the probability $\frac{D(x)^2}{\sum_{x \in X} D(x)^2}$. Here, $D(x)$ is a distance between an object x and the nearest cluster center which is already selected.

The ECBO++ uses the above procedure as selection of initial cluster centers. The objective function and constraints of The ECBO++ are the same as ones of the ECBO. We present the ECBO++ algorithm as Algorithm 2.

Algorithm 2. ECBO++

Step 0. Give the constants K or c.
Step 1. Select the initial cluster centers by following process:
  Step 1a. Select an object $x \in X$ randomly as a cluster center.
  Step 1b. Select $x \in X$ with the probability $D(x)^2 / \sum_{x \in X} D(x)^2$ as a new center v_i.
  Step 1c. Iterate **Step 1b.** until we select c cluster centers.
Step 2. Update U by the simplex method.
Step 3. Update V by $v_i = \sum_{k=1}^n u_{ki} x_k / \sum_{k=1}^n u_{ki}$.
Step 4. If V changes from previous V, go back to **Step 2**. Otherwise, stop.

3.2 L_1ECBO

L_1-norm, also referred to as Manhattan distance, is often used on data analysis. The distance is presented as the following equation:

$$\|x - y\|_1 = \sum_{j=1}^{p} |x_j - y_j|.$$

Here, $x = (x_1, \ldots, x_p) \in \Re^p$ and $y = (y_1, \ldots, y_p) \in \Re^p$. L_1 fuzzy c-means [4] (L_1FCM) has higher robustness against outliers than FCM with squared Euclidean distance. Therefore, it is expected that the robustness of outliers by ECBO is improves by introducing the L_1-norm similar to L_1FCM. We consider an objective function with the L_1-norm instead of the squared Euclidean distance in (1) as follows:

$$J_{L_1\mathrm{ECBO}}(U, V) = \sum_{k=1}^{n} \sum_{i=1}^{c} u_{ki} \|x_k - v_i\|_1 = \sum_{k=1}^{n} \sum_{i=1}^{c} \sum_{j=1}^{p} u_{ki} |x_{kj} - v_{ij}| = \sum_{i=1}^{c} \sum_{j=1}^{p} J_{ij}.$$

The constraints are the same as ones of the ECBO. Here, J_{ij} is referred to as the semi-objective function.

When J_{ij} for all i and j is minimized, the objective function $J_{L_1\mathrm{ECBO}}$ is also minimized. Therefore, we can consider minimization of J_{ij} instead of $J_{L_1\mathrm{ECBO}}$. Because J_{ij} is a convex and piecewise linear function, the point that a sign of $\frac{\partial J_{ij}}{\partial v_{ij}}$ is changed from minus to plus is an optimal solution.

For exploring the solution, we sort the j-th coordinate of all objects $\{x_{1j}, \ldots, x_{nj}\}$ in ascending order to $x_{q_j(1)j} \leq x_{q_j(2)j} \leq \cdots \leq x_{q_j(n)j}$. Here, $q_j(k)$ ($k = 1, \ldots, n$) are a permutation of $\{1, \ldots, n\}$, and they correspond to the object numbers. By using $x_{q_j(k)}$ and J_{ij}, we can rewrite J_{ij} as follows:

$$J_{ij} = \sum_{k=1}^{n} u_{ki} |x_{kj} - v_{ij}| = \sum_{k=1}^{n} u_{q_j(k)i} |x_{q_j(k)j} - v_{ij}|.$$

Thus, $\frac{\partial J_{ij}}{\partial v_{ij}}\Big|_{x_{q(k)j}}$ is expressed as follows:

$$\frac{\partial J_{ij}}{\partial v_{ij}}\Big|_{x_{q_j(r)j}} = -\sum_{k=r+1}^{n} u_{q_j(k)i} + \sum_{k=1}^{r} u_{q_j(k)i}.$$

The point $v_{ij} = x_{q_j(r)j}$ that the sign of derivative changes from negative to positive is the optimal solution. We present the L_1ECBO algorithm as Algorithm 3.

Algorithm 3. L_1ECBO

Step 0. Give the constants K or c.
Step 1. Give the initial cluster centers V randomly.
Step 2. Update U by the simplex method.
Step 3. Update V by minimizing J_{ij} for v_{ij}.
Step 4. If V changes from previous V, go back to **Step 2**. Otherwise, stop.

3.3   MECBO

k-medoids clustering [5] is a method in which each cluster center are represented by an object that is the nearest to the center. Such an object is referred to as the medoid. The k-medoids clustering can classify the dataset if the distances between objects are given. It is known that the k-medoids clustering is more robust against outliers than HCM [6].

A The method proposed by Park et al. [6] minimizes the objective function by iterative optimization for the medoids V and the belongingness of each object to clusters. $x_k \in C_i$ of which the sum of distance between $x_j \in C_i$ is minimum for i is selected as a new medoid, that is,

$$v_i = \arg \min_{x_k \in X_i} \sum_{x_j \in C_i} d_{kj}. \tag{5}$$

Here, $d_{kj} = \|x_k - x_j\|$. Therefore, we apply the medoids to ECBO which is referred to as medoid-ECBO (MECBO). The objective function and constraints are the same as ones of the ECBO, but the measure of distance in the function of the MECBO is Euclidean distance. We introduce the above procedure of k-medoids clustering to ECBO to obtain V. The simplex method is used to obtain the optimal solutions to U. We present the MECBO algorithm as Algorithm 4.

Algorithm 4. MECBO

Step 0. Give the constants K or c.
Step 1. Give the initial cluster centers V randomly.
Step 2. Update U by the simplex method.
Step 3. Update V by (5).
Step 4. If V changes from previous V, go back to **Step 2**. Otherwise, stop.

3.4   KECBO

Since the ECBO is based on the HCM, it can not classify datasets which consist of clusters with nonlinear boundary. Therefore, we propose KECBO in which a kernel function is introduced to classify such datasets.

Kernel Hard c-means (KHCM) proposed by Girolami [7] is a method based on HCM to classify datasets which are not able to be classified with linear boundaries. In KHCM, a kernel function, which maps objects in the original space to a higher dimensional feature space, plays very important role. KHCM classifies all objects not in the original space but in the feature space by the kernel function. We introduce the idea of kernel functions in the ECBO.

Let $\phi : \Re^p \to F$ be mapping from the p-dimensional original space to a higher feature space F. The objective function of KHCM and KECBO are the same as follows:

$$J_{\text{KECBO}}(U, V) = \sum_{k=1}^{n} \sum_{i=1}^{c} u_{ki} d_\phi(x_k, v_i) = \sum_{k=1}^{n} \sum_{i=1}^{c} u_{ki} \|\phi(x_k) - v_i^\phi\|^2.$$

Here, v_i^ϕ is a cluster center on F. The constraints of KECBO are as same as the ECBO.

If we calculated the distance on a feature space directly, calculation cost becomes enormous. Therefore, the distance is usually calculated by using a the following κ which represents a inner product of vectors on a feature space.

$$\kappa(x, y) = \langle \phi(x), \phi(y) \rangle.$$

κ is referred to as the kernel function.

The squared distance on a feature space d_ϕ is calculated as follows:

$$d_\phi(x, y) = \|\phi(x) - \phi(y)\|^2 = \kappa(x, x) - 2\kappa(x, y), +\kappa(y, y)$$

We can calculate the inner product in F by using κ in the original space without explicit definition of ϕ. Typical kernel function is the Gaussian kernel as follows:

$$\kappa(x_i, x_j) = \exp\left(-\frac{\|x_i - x_j\|^2}{2\sigma^2}\right).$$

In general, the value of v_i^ϕ or the cluster center can not be directly calculated in the feature space. Thus, the distance between an object and a cluster center is calculated by the following formula deformation:

$$\|\phi(x_k) - v_i^\phi\|^2 = \kappa(x_k, x_k) - \frac{2}{|C_i|} \sum_{x_l \in C_i} \kappa(x_k, x_l) + \frac{1}{|C_i|^2} \sum_{x_l \in C_i} \sum_{x_m \in C_i} \kappa(x_l, x_m).$$

$$(6)$$

KECBO updates a distance matrix $\Delta = (d_\phi(x_k, v_i))_{k=1 \sim n,\ i=1 \sim c}$ instead of calculation of V. The simplex method is used to obtain the optimal solutions to U.

We present the KECBO algorithm as Algorithm 5.

Algorithm 5. Kernel ECBO

Step 0. Give the constants K or c.
Step 1. Give the initial cluster centers V randomly.
Step 2. Calculate Δ.
Step 3. Update U by the simplex method.
Step 4. Update Δ.
Step 5. If U changes from previous U, go back to **Step 3**. Otherwise, stop.

---

4 Numerical Examples

We compared the proposed methods to an conventional method of E2DCKM in some data sets. FCM was used as a method to classify one cluster into two clusters in E2DCKM.

4.1   Information Loss Function

We were not able to compare the results in each proposed method because of difference of objective function. Therefore, we compared the results using Information Loss function (IL) [8]. It is a function which measures suitability of clustering. When loss of information is small, IL is small enough and the result is able to be considered suitable. IL is expressed by the following expression:

$$IL = \sum_{i=1}^{c} |C_i| \cdot \sum_{j=1}^{p} \frac{\hat{N}_j(C_i) - \check{N}_j(C_i)}{\hat{N}_j(X) - \check{N}_j(X)}.$$

$\hat{N}_j(S)$ is the maximum value of the j-th coordinate of the object in the set $S \subseteq X$, and $\check{N}_j(S)$ is the minimum value of the j-th coordinate of the object in S.

If all objects are in a cluster, IL is the maximum value. Conversely, if each object is in different clusters, IL is the minimum value 0.

4.2   Double Circle Data

We examined the dataset which have 50 objects in the small circle and 100 objects on the big circle. We present the dataset as Fig. 1. We examined the dataset for 100 times by each method at $c = 2, 3, 5$.

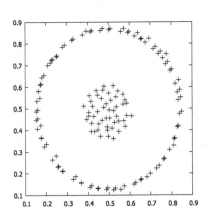

Fig. 1. Double circle data

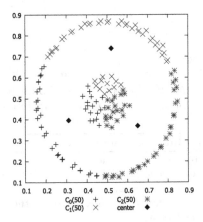

Fig. 2. The result of ECBO for double circle data (IL = 185.00)

We present the best values and average of IL values as Tables 1 and 2, respectively. Compared to E2DCKM, the average of IL values of all proposed methods had been very small. A little difference between the average and the minimum of IL values means that good results are obtained stably by proposed methods. We found little difference between IL values of ECBO and ECBO++. Further, the L_1ECBO showed the best value on $c = 2, 5$ and the second good value on

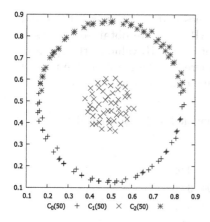

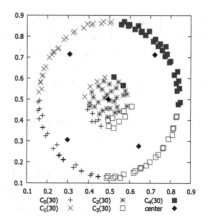

Fig. 3. The result of KECBO for double circle data (IL = 181.23)

Fig. 4. The result of ECBO for double circle data (IL = 130.65)

Table 1. The best IL value of each method in double circle data

c	2	3	5
E2DCKM	223.7	183.4	135.1
ECBO	224.6	185.0	130.7
ECBO++	224.6	185.0	130.4
L_1ECBO	224.5	181.5	127.8
MECBO	225.4	185.2	131.3
KECBO	237.3	181.2	134.3

Table 2. The average IL value of each method in double circle data

c	2	3	5
E2DCKM	241.3	201.9	151.4
ECBO	234.0	188.4	138.2
ECBO++	234.3	188.4	136.7
L_1ECBO	228.5	192.9	140.6
MECBO	238.0	198.1	140.8
KECBO	257.4	212.1	171.2

$c = 3$. KECBO showed the good value on $c = 3$, but it showed the worst value in other case.

We present the average execution time as Table 3. MECBO is the shortest execution time, and KECBO is the longest execution time.

We present the results that ECBO and KECBO classified the dataset into three clusters as Figs. 2 and 3, respectively. Against the linear classification of ECBO, the classification of KECBO is nonlinear as the small circle and the big one. Further, the IL value of KECBO is smaller than ECBO.

We also present the results that ECBO, KECBO and L_1ECBO classified the dataset into five clusters as Figs. 4, 5 and 6, respectively. In this case, the IL value of L_1ECBO is the minimum.

As mentioned above, KECBO can classify the dataset which consists of clusters with nonlinear boundary. However, according to the dataset distribution, the IL values of the classification results by KECBO are not always the minimum.

4.3 Fischer's Iris Dataset

Fischer's Iris dataset consists of 50 samples from each of three species of Iris. Each sample has four features: the length and the width of the sepals and petals.

We executed each method in $c = 3$ for 100 times and we compared the result of each method by Adjusted Rand Index (AARI) [9]. We obtained the result of HCM and FCM for comparison. We did not obtain IL value in the result of HCM and FCM because the IL value is considered in the methods We present the result of each method about AARI, IL value and execution time as Table 4.

Table 3. The average execution time of each method in double circle data [ms]

c	2	3	5
E2DCKM	0.48	0.77	0.92
ECBO	12.34	19.94	33.68
ECBO++	13.53	19.73	30.52
L_1ECBO	13.31	19.06	38.51
MECBO	10.64	16.01	29.58
KECBO	44.42	72.59	136.94

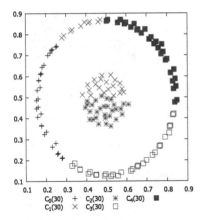

Fig. 5. The result of KECBO for double circle data (IL = 134.32)

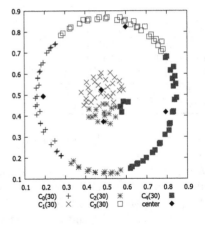

Fig. 6. The result of L_1ECBO for double circle data (IL = 127.77)

The ARI of ECBO, ECBO++, and MECBO is the same value, and the one of L_1ECBO and KECBO is larger than ECBO. In addition, ARI of each proposed method is very large in comparison with E2DCKM, HCM, and FCM. Although ARI of KECBO in $\sigma = 0.6$ is the largest, the variance of ARI is relatively large and the minimum value of ARI is relatively small. Therefore, the classification result in this case seems unstable. Furthermore, although ARI of KECBO in $\sigma = 2.0$ is relatively small, the value is larger than HCM and FCM. In addition, The variance of ARI is 0, and the result of classification in this case seems stable.

The execution time of ECBO++ is the shortest, and the one of MECBO is the second shortest in the proposed methods.

Table 4. Each value of each method in the Iris data

Method	E2DCKM	HCM	FCM ($m = 1.5$)	ECBO	ECBO++
ARI (min.)	0.216	0.420	0.422	0.893	0.893
ARI (avg.)	0.746	0.672	0.645	0.893	0.893
ARI (var.)	0.013	0.0098	0.016	0.000	0.000
IL (min.)	271.8	-	-	272.4	272.4
IL (ave.)	297.4	-	-	272.4	272.4
Time (avg.) [ms]	0.905	0.24	6.93	15.00	12.91

Method	L_1ECBO	MECBO	KECBO ($\sigma = 0.6$)	KECBO ($\sigma = 2.0$)
ARI (min.)	0.893	0.893	0.199	0.785
ARI (avg.)	0.903	0.893	0.851	0.785
ARI (var.)	5.87×10^{-5}	0.000	0.010	0.000
IL (min.)	273.2	272.4	273.7	272.4
IL (ave.)	273.7	272.4	304.5	272.4
Time (avg.) [ms]	17.20	14.76	84.00	55.42

5 Extension of ECBO for Spherical Data

We extend this ECBO to apply data on a sphere. The data on a sphere is referred to as the spherical data. Generally, each norm of spherical data is normalized as a unit. In other words, all spherical data is on a unit sphere. Consequently, each spherical datum has only direction. Therefore, cosine dissimilarity between x and y is usually used when we handle spherical data as follows:

$$d_{\mathrm{SECBO}}(x, y) = \alpha - \cos(x, y) = \alpha - \langle x, y \rangle. \quad (||x|| = ||y|| = 1)$$

$\alpha \geq 3/2$ from the reason mentioned in Ref. [10].

The objective function of the extended ECBO for spherical data (called SECBO) is as follows:

$$J_{\mathrm{SECBO}} = \sum_{k=1}^{n} \sum_{i=1}^{c} u_{ki} d_{\mathrm{SECBO}}(x_k, v_i) = \sum_{k=1}^{n} \sum_{i=1}^{c} u_{ki}(\alpha - \langle x_k, v_i \rangle).$$

The constraints are as same as ECBO. Notice that minimization of J_{SECBO} is as same as maximization of $\sum_{k=1}^{n} \sum_{i=1}^{c} u_{ki} \langle x_k, v_i \rangle$.

The algorithm of SECBO is showed in Algorithm 6.

Algorithm 6. SECBO

Step 0. Give the constants K or c.

Step 1. Give the initial cluster centers V randomly.

Step 2. Update U by the simplex method.

Step 3. Update V by calculating the centroid of the cluster as $v_i = \frac{\sum_{i=1}^{c} u_{ki} x_k}{||\sum_{i=1}^{c} u_{ki} x_k||}$.

Step 4. If V changes from previous V, go back to **Step 2.** Otherwise, stop.

We'll be able to construct SECBO++, L_1SECBO, Medoid SECBO, and KSECBO in similar way of ECBO. In particular, the idea for SECBO++ can be found in spherical k-means++ clustering proposed by one of the author in Ref. [10].

6 Conclusion

In this study, we proposed four types of even-sized clustering algorithm using an optimization method, i.e., ECBO++, L_1ECBO, MECBO, and KECBO. ECBO++, L_1ECBO, MECBO, and KECBO are solutions to solve the following problems: (1) the results strongly depend on initial values (initial-value dependence), (2) the algorithm is not robust against outliers, (3) it needs a lot of calculation cost, and (4) it is very difficult to classify datasets which consist of clusters with nonlinear boundary, respectively. Next, we estimated the effectiveness of our proposed algorithms in some numerical examples. Third, we mentioned one extension of ECBO for dealing with datasets on a sphere.

As presented in the numerical examples, the proposed methods can decrease the IL value in comparison with the conventional method. Whereas the best IL values of E2DCKM were often smaller than the proposed methods, the average of IL values of the proposed methods were always smaller than E2DCKM and good results were stably obtained by the proposed methods.

The ECBO was constructed by adding the constraints for cluster size to HCM and the membership grade was obtained by the simplex method. Thus, each cluster shape was hyperspherical and the data well gather.

The ECBO++ was constructed by introducing the selection method of initial cluster centers of k-means++ into the ECBO. The results of the minimum IL values of ECBO and ECBO++ were almost the same, but the variance of IL values and the execute time of ECBO++ were improved from ECBO.

The calculation of cluster centers in L_1ECBO and MECBO were the same as L_1 fuzzy c-means and k-medoids, respectively. In this study, we presented numerical examples on Euclidean space and with no outliers. However, It is expected that both the methods have the robustness against outliers and that MECBO can deal with the graph data.

KECBO was constructed by introducing a kernel function into ECBO. It classified the datasets which consist of clusters with nonlinear boundary, and the results of IL values were comparatively smaller.

Acknowledgment. We would like to thank gratefully and sincerely Professor Emeritus Sadaaki Miyamoto of University of Tsukuba, Japan, Professor Vicenç Torra of University of Skövde, Sweden, and Associate Professor Yuchi Kanzawa of Shibaura Institute of Technology, Japan, for their advice. This study was supported by JSPS KAKENHI Grant Numbers JP26330270, JP26330271, and JP16K16128.

References

1. Ogata, Y., Endo, Y.: A note on the K-member clustering problem. In: The 29th Fuzzy System Symposium (FSS), MB2-2 (2013). (in Japanese)

2. Hirano, T., Endo, Y., Kinoshita, N., Hamasuna, Y.: On even-sized clustering algorithm based on optimization. In: Proceedings of Joint 7rd International Conference on Soft Computing and Intelligent Systems and 15th International Symposium on advanced Intelligent Systems (SCIS & ISIS), TP4-3-5-(3), #69 (2014)
3. Arthur, D., Vassilvitskii, S.: k-means++: the advantages of careful seeding. In: Proceedings of the Eighteenth Annual ACM-SIAM Symposium on Discrete Algorithms, Society for Industrial and Applied Mathematics Philadelphia, PA, USA (2007)
4. Miyamoto, S., Agusta, Y.: An efficient algorithm for l_1 fuzzy c-means and its termination. Control Cybern. **24**, 421–436 (1995)
5. Kaufman, L., Rousseeuw, P.J.: Finding Groups in Data: An Introduction to Cluster Analysis. Wiley, New York (1990)
6. Park, H.-S., Jun, C.-H.: Simple and fast algorithm for k-medoids clustering. Expert Syst. Appl. **36**(2), 3336–3341 (2009)
7. Girolami, M.: Mercer kernel-based clustering in feature space. IEEE Trans. Neural Netw. **13**(3), 780–784 (2002)
8. Byun, J.-W., Kamra, A., Bertino, E., Li, N.: Efficient k-anonymization using clustering techniques. In: Kotagiri, R., Radha Krishna, P., Mohania, M., Nantajeewarawat, E. (eds.) DASFAA 2007. LNCS, vol. 4443, pp. 188–200. Springer, Heidelberg (2007)
9. Hubert, L., Arabie, P.: Comparing partitions. J. Classif. **2**, 193–218 (1985)
10. Endo, Y., Miyamoto, S.: Spherical k-means++ clustering. In: Torra, V., Narukawa, T. (eds.) MDAI 2015. LNCS, vol. 9321, pp. 103–114. Springer, Heidelberg (2015)

On Bezdek-Type Possibilistic Clustering for Spherical Data, Its Kernelization, and Spectral Clustering Approach

Yuchi Kanzawa[✉]

Shibaura Institute of Technology, Koto, Tokyo 135-8548, Japan
kanzawa@sic.shibaura-it.ac.jp

Abstract. In this study, a Bezdek-type fuzzified possibilistic clustering algorithm for spherical data (bPCS), its kernelization (K-bPCS), and spectral clustering approach (sK-bPCS) are proposed. First, we propose the bPCS by setting a fuzzification parameter of the Tsallis entropy-based possibilistic clustering optimization problem for spherical data (tPCS) to infinity, and by modifying the cosine correlation-based dissimilarity between objects and cluster centers. Next, we kernelize bPCS to obtain K-bPCS, which can be applied to non-spherical data with the help of a given kernel, e.g., a Gaussian kernel. Furthermore, we propose a spectral clustering approach to K-bPCS called sK-bPCS, which aims to solve the initialization problem of bPCS and K-bPCS. Furthermore, we demonstrate that this spectral clustering approach is equivalent to kernelized principal component analysis (K-PCA). The validity of the proposed methods is verified through numerical examples.

Keywords: Possibilistic clustering · Spherical data · Bezdek-type fuzzification · Kernel clustering · Spectral clustering

1 Introduction

Fuzzy c-means (FCM), proposed by Bezdek [1], is the most popular algorithm for performing fuzzy clustering on Euclidean data, which is fuzzified through its membership in the hard c-means objective function [2]. Other hard c-means fuzzification methods include entropy-regularized FCM (eFCM) [3] and Tsallis entropy-based FCM (tFCM) [4].

The FCM family is a useful family of clustering methods; however, their memberships do not always correspond well to the degree of belonging of the data. To address this weakness of the FCM family, Krishnapuram and Keller [5] proposed a possibilistic c-means (PCM) algorithm that uses a possibilistic membership function. Krishnapuram and Keller [6] and Ménard et al. [4] proposed other possibilistic clustering techniques that employ Shannon entropy and Tsallis entropy, respectively referred to as entropy-regularized PCM (ePCM) and Tsallis-entropy-regularized PCM (tPCM).

© Springer International Publishing Switzerland 2016
V. Torra et al. (Eds.): MDAI 2016, LNAI 9880, pp. 178–190, 2016.
DOI: 10.1007/978-3-319-45656-0_15

All these clustering methods are designed for Euclidean data. However, in some application domains, Euclidean data clustering methods may yield poor results. For example, information retrieval applications show that cosine similarity is a more accurate measure for clustering text documents than a Euclidean distortion of dissimilarity [7]. Such domains require spherical data and only consider the directions of the unit vectors. In particular, spherical K-means [8] and its fuzzified variants [9–13] are designed to process spherical data. Furthermore, two possibilistic clustering algorithms for spherical data have been proposed, which are based on the Shannon and Tsallis entropies and respectively referred to as entropy-based possibilistic clustering for spherical data (ePCS) and Tsallis-entropy-based possibilistic clustering for spherical data (tPCS) [14]. However, a Bezdek-type fuzzified possibilistic approach for clustering spherical data has not been proposed in the literature; this was a motivation for this work.

In this study, a Bezdek-type fuzzified possibilistic clustering algorithm for spherical data (bPCS), its kernelization (K-bPCS), and spectral clustering approach (sK-bPCS) are proposed. First, we propose bPCS by setting the fuzzification parameter of tPCS to infinity and modifying the cosine correlation-based dissimilarity between the objects and cluster centers. Next, we kernelize bPCS to obtain K-bPCS, which can be applied to non-spherical data with the help of a given kernel, e.g., a Gaussian kernel. Furthermore, we propose a spectral clustering approach to K-bPCS called sK-bPCS, which aims to solve the initialization problem of K-bPCS. Furthermore, we see that this spectral clustering approach is equivalent to kernelized principal component analysis (K-PCA) [15]. The validity of the proposed methods is verified through numerical examples.

The rest of this paper is organized as follows. In Sect. 2, the notation and the conventional methods are introduced. Section 3 presents the proposed methods, and Sect. 4 provides some numerical examples. Section 5 contains our concluding remarks.

2   Preliminaries

Let $X = \{x_k \in \mathbb{S}^{p-1} \mid k \in \{1, \ldots, N\}\}$ be a dataset of points on the surface of a p-dimensional unit hypersphere $\mathbb{S}^{p-1} = \{x \in \mathbb{R}^p \mid \|x\|_2 = 1\}$, which is referred to as spherical data. The membership of an x_k that belongs to the i-th cluster is denoted by $u_{i,k}$ ($i \in \{1, \ldots, C\}, k \in \{1, \ldots, N\}$) and the set of $u_{i,k}$ is denoted by u, which is also known as the partition matrix. The cluster center set is denoted by $v = \{v_i \mid v_i \in \mathbb{S}^{p-1}, i \in \{1, \ldots, C\}\}$. The value $1 - x_k^\mathsf{T} v_i$ can be used as the dissimilarity between object x_k and cluster center v_i.

Three methods of fuzzy clustering for spherical data, Bezdek-type, entropy-regularized, and Tsallis-entropy-based, can be obtained for the following optimization problems:

$$\underset{u,v}{\text{minimize}} \sum_{i=1}^{C} \sum_{k=1}^{N} (u_{i,k})^m (1 - x_k^\mathsf{T} v_i), \tag{1}$$

$$\underset{u,v}{\text{minimize}} \sum_{i=1}^{C} \sum_{k=1}^{N} u_{i,k}(1 - x_k^{\mathsf{T}} v_i) + \lambda^{-1} \sum_{i=1}^{C} \sum_{k=1}^{N} u_{i,k} \log(u_{i,k}), \qquad (2)$$

$$\underset{u,v}{\text{minimize}} \sum_{i=1}^{C} \sum_{k=1}^{N} u_{i,k}^m(1 - x_k^{\mathsf{T}} v_i) + \frac{\lambda^{-1}}{m-1} \sum_{i=1}^{C} \sum_{k=1}^{N} [u_{i,k}^m - u_{i,k}], \qquad (3)$$

subject to

$$\sum_{i=1}^{C} u_{i,k} = 1, \qquad (4)$$

$$\|v_i\|_2 = 1, \qquad (5)$$

and referred to as bFCS [16], eFCS [9] and tFCS [13], respectively.

Two possibilistic clustering objective functions for spherical data, entropy-based possibilistic clustering for spherical data (ePCS) [14] and Tsallis entropy-based possibilistic clustering for spherical data (tPCS) [14], are obtained by using the possibilistic constraint in place of the probabilistic constraint in Eq. (4) in the fuzzy clustering objective functions for spherical data, as

$$\underset{u,v}{\text{minimize}} \sum_{i=1}^{C} \sum_{k=1}^{N} u_{i,k}(1 - x_k^{\mathsf{T}} v_i) + \lambda^{-1} \sum_{i=1}^{C} \sum_{k=1}^{N} u_{i,k} \log(u_{i,k}) - \alpha \sum_{i=1}^{C} \sum_{k=1}^{N} u_{i,k}, \quad (6)$$

$$\underset{u,v}{\text{minimize}} \sum_{i=1}^{C} \sum_{k=1}^{N} u_{i,k}^m(1 - x_k^{\mathsf{T}} v_i) + \frac{\lambda^{-1}}{m-1} \sum_{i=1}^{C} \sum_{k=1}^{N} [u_{i,k}^m - u_{i,k}] - \alpha \sum_{i=1}^{C} \sum_{k=1}^{N} u_{i,k},$$
$$\qquad (7)$$

respectively, subject to Eq. (5), where α is a scale parameter and determined such that the maximal membership value is one. However, to the best of our knowledge, a possibilistic approach to Bezdek-type spherical clustering has not yet been investigated.

3 Proposed Method

3.1 Basic Concept

Comparing the bFCS and tFCS optimization problems in Eqs. (1) and (3), we find that the bFCS optimization problem is obtained by setting $\lambda \to +\infty$ in tFCS. Therefore, the bPCS optimization problem, the basis of the proposed method in this study, could also be obtained by setting $\lambda \to +\infty$ in tPCS, as

$$\underset{u,v}{\text{minimize}} \sum_{i=1}^{C} \sum_{k=1}^{N} u_{i,k}^m(1 - x_k^{\mathsf{T}} v_i) - \alpha \sum_{i=1}^{C} \sum_{k=1}^{N} u_{i,k} \qquad (8)$$

subject to Eq. (5). However, the membership equation obtained by solving this optimization problem is singular at $x_k = v_i$. A typical way of addressing singularities is regularization. Therefore, modifying the object-cluster dissimilarity from $1 - x_k^\mathsf{T} v_i$ to $\eta - x_k^\mathsf{T} v_i$ with parameter $\eta(\neq 1)$, the singularity is removed, and we obtain the bPCS optimization problem described as

$$\underset{u,v}{\text{minimize}} \sum_{i=1}^{C} \sum_{k=1}^{N} u_{i,k}^m (\eta - x_k^\mathsf{T} v_i) - \alpha \sum_{i=1}^{C} \sum_{k=1}^{N} u_{i,k} \qquad (9)$$

subject to Eq. (5).

Here, we consider the condition of the pair of parameters (m, η) in bPCS such that the bPCS optimization problem has a valid solution in situations such that the data are on the first quadrant of sphere, and reserve the case in which data are distributed over the whole sphere for future work. Data on the first quadrant of sphere arises in document data, where the document-term frequency is normalized on to the first quadrant of a unit sphere of dimension equal to the number of terms. Through this consideration, we find that bPCS has a valid parameter range such that not only $m > 1, \eta > 1$ but also $0 < m < 1, \eta \leq 0$. We focus on the case where $0 < m < 1, \eta = 0$, and will address the other cases in future work. This is because the case with $0 < m < 1, \eta = 0$ includes its extension to the spectral clustering approach discussed in later.

We subsequently kernelize bPCS to obtain K-bPCS. The original bPCS can only be applied to objects on the first quadrant of the unit hypersphere, whereas its kernelized algorithm can be applied to wider classes of objects. Here, we consider objects $\{x_k\}_{k=1}^N$ not on the first quadrant of the unit hypersphere. The elements of the kernel matrix obtained from the Gaussian kernel are described as

$$K_{k,\ell} = \exp(-\sigma^2 \|x_k - x_\ell\|_2^2). \qquad (10)$$

Note that

$$K_{k,k} = 1 \text{ and } K_{k,\ell} \in [0, 1], \qquad (11)$$

that is, the norm induced by the inner product of each feature vector is 1, and the inner product of a pair of feature vectors ranges from zero to one. This implies that the feature vectors corresponding to the Gaussian kernel are on the first quadrant of the unit hypersphere. Therefore, using an adequate kernel does not restrict the dataset that can be applied to K-bPCS.

Finally, we derive a spectral clustering approach for K-bPCS. We can obtain an equivalent maximization problem for the cluster centers by substituting the membership update equation into the original problem. Next, assuming orthogonality among cluster centers, an eigenproblem is obtained from the optimization problem with the fuzzification parameter $m = 0.5$, using the coefficients by which cluster centers are expressed as linear combinations of objects. This implies that the globally optimal solution can be obtained by solving that eigenproblem. Thus, no specific initial value setting is needed, so we expect to overcome the local convergence problem of the original algorithm. Because the actual cluster centers are not orthogonal, some iterations are executed to update the membership values. Therefore, similar to spectral clustering, this algorithm consists of two stages: (1) solving the eigenproblem, and (2) updating the optimal solutions. Furthermore, we see that this first stage is equivalent to K-PCA [15].

3.2   bPCS

In this subsection, the optimal solution of (u, v) for the bPCS optimization problem in Eqs. (8) and (5) is derived and the bPCS algorithm is proposed. Let us begin by considering the conditions on the pairs of parameters (m, η) in bPCS such that the bPCS optimization problem has a valid solution in situations such that the data are on the first quadrant of sphere. The Lagrange function $L(u, v)$ for the optimization problem is described as

$$L(u, v) = \sum_{i=1}^{C} \sum_{k=1}^{N} (u_{i,k})^m (\eta - x_k^\mathsf{T} v_i) - \alpha \sum_{i=1}^{C} \sum_{k=1}^{N} u_{i,k} + \sum_{i=1}^{C} \nu_i (1 - \|v_i\|_2^2) \quad (12)$$

with Lagrange multiplier $\nu = (\nu_1, \ldots, \nu_C)$. For $L(u, v)$ to have the optimal solution of (u, v), it is necessary for $L(u, v)$ to be convex for u and v. The second derivative of $L(u, v)$ with respect to v is

$$\frac{\partial^2 L(u, v)}{\partial^2 v_i} = 2\nu_i \mathbf{1}, \quad (13)$$

where $\mathbf{1}$ is the p-dimensional vector of all ones. Then, $L(u, v)$ is convex for v with positive values of ν. The fact that the value of ν is determined as a positive value will be discussed later. The second derivative of $L(u, v)$ with respect to u is

$$\frac{\partial^2 L(u, v)}{\partial^2 u_{i,k}} = m(m - 1) u_{i,k}^{m-2} (\eta - x_k^\mathsf{T} v_i). \quad (14)$$

Because memberships are non-negative, the necessary condition for $L(u, v)$ convexity is

$$m(m - 1)(\eta - x_k^\mathsf{T} v_i) \geq 0, \quad (15)$$

from which we have three possible cases:(1) $1 < m$, $\eta \geq x_k^\mathsf{T} v_i$, (2) $0 < m < 1$, $\eta \leq x_k^\mathsf{T} v_i$, and (3) $m < 0$, $\eta \geq x_k^\mathsf{T} v_i$, where the last case is useless because $u_{i,k}^m$ in the objective function decreases for $u_{i,k}$ with $m < 0$. Based on the fact that the data are on the first quadrant of the unit sphere, and assuming naturally that clusters are also on the first quadrant of the unit sphere, we have

$$x_k^\mathsf{T} v_i \in [0, 1], \quad (16)$$

and hence, $1 < m$ and $\eta \geq 1$ covers the first case, and $0 < m < 1$ and $\eta \geq 0$ covers the second case.

Hereafter, we focus on the case in which $0 < m < 1$ and $\eta = 0$, that is, the optimization problem is described as

$$\underset{u,v}{\text{minimize}} \sum_{i=1}^{C} \sum_{k=1}^{N} (u_{i,k})^m (-x_k^\mathsf{T} v_i) - \alpha \sum_{i=1}^{C} \sum_{k=1}^{N} u_{i,k}$$

$$\Leftrightarrow \underset{u,v}{\text{maxmize}} \sum_{i=1}^{C} \sum_{k=1}^{N} (u_{i,k})^m x_k^\mathsf{T} v_i + \alpha \sum_{i=1}^{C} \sum_{k=1}^{N} u_{i,k} \tag{17}$$

subject to Eq. (5). The necessary conditions for optimality can be written as

$$\frac{\partial L(u,v)}{\partial u_{i,k}} = 0, \tag{18}$$

$$\frac{\partial L(u,v)}{\partial v_i} = 0, \tag{19}$$

$$\frac{\partial L(u,v)}{\partial \nu_i} = 0. \tag{20}$$

Optimal membership is given by Eq. (18) in the form

$$u_{i,k} = (s_{i,k})^{1/(1-m)}, \tag{21}$$

where

$$s_{i,k} = x_k^\mathsf{T} v_i, \tag{22}$$

and the scale parameter α is set to $-m$. The optimal cluster center is obtained using Eq. (19) as

$$v_i = \frac{1}{2\nu_i} \sum_{k=1}^{N} (u_{i,k})^m x_k \tag{23}$$

with Lagrange multiplier ν_i. By considering the squared norm and taking Eq. (20) $\Leftrightarrow$ Eq. (5) into account, we have

$$\frac{1}{(2\nu_i)^2} \left\| \sum_{k=1}^{N} (u_{i,k})^m x_k \right\|_2^2 = 1 \Leftrightarrow \frac{1}{2\nu_i} = \frac{1}{\left\| \sum_{k=1}^{N} (u_{i,k})^m x_k \right\|_2}. \tag{24}$$

Inserting Eq. (24) into Eq. (23) and eliminating ν_i, we have

$$v_i = \frac{\sum_{k=1}^{N} (u_{i,k})^m x_k}{\left\| \sum_{k=1}^{N} (u_{i,k})^m x_k \right\|_2}. \tag{25}$$

These equations are alternatively iterated until convergence as shown in Algorithm 1.

Algorithm 1 (bPCS)

STEP 1. Given the number of clusters C, specify the fuzzification parameter m and set the initial cluster centers as v.
STEP 2. Calculate s using Eq. (22)
STEP 3. Calculate u using Eq. (21).
STEP 4. Calculate v using Eq. (25).
STEP 5. Check the stopping criterion for (u, v). If the criterion is not satisfied, go to STEP 2.   □

3.3   K-bPCS

For a given set of objects $X = \{x_k \mid k \in \{1, \ldots, N\}\}$, K-bPCS assumes that the kernel matrix $K \in \mathbb{R}^{N \times N}$ is known. Let $\mathbb{H}$ be a higher-dimensional feature space, $\Phi : X \to \mathbb{H}$ be a map from the data set X to the feature space $\mathbb{H}$, and $W = \{W_i \in \mathbb{H} \mid i \in \{1, \ldots, C\}\}$ be a set of cluster centers in the feature space.

K-bPCS solves the following optimization problem:

$$\underset{u,W}{\text{maximize}} \sum_{i=1}^{C} \sum_{k=1}^{N} (u_{i,k})^m s_{i,k} + \alpha \sum_{i=1}^{C} \sum_{k=1}^{N} u_{i,k} \tag{26}$$

$$\text{subject to } \langle W_i, W_i \rangle = 1, \tag{27}$$

where $s_{i,k} = \langle \Phi(x_k), W_i \rangle$. Lagrangian $L(u, v)$ is described as

$$L(u, v) = \sum_{i=1}^{C} \sum_{k=1}^{N} (u_{i,k})^m s_{i,k} + \alpha \sum_{i=1}^{C} \sum_{k=1}^{N} u_{i,k} + \sum_{i=1}^{C} \nu_i \left(1 - \|W_i\|_{\mathbb{H}}^2\right) \tag{28}$$

with Lagrange multipliers ν. Following a derivation similar to that of bPCS, the optimal solutions of membership and cluster centers are described as

$$u_{i,k} = (s_{i,k})^{1/(1-m)}, \tag{29}$$

$$W_i = \frac{\sum_{k=1}^{N} (u_{i,k})^m \Phi(x_k)}{\left\| \sum_{k=1}^{N} (u_{i,k})^m \Phi(x_k) \right\|_{\mathbb{H}}}, \tag{30}$$

where the scale parameter α is set to $-m$. Generally, Φ cannot be given explicitly, so the K-bPCS algorithm assumes that a kernel function $\mathcal{K} : \mathbb{R}^p \times \mathbb{R}^p \to \mathbb{R}$ is given. This function describes the inner product value of pairs of the objects in the feature space as $\mathcal{K}(x_k, x_j) = \langle \Phi(x_k), \Phi(x_j) \rangle$. However, it can be interpreted that Φ is given explicitly by allowing $\mathbb{H} = \mathbb{R}^N$, $\Phi(x_k) = e_k$, where e_k is the N-dimensional unit vector whose ℓ-th element is the Kronecker delta $\delta_{k,\ell}$, and by introducing $K \in \mathbb{R}^{N \times N}$ such that

$$K_{k,j} = \langle \Phi(x_k), \Phi(x_j) \rangle. \tag{31}$$

Using this kernel matrix K, $s_{i,k}$ is described as

$$s_{i,k} = \left\langle \Phi(x_k), \frac{\sum_{\ell=1}^{N} u_{i,\ell}^{\frac{1}{m}} \Phi(x_\ell)}{\left\| \sum_{\ell=1}^{N} u_{i,\ell}^{\frac{1}{m}} \Phi(x_\ell) \right\|_{\mathbb{H}}} \right\rangle = \frac{\sum_{\ell=1}^{N} u_{i,\ell}^{\frac{1}{m}} \langle \Phi(x_k), \Phi(x_\ell) \rangle}{\sqrt{\left\langle \sum_{\ell=1}^{N} u_{i,\ell}^{\frac{1}{m}} \Phi(x_\ell), \sum_{r=1}^{N} u_{i,r}^{\frac{1}{m}} \Phi(x_r) \right\rangle}}$$

$$= \frac{\sum_{\ell=1}^{N} u_{i,\ell}^{\frac{1}{m}} K_{k,\ell}}{\sqrt{\sum_{\ell=1}^{N} \sum_{r=1}^{N} u_{i,\ell}^{\frac{1}{m}} u_{i,r}^{\frac{1}{m}} \langle \Phi(x_\ell), \Phi(x_r) \rangle}} = \frac{\sum_{\ell=1}^{N} u_{i,\ell}^{\frac{1}{m}} K_{k,\ell}}{\sqrt{\sum_{\ell=1}^{N} \sum_{r=1}^{N} u_{i,\ell}^{\frac{1}{m}} u_{i,r}^{\frac{1}{m}} K_{\ell,r}}}. \quad (32)$$

Therefore, the K-bPCS algorithm consists of updating (u, s) as follows:

Algorithm 2 (K-bPCS)

STEP 1. Given the number of clusters C and fuzzification parameter m, and set the initial partition matrix to u.

STEP 2. Calculate s using Eq. (32).

STEP 3. Calculate u using Eq. (29).

STEP 4. Check the stopping criterion for (u, s). If the criterion is not satisfied, go to Step 2.   □

This algorithm can be applied to any kernel matrix, satisfying Eq. (11) such as the Gaussian kernel and its variants (e.g., [17]).

3.4 Spectral Clustering Approach to K-bPCS

In this subsection, we propose a spectral clustering approach to K-bPCS. First, we obtain an equivalent objective function to Eq. (26) with $m = 0.5$ as

$$\sum_{i=1}^{C} \sum_{k=1}^{N} (u_{i,k})^{0.5} s_{i,k} - 0.5 \sum_{i=1}^{C} \sum_{k=1}^{N} u_{i,k} = \sum_{i=1}^{C} \sum_{k=1}^{N} (s_{i,k})^2 - 0.5 \sum_{i=1}^{C} \sum_{k=1}^{N} (s_{i,k})^2$$

$$= 0.5 \sum_{i=1}^{C} \sum_{k=1}^{N} (s_{i,k})^2 \quad (33)$$

by substituting the membership update equation Eq. (29) into the original problem in Eq. (26). Furthermore, we rewrite cluster center W_i as a linear combination, similar to Eq. (30), i.e.,

$$W_i = \sum_{\ell=1}^{N} a_{i,\ell} \Phi(x_\ell) \quad (34)$$

with coefficients $a_{i,\ell}$, $s_{i,k}$ can be written as

$$s_{i,k} = \langle \Phi(x_k), W_i \rangle = \sum_{\ell=1}^{N} a_{i,\ell} \langle \Phi(x_k), \Phi(x_\ell) \rangle = \sum_{\ell=1}^{N} a_{i,\ell} K_{k,\ell}. \quad (35)$$

Therefore, the objective function for Eq. (33) that is equivalent to Eq. (26) is given by

$$\sum_{k=1}^{N}\sum_{i=1}^{C} s_{i,k}^2 = \sum_{k=1}^{N}\sum_{i=1}^{C}\left(\sum_{\ell=1}^{N} a_{i,\ell} K_{k,\ell}\right)^2 = \sum_{k=1}^{N}\sum_{i=1}^{C}\sum_{\ell=1}^{N}\sum_{r=1}^{N} a_{i,\ell} a_{i,r} K_{k,\ell} K_{k,r}$$
$$= \operatorname{trace}(A^{\mathsf{T}} K^2 A), \tag{36}$$

where the (ℓ, i)-th element of $A \in \mathbb{R}^{N \times C}$ is $a_{i,\ell}$. Additionally, we assume orthogonality among cluster centers, where

$$\langle W_i, W_j \rangle = \delta_{i,j}, \tag{37}$$

and (34) implies that

$$\langle W_i, W_j \rangle = \left\langle \sum_{\ell=1}^{N} a_{i,\ell} \Phi(x_\ell), \sum_{r=1}^{N} a_{j,r} \Phi(x_r) \right\rangle = \sum_{\ell=1}^{N}\sum_{r=1}^{N} a_{i,\ell} a_{j,r} \langle \Phi(x_\ell), \Phi(x_r) \rangle$$
$$= \sum_{\ell=1}^{N}\sum_{r=1}^{N} a_{i,\ell} a_{j,r} K_{\ell,r} = \delta_{i,j}, \tag{38}$$

that is,

$$A^{\mathsf{T}} K A = E, \tag{39}$$

where E is the N-dimensional unit matrix. Therefore, the optimization problem of K-bPCS under the assumption in Eq. (37) is simply

$$\operatorname*{maximize}_{A} \operatorname{trace}(A^{\mathsf{T}} K^2 A) \tag{40}$$
$$\text{subject to } A^{\mathsf{T}} K A = E. \tag{41}$$

Using $B = K^{\frac{1}{2}} A$, the above problem can be described as

$$\operatorname*{maximize}_{B} \operatorname{trace}(B^{\mathsf{T}} K B) \tag{42}$$
$$\text{subject to } B^{\mathsf{T}} B = E,$$

whose globally optimal solution can be obtained from the first C eigenvectors $\{b_i\}_{i=1}^{C}$ of K, written in descending order as $B = (b_1, \ldots, b_C)$, from which we have $A = K^{-\frac{1}{2}} B$. Then, $s_{i,k}$ is given by

$$s_{i,k} = \sum_{\ell=1}^{N} a_{i,\ell} K_{k,\ell} = e_k^{\mathsf{T}} K a_i = e_k^{\mathsf{T}} K K^{-\frac{1}{2}} b_i = e_k^{\mathsf{T}} K^{\frac{1}{2}} b_i = \sqrt{\lambda_i} e_k^{\mathsf{T}} b_i = \sqrt{\lambda_i} b_{i,k}, \tag{43}$$

where λ_i is the eigenvalue corresponding to b_i and e_k is the k-th unit vector. With this expression for $s_{i,k}$, the membership $u_{i,k}$ is described as

$$u_{i,k} = s_{i,k}^2 = \lambda_i b_{i,k}^2. \tag{44}$$

Because the actual situation is not well-separated, some iterations must be executed to update the memberships according to Eqs. (29) and (32). The above analysis suggests the following algorithm:

Algorithm 3 (Spectral Clustering Approach to K-bPCS)

STEP 1. Given the number of clusters C, obtain the first C eigenpairs $\{(\lambda_i, b_i)\}_{i=1}^{C}$ of K in descending order, and set the initial partition according to Eq. (44).
STEP 2. Calculate s using Eq. (32).
STEP 3. Calculate u using Eq. (29).
STEP 4. Check the stopping criterion for (u, s). If the criterion is not satisfied, go to Step. 2.   □

In this algorithm, a random initial value setting is not needed to solve the local convergence problem of the original algorithm. Similar to other spectral clustering techniques, this algorithm consists of two stages: (1) solving the eigenproblem, and (2) updating the optimal solutions.

Here, we see that solving the trace maximization problem in Eqs. (40) and (41) is equal to K-PCA [15], that is, obtaining the subspace $\mathrm{span}\{W_i \in \mathbb{H}\}_{i=1}^{C}$ of $\mathbb{H}$ such that the sum of squared distance between the original object $\Phi(x_k)$ and its projection $\sum_{i=1}^{C}\langle\Phi(x_k), W_i\rangle W_i$ to this subspace is minimal such that the value is described as

$$\sum_{k=1}^{N} \|\Phi(x_k)\|_{\mathbb{H}}^2 - \|\sum_{i=1}^{C}\langle\Phi(x_k), W_i\rangle W_i\|_{\mathbb{H}}^2. \tag{45}$$

From the orthogonality of basis $\{W_i \in \mathbb{H}\}_{i=1}^{C}$ as in Eq. (37), Eq. (45) is rewritten as

$$\sum_{k=1}^{N} \|\Phi(x_k)\|_{\mathbb{H}}^2 - \sum_{i=1}^{C}\langle\Phi(x_k), W_i\rangle^2. \tag{46}$$

Rewriting the basis W_i as a linear combination of objects $\{\Phi(x_k)\}_{k=1}^{N}$, as in Eq. (34), Eq. (46) can be described as

$$\sum_{k=1}^{N}\langle\Phi(x_k), \Phi(x_k)\rangle - \sum_{i=1}^{C}\langle\Phi(x_k), \sum_{\ell=1}^{N} a_{i,\ell}\Phi(x_\ell)\rangle^2$$

$$= \sum_{k=1}^{N} K_{k,k} - \sum_{i=1}^{C}\left(\sum_{\ell=1}^{N} a_{i,\ell}K_{k,\ell}\right)^2$$

$$= \sum_{k=1}^{N} K_{k,k} - \sum_{i=1}^{C} \sum_{\ell=1}^{N} \sum_{r=1}^{N} a_{i,\ell} a_{i,r} K_{k,\ell} K_{k,r}$$
$$= \text{trace}(K - AK^2 A), \tag{47}$$

and the orthogonality of basis $\{W_i \in \mathbb{H}\}_{i=1}^{C}$ is described as in Eq. (39). This minimization problem is equivalent to the trace maximization problem in Eqs. (40) and (41). Therefore, the initialization step in Algorithm 3 obtains the basis of the feature space induced from the kernel K such that the sum of squared distance between the original object and its projection to the subspace span$\{W_i \in \mathbb{H}\}_{i=1}^{C}$ is minimal.

4   Numerical Examples

This section provides numerical examples based on artificial and actual datasets. We use the adjusted Rand index (ARI) [18] to evaluate the clustering results. ARI takes a value not greater than one, and higher values are preferred.

The first example illustrates the performance of bPCS (Algorithm 1) using a dataset containing three clusters, each of which contains 50 points in the first quadrant of the unit sphere (Fig. 1). Using the parameter setting $m = 0.9$, bPCS achieves the highest ARI value, ARI $= 1$, and partitions this dataset adequately, as shown in Fig. 2, where squares, circles, and triangles indicate the maximal memberships generated by the algorithm during the test.

The second example shows the performance of bPCS (Algorithm 1) using a dataset of the Oz books [16]; this is a corpus containing 39,353 terms from 21 Oz-related books, 14 authored by L. Frank Baum and seven authored by Ruth Plumly Thompson. In this example, the dataset should be partitioned into two clusters using the proposed methods. With the parameter setting $m = 0.9$, bPCS achieves ARI $= 1$ and partitions this dataset adequately.

The third example illustrates the validity of K-bPCS and sK-bPCS (Algorithms 2 and 3, respectively) using the artificial dataset shown in Fig. 3, which consists of two nonlinearly bordered clusters. In both algorithms, the Gaussian kernel

$$K_{k,\ell} = \exp(-\sigma \|x_k - x_\ell\|_2^2) \tag{48}$$

is used, where the parameter $\sigma = 0.1$ for this dataset. K-bPCS was applied to this dataset with $C = 2$, $m = 0.5$, and 100 different initial settings, and the result with the minimal objective function value was selected. sK-bPCS was also applied to the same dataset with $C = 2$, and no initial setting was needed. Both algorithms achieves ARI $= 1$, and partition this dataset adequately as shown in Fig. 4. However, K-bPCS partitions the data correctly using 57 initial settings and fails for the other 43 initial settings. This implies that some initial settings risk partitioning failure. In contrast, sK-bPCS partitions the data correctly without initial settings. Therefore, sK-bPCS seems to outperform K-bPCS; however, this finding will be more thoroughly investigated with additional datasets in future work.

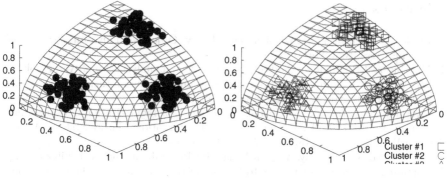

Fig. 1. Artificial dataset #1   **Fig. 2.** Result for artificial dataset #1

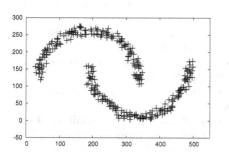
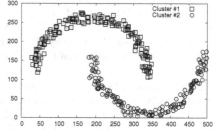

Fig. 3. Artificial dataset #2   **Fig. 4.** Result for artificial dataset #2

5   Conclusions

In this study, a Bezdek-type fuzzified possibilistic clustering algorithm for spherical data bPCS, its kernelization K-bPCS, and spectral clustering approach sK-bPCS were proposed. In the theoretical discussion, the proposed spectral clustering approach was shown to be identical to K-PCA. The validity of the proposed methods was confirmed through numerical examples.

In this study, data are assumed on the first quadrant of the unit sphere; however, in future work, we will consider the case in which data are distributed over the whole sphere. In addition, we plan to consider cases with other parameter values of (m, η) than those used in this study (i.e., $0 < m < 1$ and $\eta = 0$). We will also test the proposed algorithms on many additional datasets and compare with conventional methods. Finally, we will apply sequential cluster extraction [19], which is another algorithm for possibilistic clustering and develop a possibilistic clustering approach for other data types.

References

1. Bezdek, J.: Pattern Recognition with Fuzzy Objective Function Algorithms. Plenum Press, New York (1981)
2. MacQueen, J.B.: Some methods of classification and analysis of multivariate observations. In: Proceedings of the 5th Berkeley Symposium on Mathematical Statistics and Probability, vol. 1, pp. 281–297 (1967)
3. Miyamoto, S., Mukaidono, M.: Fuzzy c-means as a regularization and maximum entropy approach. In: Proceedings of the 7th International Fuzzy Systems Association World Congress (IFSA 1997), vol. 2, pp. 86–92 (1997)
4. Ménard, M., Courboulay, V., Dardignac, P.: Possibilistic and probabilistic fuzzy clustering: unification within the framework of the non-extensive thermostatistics. Pattern Recogn. **36**, 1325–1342 (2003)
5. Krishnapuram, R., Keller, J.M.: A possibilistic approach to clustering. IEEE Trans. Fuzzy Syst. **1**, 98–110 (1993)
6. Krishnapuram, R., Keller, J.M.: The possibilistic c-means algorithm: insights and recommendations. IEEE Trans. Fuzzy Syst. **4**, 393–396 (1996)
7. Strehl, A., Ghosh, J., Mooney, R.: Impact of similarity measures on web-page clustering. In: Proceedings of the AAAI2000, pp. 58–64 (2000)
8. Dhillon, I.S., Modha, D.S.: Concept decompositions for large sparse text data using clustering. Mach. Learn. **42**, 143–175 (2001)
9. Miyamoto, S., Mizutani, K.: Fuzzy multiset model and methods of nonlinear document clustering for information retrieval. In: Torra, V., Narukawa, Y. (eds.) MDAI 2004. LNCS (LNAI), vol. 3131, pp. 273–283. Springer, Heidelberg (2004)
10. Mizutani, K., Inokuchi, R., Miyamoto, S.: Algorithms of nonlinear document clustering based on fuzzy set model. Int. J. Intell. Syst. **23**(2), 176–198 (2008)
11. Kanzawa, Y.: A maximizing model of Bezdek-like spherical fuzzy c-means. Int. J. Intell. Syst. **19**(5), 662–669 (2015)
12. Kanzawa, Y.: On kernelization for a maximizing model of bezdek-like spherical fuzzy c-means clustering. In: Torra, V., Narukawa, Y., Endo, Y. (eds.) MDAI 2014. LNCS, vol. 8825, pp. 108–121. Springer, Heidelberg (2014)
13. Kanzawa, Y.: Fuzzy clustering based on α-divergence for spherical data and for categorical multivariate data. In: Proceedings of the FUZZ-IEEE2015, #15091 (2015)
14. Kanzawa, Y.: On possibilistic clustering methods based on Shannon/Tsallis-entropy for spherical data and categorical multivariate data. In: Torra, V., Narukawa, T. (eds.) MDAI 2015. LNCS, vol. 9321, pp. 115–128. Springer, Heidelberg (2015)
15. Scholkopf, B., Smola, A., Muller, K.: Nonlinear component analysis as a kernel eigenvalue problem. Neural Comput. **10**, 1299–1319 (1998)
16. Hornik, K., Feinerer, I., Kober, M., Buchta, C.: Spherical k-means clustering. J. Stat. Softw. **50**(10), 1–22 (2012)
17. Zelnik-Manor, L., Perona, P.: Self-tuning spectral clustering. Adv. Neural Inf. Process. Syst. **17**, 1601–1608 (2005)
18. Hubert, L., Arabie, P.: Comparing partitions. J. Classif. **2**, 193–218 (1985)
19. Kanzawa, Y.: Sequential cluster extraction using power-regularized possibilistic c-means. JACIII **19**(1), 67–73 (2015)

Hierarchical Clustering via Penalty-Based Aggregation and the Genie Approach

Marek Gagolewski[1,2]($\boxtimes$), Anna Cena[1], and Maciej Bartoszuk[2]

[1] Systems Research Institute, Polish Academy of Sciences,
ul. Newelska 6, 01-447 Warsaw, Poland
`gagolews@ibspan.waw.pl`
[2] Faculty of Mathematics and Information Science,
Warsaw University of Technology, ul. Koszykowa 75, 00-662 Warsaw, Poland

Abstract. The paper discusses a generalization of the nearest centroid hierarchical clustering algorithm. A first extension deals with the incorporation of generic distance-based penalty minimizers instead of the classical aggregation by means of centroids. Due to that the presented algorithm can be applied in spaces equipped with an arbitrary dissimilarity measure (images, DNA sequences, etc.). Secondly, a correction preventing the formation of clusters of too highly unbalanced sizes is applied: just like in the recently introduced *Genie* approach, which extends the single linkage scheme, the new method averts a chosen inequity measure (e.g., the Gini-, de Vergottini-, or Bonferroni-index) of cluster sizes from raising above a predefined threshold. Numerous benchmarks indicate that the introduction of such a correction increases the quality of the resulting clusterings significantly.

Keywords: Hierarchical clustering · Aggregation · Centroid · Gini-index · Genie algorithm

1 Introduction

A data analysis technique called clustering or data segmentation (see, e.g., [10]) aims at grouping – in an unsupervised manner – a family of objects into a number of subsets in such a way that items within each cluster are more similar to each other than to members of different clusters. The focus of this paper is on hierarchical clustering procedures, i.e., on algorithms which do not require the number of output clusters to be fixed a priori. Instead, each method of this sort results in a sequence of nested partitions that can be cut at an arbitrary level.

Recently, we proposed a new algorithm, named *Genie* [8]. Its reference implementation has been included in the `genie` package for R [15] see http://cran.r-project.org/web/packages/genie/. In short, the method is based on the single linkage criterion: in each iteration, the pair of closest data points from two different clusters is looked up in order to determine which subsets are to be merged. However, if an economic inequity measure (e.g., the Gini-index) of current cluster sizes raises above a given threshold, a forced merge of low-cardinality clusters

© Springer International Publishing Switzerland 2016
V. Torra et al. (Eds.): MDAI 2016, LNAI 9880, pp. 191–202, 2016.
DOI: 10.1007/978-3-319-45656-0_16

occurs so as to prevent creating a few very large clusters and many small ones. Such an approach has many advantages:

- By definition, the *Genie* clustering is more resistant to outliers.
- A study conducted on 29 benchmark sets revealed that the new approach reflects the underlying data structure better than not only when the average, single, complete, and Ward linkages are used, but also when the k-means and BIRCH algorithms are applied.
- It relies on arbitrary dissimilarity measures and thus may be used to cluster not only points in the Euclidean space, but also images, DNA or protein sequences, informetric data, etc., see [6].
- Just like the single linkage, it may be computed based on the minimal spanning tree. A modified, parallelizable Prim-like algorithm (see [14]) can be used so as to guarantee that a chosen dissimilarity measure is computed exactly once for each unique pair of data points. In such a case, its memory use is linear and thus the algorithm can be used to cluster much larger data sets than with the Ward, complete, or average linkage.

The current contribution is concerned with a generalization of the centroid linkage scheme, which merges two clusters based on the proximity of their centroids. We apply, analyze, and test the performance of the two following extensions:

- First of all, we note that – similarly as in the case of the generalized fuzzy (weighted) k-means algorithm [5] – the linkage can take into account arbitrary distance-based penalty minimizers which are related to idempotent aggregation functions on spaces equipped with a dissimilarity measure.
- Secondly, we incorporate the *Genie* correction for cluster sizes in order to increase the quality of the resulting data subdivision schemes.

The paper is set out as follows. The new linkage criterion is introduced in Sect. 2. In Sect. 3 we test the quality of the resulting clusterings on benchmark data of different kinds. A possible algorithm to employ the discussed linkage criterion is given in Sect. 4. The paper is concluded in Sect. 5.

2   New Linkage Criterion

For some set $\mathcal{X}$, let $\{\mathbf{x}^{(1)}, \mathbf{x}^{(2)}, \ldots, \mathbf{x}^{(n)}\} \subseteq \mathcal{X}$ be an input data sequence and $\eth$ be a pairwise dissimilarity measure (distance, see [6]), i.e., a function $\eth :$ $\mathcal{X} \times \mathcal{X} \to [0, \infty]$ such that (a) $\eth$ is symmetric, i.e., $\eth(\mathbf{x}, \mathbf{y}) = \eth(\mathbf{y}, \mathbf{x})$ and (b) $(\mathbf{x} = \mathbf{y}) \implies \eth(\mathbf{x}, \mathbf{y}) = 0$ for any $\mathbf{x}, \mathbf{y} \in \mathcal{X}$.

Example 1. Numerous practically useful examples of spaces like $(\mathcal{X}, \eth)$ can be found very easily. The clustered data sets may consist of points in $\mathbb{R}^d$, character strings (DNA and protein sequences in particular), rankings, graphs, equivalence relations, intervals, fuzzy numbers, citation sequences, images, time series, and so on. For each such $\mathcal{X}$, many popular distances can be utilized, e.g., respectively, the Euclidean, Levenshtein, Kendall, etc. ones, see, e.g., [5,6].   ⊡

Each hierarchical clustering procedure works in the following way. At the j-th step, $j = 0, \ldots, n-1$, there are $n-j$ clusters. It is always true that $\mathcal{C}^{(j)} = \{C_1^{(j)}, \ldots, C_{n-j}^{(j)}\}$ is a partition of the input data set. Formally, $C_u^{(j)} \cap C_v^{(j)} = \emptyset$ for $u \neq v$, $C_u^{(j)} \neq \emptyset$, and $\bigcup_{u=1}^{n-j} C_u^{(j)} = \{\mathbf{x}^{(1)}, \mathbf{x}^{(2)}, \ldots, \mathbf{x}^{(n)}\}$. Initially, the first partitioning consists solely of singletons, i.e., we have that $C_i^{(0)} = \{\mathbf{x}^{(i)}\}$ for $i = 1, \ldots, n$. When proceeding from step $j-1$ to j, a predefined linkage scheme determines which of the two clusters $C_u^{(j-1)}$ and $C_v^{(j-1)}$, $u < v$, are to be merged so as to we get $C_i^{(j)} = C_i^{(j-1)}$ for $u \neq i < v$, $C_u^{(j)} = C_u^{(j-1)} \cup C_v^{(j-1)}$, and $C_i^{(j)} = C_{i+1}^{(j-1)}$ for $i > v$. For instance, the single (minimum) linkage scheme assumes that u and v are such that:

$$\arg\min_{(u,v),u<v} \left(\min_{\mathbf{a} \in C_u^{(j-1)}, \mathbf{b} \in C_v^{(j-1)}} \mathfrak{d}(\mathbf{a}, \mathbf{b}) \right),$$

the complete (maximum) linkage is based on:

$$\arg\min_{(u,v),u<v} \left(\max_{\mathbf{a} \in C_u^{(j-1)}, \mathbf{b} \in C_v^{(j-1)}} \mathfrak{d}(\mathbf{a}, \mathbf{b}) \right),$$

and the average linkage on:

$$\arg\min_{(u,v),u<v} \left(\frac{1}{|C_u^{(j-1)}||C_v^{(j-1)}|} \sum_{\mathbf{a} \in C_u^{(j-1)}, \mathbf{b} \in C_v^{(j-1)}} \mathfrak{d}(\mathbf{a}, \mathbf{b}) \right),$$

see, e.g., [10] for a discussion.

Moreover, assuming that $\mathcal{X} = \mathbb{R}^d$ for some $d \geq 1$ and that $\mathfrak{d}$ is the Euclidean metric, we may consider the *centroid linkage criterion*:

$$\arg\min_{(u,v),u<v} \mathfrak{d}\left(\boldsymbol{\mu}(C_u^{(j-1)}), \boldsymbol{\mu}(C_v^{(j-1)}) \right),$$

where $\boldsymbol{\mu}(C)$, $C = \{\mathbf{x}^{(i_1)}, \ldots, \mathbf{x}^{(i_m)}\}$, denotes the *centroid* of C given by:

$$\boldsymbol{\mu}(\{\mathbf{x}^{(i_1)}, \ldots, \mathbf{x}^{(i_m)}\}) = \left(\frac{1}{m} \sum_{j=1}^m x_1^{(i_j)}, \ldots, \frac{1}{m} \sum_{j=1}^m x_d^{(i_j)} \right),$$

that is, the componentwise arithmetic mean of points in C. It can easily be shown that in such a setting we have that:

$$\boldsymbol{\mu}(\{\mathbf{x}^{(i_1)}, \ldots, \mathbf{x}^{(i_m)}\}) = \arg\min_{\mathbf{y} \in \mathbb{R}^d} \sqrt{\frac{1}{m} \sum_{j=1}^m \mathfrak{d}^2(\mathbf{x}^{(i_j)}, \mathbf{y})}.$$

Just as in the case of the fuzzy k-means algorithm [5], again for an arbitrary $\mathcal{X}$ and $\mathfrak{d}$, we may generalize the above cluster aggregation method as follows:

$$\boldsymbol{\mu}_\varphi(\{\mathbf{x}^{(i_1)}, \ldots, \mathbf{x}^{(i_m)}\}) = \arg\min_{\mathbf{y} \in \mathcal{X}'} \varphi^{-1} \left(\frac{1}{m} \sum_{j=1}^m \varphi\left(\mathfrak{d}(\mathbf{x}^{(i_j)}, \mathbf{y}) \right) \right), \quad (1)$$

where $\mathcal{X}' \subseteq \mathcal{X}$ and $\varphi : [0, \infty] \to [0, \infty]$ is a strictly increasing continuous function such that $\varphi(0) = 0$. In other words, μ_φ determines a minimizer of a distance-based penalty function given via a quasi-arithmetic mean. Let us observe that it is an idempotent fusion function, see [3, 7]. Assuming that the solution to (1) exists and is unique, the incorporation of μ_φ leads us to a generalized centroid linkage scheme that can work in arbitrary spaces equipped with a dissimilarity measure.

Remark 1. Most commonly, φ is set to be a power fuction, i.e., $\varphi(\delta) = \delta^p$ for some $p \geq 1$. For $\mathcal{X}' = \mathcal{X}$, the power-mean-based penalty minimizer corresponding to $p = 1$ is usually called the *1-median*, for $p = 2$ – *centroid*, and $p = \infty$ – *1-center*. However, special attention should be paid to whether a chosen fusion function can be computed sufficiently easily. In particular, if $\mathcal{X} = \mathbb{R}^d$, $p = 2$, and $\mathfrak{d}$ is the Euclidean distance, then we noted that the solution is the componentwise arithmetic mean. Moreover, if $p = 1$ and $\mathfrak{d}$ is the Manhattan distance, then we shall compute the componentwise median. On the other hand, for $p = 1$ or $p = \infty$ and $\mathfrak{d}$ being the Euclidean distance, there is no open-form solution (but, e.g., the Weiszfeld algorithm or a quadratic programming task can be applied, see, e.g., [7]). However, e.g., the search for 1-median with respect to the Levenshtein distance on the space of non-trivial character strings yields an NP-complete problem. ⊡

Remark 2. We can also set $\mathcal{X}' = \{\mathbf{x}^{(i_1)}, \ldots, \mathbf{x}^{(i_m)}\}$ which leads to the concept of a *set exemplar*. In particular, if $\varphi(d) = d$, then the corresponding distance-based penalty minimizer is named *medoid*. The computation of such a fusion function is always relatively easy $(O(m^2)$-time is needed). Yet, we should note that if the set of penalty minimizers is non-unique, some tie breaking rule (e.g., point index-based one) should be additionally introduced. ⊡

In order to increase the clustering quality in the presence of potential outliers (at least if the true underlying cluster structure is not heavily unbalanced), we can also incorporate a correction used in the single-linkage-based Genie [8] algorithm. In order to do so, firstly, let us recall the notion of an inequity index, see [2, 4, 9].

Definition 1. *For a fixed $n \in \mathbb{N}$, let $\mathcal{G}$ denote the set of all non-increasingly ordered n-tuples with elements in the set of non-negative integers, i.e., $\mathcal{G} = \{(x_1, \ldots, x_n) \in \mathbb{N}_0^n : x_1 \geq \cdots \geq x_n\}$. Then $\mathsf{F} : \mathcal{G} \to [0, 1]$ is an inequity index, whenever:*

(a) it is Schur-convex, i.e., for any $\mathbf{x}, \mathbf{y} \in \mathcal{G}$ with $\sum_{i=1}^n x_i = \sum_{i=1}^n y_i$, if it holds for all $i = 1, \ldots, n$ that $\sum_{j=1}^i x_j \leq \sum_{j=1}^i y_j$, then $\mathsf{F}(\mathbf{x}) \leq \mathsf{F}(\mathbf{y})$,
(b) $\inf_{\mathbf{x} \in \mathcal{G}} \mathsf{F}(\mathbf{x}) = 0$,
(c) $\sup_{\mathbf{x} \in \mathcal{G}} \mathsf{F}(\mathbf{x}) = 1$.

Example 2. Noteworthy instances of inequity indices, see [2], include the normalized Gini-index:

$$G(\mathbf{x}) = \frac{\sum_{i=1}^{n-1} \sum_{j=i+1}^n |x_i - x_j|}{(n-1) \sum_{i=1}^n x_i}, \tag{2}$$

the normalized Bonferroni-index:

$$B(\mathbf{x}) = \frac{n}{n-1}\left(1 - \frac{\sum_{i=1}^{n}\frac{1}{n-i+1}\sum_{j=i}^{n}x_j}{\sum_{i=1}^{n}x_i}\right), \tag{3}$$

or the normalized de Vergottini-index:

$$V(\mathbf{x}) = \frac{1}{\sum_{i=2}^{n}\frac{1}{i}}\left(\frac{\sum_{i=1}^{n}\frac{1}{i}\sum_{j=1}^{i}x_j}{\sum_{i=1}^{n}x_i} - 1\right). \tag{4}$$

It may be shown that all the indices may be computed in $O(n)$-time given a sorted $\mathbf{x}$.   ⊡

Now let F be a fixed inequity index and $g \in (0,1]$ be some threshold. The Genie-based generalized centroid linkage criterion proceeds as follows. At the j-th step let $c_i = |C_i^{(j)}|$ and denote with $c_{(i)}$ the i-th smallest value in $(c_1, \ldots, c_{n-j})$. Now:

1. if $F(c_{(n-j)}, \ldots, c_{(1)}) \le g$, then apply the standard generalized centroid linkage criterion:
$$\arg\min\nolimits_{(u,v),u<v}\eth\left(\boldsymbol{\mu}_\varphi(C_u^{(j-1)}), \boldsymbol{\mu}_\varphi(C_v^{(j-1)})\right),$$

2. otherwise, i.e., if $F(c_{(n-j)}, \ldots, c_{(1)}) > g$, restrict the search domain only to pairs of clusters such that one of them is of the smallest size:
$$\arg\min\nolimits_{\substack{(u,v),u<v,\\ c_u=c_{(1)}\text{ or }c_v=c_{(1)}}}\eth\left(\boldsymbol{\mu}_\varphi(C_u^{(j-1)}), \boldsymbol{\mu}_\varphi(C_v^{(j-1)})\right).$$

Such a linkage scheme prevents drastic increases of the selected inequity measure and guarantees that small clusters are linked to some other ones much earlier. Whatever the choice of F, for $g = 1$ we obtain the ordinary generalized centroid linkage scheme. To recall, the original Genie algorithm [8] minimizes the value of $\min_{\mathbf{a}\in C_u^{(j-1)}, \mathbf{b}\in C_v^{(j-1)}}\eth(\mathbf{a},\mathbf{b})$ instead of $\eth\left(\boldsymbol{\mu}_\varphi(C_u^{(j-1)}), \boldsymbol{\mu}_\varphi(C_v^{(j-1)})\right)$, which for $g = 1$ reduces itself to the single linkage criterion.

3   Benchmarks

For testing purposes we use the benchmark data sets already studied in [8]. They are described in more detail and available for download at http://www.gagolewski.com/resources/data/clustering/. These include 21 data sets in the Euclidean space: *iris, iris5, s1, s2, s3, s4, a1, a2, a3, g2-2-100, g2-16-100, g2-64-100, unbalance, spiral, D31, R15, flame, jain, Aggregation, Compound, pathbased* as well as 6 non-Euclidean ones: strings over the $\{\mathsf{a}, \mathsf{c}, \mathsf{t}, \mathsf{g}\}$ alphabet (*actg1, actg2, actg3*, for use with the Levenshtein distance) and 0–1 vectors of fixed lengths (*binstr1, binstr2, binstr3*, for use with the Hamming distance). *digits2k_pixels* and *digits2k_points* were omitted from the analysis, as the performance of all the clustering algorithms is very weak in their case.

It is worth emphasizing that each data set comes with a vector of reference labels, which can be used to assess the performance of a clustering algorithm. For this purpose, we rely on the well-known notion of the FM-index, which gives the value 1 if a computed clustering (a dendrogram should be cut at an appropriate level) fully agrees with the reference one and 0 if it is totally discordant.

3.1   Choosing Different Inequity Measures

Firstly, let us study the effects of choosing different inequity measures on the original Genie algorithm (in [8], only the Gini-index was considered, but the algorithm's performance was already outstanding).

Figure 1 depicts the average FM-index computed over 21 Euclidean benchmark sets as a function of an inequity index threshold, g. Three measures of inequity are taken into account: the ones by Gini, Bonferroni, and de Vergottini. Additionally, the shaded regions span from the 1st to the 3rd quartiles of the empirical FM-index distributions.

The thresholds yielding the highest average FM-scores are equal to 0.2 in the case of the Bonferroni- and the Gini-index and 0.1 for the de Vergottini-index. Notably, in such cases the empirical FM-index distributions do not significantly differ from each other (as measured by the Wilcoxon (paired) signed rank test, all p-values > 0.1). This suggests that the actual choice of an inequity measure

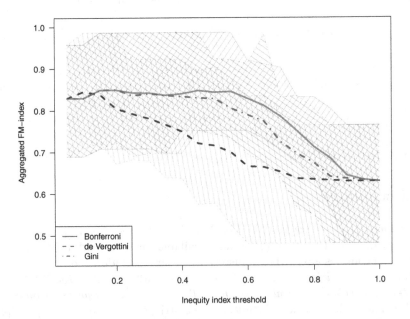

Fig. 1. The original Genie algorithm's performance depending on the choice of an inequity measure. The bold lines represent averaged FM-indices (21 benchmark sets), while the filled areas span from the 1st to the 3rd quartile of the empirical FM-index distribution.

(at least, as far as indices given by (2)–(4) are concerned) is not so important – special attention should rather be paid to the threshold selection.

For a better understanding of the reasons why it is so, let us focus on the *Aggregation* dataset, which consists of 788 observations. Firstly, we determine the dendrogram using the average linkage method with respect to the Euclidean distance. Next, we cut the dendrogram so as to obtain $2, 3, \ldots, 788$ clusters. Then, for each data set partition obtained in this manner, we compute the three inequity indices for the cluster size distribution. Figure 2 depicts the relationships between values of the three inequity indices. Of course, we observe that each index is not a 1-to-1 function of another one, but the data points are highly correlated (pairwise correlation coefficients – Gini vs Bonferroni: Pearson's $r = 0.98$, Spearman's $\varrho = 0.97$; Gini vs de Vergottini: $r = 0.77$, $\varrho = 0.83$; Bonferroni vs de Vergottini: $r = 0.71$, $\varrho = 0.87$). As for the other data sets similar regularities are detected, we deduce that the actual choice of an inequity index is not as important as choosing the right threshold. However, as far as the current benchmark sets are concerned, from Fig. 1 it seems that such a threshold can be found much more easily in the case of the Gini- or Bonferroni-index than while the de Vergottini-index is in use.

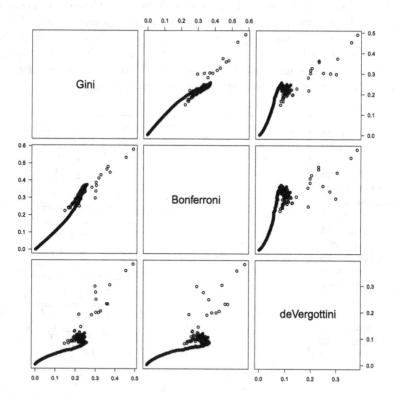

Fig. 2. Pairwise relationships between three inequity indices for the cluster size distributions as a function of the number of clusters in the case of the *Aggregation* data set.

3.2   Choosing Different Penalty Minimizers

Let us now compare the effects of choosing different penalty minimizers. Again, 21 data sets and the Euclidean metric is taken into account. We consider 5 different distance-based penalty minimizers: the *centroid* ($\varphi(\delta) = \delta^2, \mathcal{X}' = \mathbb{R}^d$), *median* ($\varphi(\delta) = \delta, \mathcal{X}' = \mathbb{R}^d$), *medoid* ($\varphi(\delta) = \delta, \mathcal{X}' = \{\mathbf{x}^{(i_1)}, \ldots, \mathbf{x}^{(i_m)}\}$), *medoid2* ($\varphi(\delta) = \delta^2, \mathcal{X}' = \{\mathbf{x}^{(i_1)}, \ldots, \mathbf{x}^{(i_m)}\}$), *medoid3* ($\varphi(\delta) = \delta^3, \mathcal{X}' = \{\mathbf{x}^{(i_1)}, \ldots, \mathbf{x}^{(i_m)}\}$). The three latter fusion functions are instances of set exemplars. Moreover, the Gini-index is used.

Figure 3 depicts the box-and-whisker plots for the FM-score distributions. Please note that the FM-indices may vary depending on the permutation of observations in a data set, because the distance matrix may consist of non-unique elements. Due to that, the median of 10 trials is computed (for different random rearrangements of the input points). For each generalized centroid, we report the results generated by considering two different Gini-index thresholds: 1.0 (no Genie correction applied at all) and the one maximizing the median among the 21 FM-index measurements.

We see that in each case the application of the Genie correction has a positive impact on the median FM-score. Among the considered distance-based penalty minimizers, *medoid2* yields the best results. When the Genie correction is in use, please observe the similarity between the results generated by relying on the 4 other fusion functions, especially between *medoid* and *median* (it should be noted that the latter is much more difficult to compute). The new linkage's performance is comparable with the Ward one. On the other hand, if $g = 1.0$, then the results are much more dependent on the choice of $\boldsymbol{\mu}$.

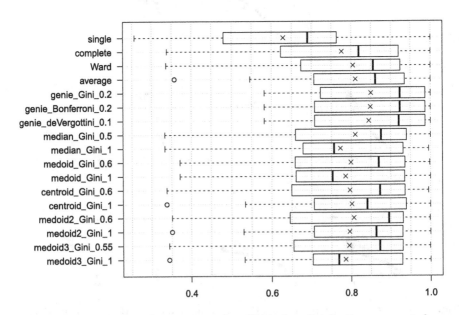

Fig. 3. Box-and-whisker plots representing FM-index distributions computed over 21 Euclidean benchmark sets for different clustering algorithms.

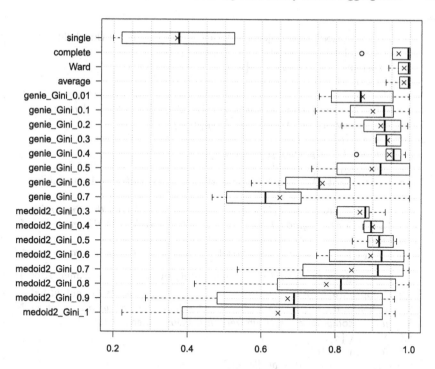

Fig. 4. Box-and-whisker plots representing FM-index distributions computed over 6 non-Euclidean benchmark sets for different clustering algorithms.

3.3 Non-Euclidean Benchmark Sets

As an example of the usefulness of the introduced algorithm in non-Euclidean spaces, let us now consider 6 different benchmark sets: *binstr1,2,3* (fixed-length 0–1 strings, the Hamming distance) and *actg1,2,3* (variable-length strings with elements in $\{a, c, t, g\}$, the Levenshtein distance).

Figure 4 depicts the FM-index distribution in the case of the single, complete, Ward, average, original Genie, and *medoid2*-based ($\varphi(\delta) = \delta^2, \mathcal{X}' = \{\mathbf{x}^{(i_1)}, \ldots, \mathbf{x}^{(i_m)}\}$) linkages. First of all, we observe that not only a too large inequity index threshold, but also a too small one may lead to unsatisfying results. Secondly, again, the Genie correction has a positive impact on the aggregated FM-index.

4   An Algorithm to Compute the New Linkage

The pseudocode of an $O(n^3)$-time and $O(n^2)$-space algorithm to compute the introduced type of clustering task is given in Fig. 5 (the cost of computing a selected penalty minimizer and inequity index is not included). The core of the routine is a quite straightforward modification of Anderberg's algorithm [1] as given in [13]. Hence, we omit a detailed discussion on the role of the *minidx*, *mindist*, etc. objects. The applied modifications include the Genie-like correction

---

0. Input: $\mathbf{x}^{(1)}, \ldots, \mathbf{x}^{(n)}$ – n objects, $g \in (0,1]$ – inequity index threshold, $\eth$ – a dissimilarity measure;
1. $ds = \text{DisjointSets}(\{1\}, \{2\}, \ldots, \{n\})$;
2. $S = \{1, \ldots, n\}$;
3. $L = \text{List}(\emptyset)$;   /* output list */
4. Let μ be an array such that $\mu[i] = \mathbf{x}^{(i)}$ for all $i \in S$;   /* by idempotence */
5. Let d be a matrix such that $d[i,j] = \eth(\mu[i], \mu[j])$;
6. **for** i in $S \setminus \{n\}$:
   6.1. $minidx[i] = \arg\min_{j=i+1,\ldots,n} d[i,j]$;
   6.2. $mindist[i] = d[i, minidx[i]]$;
7. $pq = \text{MinPriorityQueue}(minidx, mindist)$;   /* add $n-1$ point pairs */
8. **for** $i = 1, 2, \ldots, n-1$:
   8.1. **if** ds.compute_inequity() $\leq g$:   /* e.g., the Gini-index */
      8.1.1. $(a, b, \delta) = pq$.**pop()**;   /* the triple with the smallest δ */
     else:
      8.1.2. $(a, b, \delta) = pq$.**pop_conditional** $\Big($ /* Genie-like correction */

          (a, b, δ): ds.size$(a) = ds$.min_size() **or** ds.size$(b) = ds$.min_size()$\Big)$;
   8.2. L.append(a, b, δ);   /* update the output list */
   8.3. ds.link(a, b);   /* extend cluster b by a's members */
   8.4. $\mu[b] = \text{computePenaltyMinimizer}_\eth(ds$.getClusterMembers$(b))$;
   8.5. $S = S \setminus \{a\}$;
   8.6. **for** j in $S \setminus \{b\}$:
      8.6.1. $d[j,b] = d[b,j] = \eth(\mu[b], \mu[j])$;
   8.7. **for** j in S such that $j < a$ **and** $minidx[j] = a$:
      8.7.1. $minidx[j] = b$;
   8.8. **for** j in S such that $j < b$ **and** $d[j,b] < mindist[j]$:
      8.8.1. $minidx[j] = b$;
      8.8.2. $mindist[j] = d[j,b]$;
      8.8.3. pq.update$(j, minidx[j], mindist[j])$; /* update existing $(j, \cdot, \cdot)$ triple */
   8.9. $minidx[b] = \arg\min_{j=b+1,\ldots,n} d[b,j]$;
   8.a. $mindist[b] = d[b, minidx[b]]$;
   8.b. pq.update$(b, minidx[b], mindist[b])$;   /* update existing $(b, \cdot, \cdot)$ triple */
9. **return** L;

Fig. 5. A pseudocode for the introduced clustering algorithm.

(step 8.1.2, compare [8]) as well as a generic distance-based penalty minimizer instead of the Lance and Williams formula [11] in step 8.4. Note that the original Genie Algorithm [8] runs in $O(n^2)$-time and $O(n)$-space.

5 Conclusion

We have proposed a generalization of the nearest centroid linkage scheme. First of all, generic distance-based penalty minimizers may be taken into account. Due to that, the algorithm can be computed in arbitrary spaces equipped with dissimilarity measures. Secondly, the clustering quality can be improved by using a correction for the inequity of cluster size distribution, as known from the original Genie algorithm.

We noted that the actual choice of an inequity index has no significant impact on the benchmark FM-measures (at least as far as the Gini-, Bonferroni-, and de Vergottini-indices are concerned). Interestingly, if the Genie correction is in use, the choice of a distance-based penalty minimizer is not very important too. If this is not the case, we observed that μ based on the quadratic mean (*centroid, medoid2*) leads to more favorable results.

Finally, please note that, just like in the case of the original centroid and Genie linkage criteria, the "heights" at which clusters are merged are not necessarily being output in a nondecreasing order – the so-called reversals (inversions, departures from ultrametricity, see [12]) may occur. Therefore, they should be adjusted somehow when drawing corresponding dendrograms.

Acknowledgments. This study was supported by the National Science Center, Poland, research project 2014/13/D/HS4/01700.

References

1. Anderberg, M.R.: Cluster Analysis for Applications. Academic Press, New York (1973)
2. Aristondo, O., García-Lapresta, J., de la Vega, C.L., Pereira, R.M.: Classical inequality indices, welfare and illfare functions, and the dual decomposition. Fuzzy Sets Syst. **228**, 114–136 (2013)
3. Beliakov, G., Bustince, H., Calvo, T.: A Practical Guide to Averaging Functions. Springer, Heidelberg (2016)
4. Bortot, S., Marques Pereira, R.: On a new poverty measure constructed from the exponential mean. In: Proceedings of IFSA/EUSFLAT'15, pp. 333–340. Atlantis Press (2015)
5. Cena, A., Gagolewski, M.: Fuzzy K-minpen clustering and K-nearest-minpen classification procedures incorporating generic distance-based penalty minimizers. In: Carvalho, J.P., Lesot, M.-J., Kaymak, U., Vieira, S., Bouchon-Meunier, B., Yager, R.R. (eds.) IPMU 2016. CCIS, vol. 611, pp. 445–456. Springer, Heidelberg (2016). doi:10.1007/978-3-319-40581-0_36
6. Deza, M.M., Deza, E.: Encyclopedia of Distances. Springer, Heidelberg (2013)
7. Gagolewski, M.: Data Fusion: Theory, Methods, and Applications. Institute of Computer Science, Polish Academy of Sciences, Warsaw, Poland (2015)
8. Gagolewski, M., Bartoszuk, M., Cena, A.: Genie: a new, fast, and outlier-resistant hierarchical clustering algorithm. Inf. Sci. **363**, 8–23 (2016)
9. García-Lapresta, J., Lasso de la Vega, C., Marques Pereira, R., Urrutia, A.: A new class of fuzzy poverty measures. In: Proceedings of IFSA/EUSFLAT 2015, pp. 1140–1146. Atlantis Press (2015)
10. Hastie, T., Tibshirani, R., Friedman, J.: The Elements of Statistical Learning: Data Mining, Inference, and Prediction. Springer, Heidelberg (2013)
11. Lance, G.N., Williams, W.T.: A general theory of classificatory sorting strategies. Comput. J. **9**(4), 373–380 (1967)

12. Legendre, P., Legendre, L.: Numerical Ecology. Elsevier Science BV, Amsterdam (2003)
13. Müllner, D.: Modern hierarchical, agglomerative clustering algorithms. arXiv:1109.2378 [stat.ML] (2011)
14. Olson, C.F.: Parallel algorithms for hierarchical clustering. Parallel Comput. **21**, 1313–1325 (1995)
15. R Development Core Team: R: A Language and Environment for Statistical Computing. R Foundation for Statistical Computing, Vienna (2016). http://www.R-project.org

Data Privacy and Security

Complicity Functions for Detecting Organized Crime Rings

E. Vicente, A. Mateos[(✉)], and A. Jiménez-Martín

Decision Analysis and Statistics Group, Departamento de Inteligencia Artificial,
Universidad Politécnica de Madrid, Madrid, Spain
{e.vicentecestero,alfonso.mateos,antonio.jimenez}@upm.es

Abstract. Graph theory is an evident paradigm for analyzing social networks, which are the main tool for collective behavior research, addressing the interrelations between members of a more or less well-defined community. Particularly, social network analysis has important implications in the fight against organized crime, business associations with fraudulent purposes or terrorism.

Classic *centrality functions* for graphs are able to identify the key players of a network or their intermediaries. However, these functions provide little information in large and heterogeneous graphs. Often the most central elements of the network (usually too many) are not related to a collective of actors of interest, such as be a group of drug traffickers or fraudsters. Instead, its high centrality is due to the good relations of these central elements with other honorable actors.

In this paper we introduce *complicity functions*, which are capable of identifying the intermediaries in a group of actors, avoiding core elements that have nothing to do with this group. These functions can classify a group of criminals according to the strength of their relationships with other actors to facilitate the detection of organized crime rings.

The proposed approach is illustrated by a real example provided by the Spanish Tax Agency, including a network of 835 companies, of which eight were fraudulent.

1 Introduction

Graph theory is able to represent the relationships between all kinds of objects, such as human relationships, probabilistic relationships between events, electronic circuit components, computer networks, atomic networks... The analysis of these networks can be useful for valuable tasks such as detecting business associations with fraudulent purposes, counterterrorism investigation or political communication.

A *graph* [20] is a pair $G = (V, E)$, where V is the set of *vertices* or *nodes* and set E includes the *edges* or *arcs* of the graph. We denote a vertex or node by $v \in V$, and an arc by a pair of (not necessarily different) nodes (u, v). In that case, both nodes, u and v, are *neighboring nodes*. If the arc (u, v) is different to (v, u), then the graph is directed (a *digraph*).

© Springer International Publishing Switzerland 2016
V. Torra et al. (Eds.): MDAI 2016, LNAI 9880, pp. 205–216, 2016.
DOI: 10.1007/978-3-319-45656-0_17

We define the *arc-node incidence* of v in an undirected graph as the number of arcs incident to or from node v, i.e., the number of arcs starting or ending at v. If the graph is a digraph then we define the *indegree* of a node v as the number of arcs ending at (or incident to) v and the *outdegree* as the number of arcs starting at v. Therefore, the incidence degree is the sum of both.

The arcs of a graph can be assigned a weight, which denotes the intensity of the relation between the endpoint nodes of the arc. In this case, the graph is called a *weighted graph*.

The *adjacency matrix* of a given graph G with n nodes is an $n \times n$ matrix $A_G = (a_{u,v}^G)$, where $a_{u,v}^G = 1$ if $(u,v) \in E$ and $a_{u,v}^G = 0$ otherwise. In a weighted graph, this value will be the weight of the arc (u,v).

A *path* from u to v in a graph is a sequence of arcs, starting at node u and ending at node v. If such a path exists, u and v are *connected*. A *geodesic path* between two nodes of the graph is the shortest path between both nodes and the *geodesic distance* is the length of the geodesic path.

A graph is *connected* when there is a path between each pair of nodes and *complete* when all pairs of nodes are connected by an arc.

An important aspect in *social network analysis* is its *visualization*. A graph can be displayed in different ways. Classic visualization algorithms are studied in [4,8–10,19]. Secondly, the detection of communities, i.e., groups of nodes that are clearly differentiated in the graph [6,11,17], is interesting in social network analysis. Finally, another key feature of social network analysis is to identify different roles depending on the position of each actor in the network. For example, a social network may have key players, communication or action leaders that attract many other actors. The detection of these leaders can be critical in the development of information campaigns. We can also observe intermediate actors connecting different communities or other key players. All these actors can be identified by means of *centrality and intermediation indicators*.

The first studies on centrality in graphs date back to the 1950s and were conducted at [7] the Massachusetts Institute of Technology Group Networks Laboratory, in the field of sociology. Several works [1,16,21] report research into the the centrality characteristics of a group and the efficiency in the development of several more or less cooperative tasks. But the centrality concept in networks was later proven to be useful for explaining and analyzing geographical [18], political [2], economics or business [3] questions.

According to Freeman [7], it is difficult to define a centrality notion for graphs that accounts for different situations and interests. Instead the central elements are inspected to check that they have an intuitively evident set of properties. These properties are as follows:

1. The elements of a network with a higher adjacency degree, i.e., nodes that are more related with others nodes, are its central elements.
2. The elements of a network belonging to the highest possible number of geodesics between any two nodes of the network, are its central elements. These nodes are good intermediaries because they have to be traversed along the shortest path from one node to another.

3. The elements of a network closest to the other nodes, i.e., nodes that minimize the sum of geodesic distances to the other nodes, are its central elements.

Thus, the centrality function with the best trade-off between these three, not necessarily conflicting but not strictly correlated, properties will be the best, as we shall see later.

The first centrality functions in the literature were, predictably, functions that primarily satisfied the first of the three properties. However, the adjacency degree of the node it is not, by itself, a good centrality indicator. Consequently, centrality measures should be collective, where the centrality of a node depends not only on the local individual structure of the node, but also on the neighborhood structure, i.e., the centrality of a node is a function not only of the number of nodes connected to it, but also of the centrality of these nodes. This idea directly leads to two major centrality indicators: *eigenvector centrality* [14] and the *PageRank coefficient* [15], which is the basis of the Google engine search.

Besides, *hub/authority centrality* is based on the correlation between nodes with a high indegree (*authority nodes*) and nodes with a high outdegree (*hub nodes*). The HITS (*Hiperlink-Induced Topic Search*) [13] algorithm computes a membership value from each node to each one of these classes.

In this paper we propose a five-phases procedure for organized crime ring detection. The first phase computes the complicity of each actor with each fraudulent actor. Section 2 introduces the concept of complicity or suspicion of collaboration between one actor and another previously identified as fraudulent. Section 3 defines the strength of attraction between fraudulent actors, shown as the set of fraudulent nodes projected using multidimensional scaling in a plane (phase 3). These points are then grouped according to the DBSCAN algorithm (phase 4). Finally each identified actor is added to the group of fraudulent actors that maximizes its complicity (phase 5). Section 4 illustrates our approach with a real example including 835 Spanish companies, eight of which are fraudulent. Finally, some conclusions are provided in Sect. 5.

2   Dangerous Liaisons

The classic centrality measures are designed for graphs representing uniform collectives. However, the fight against organized crime requires an analysis that goes beyond the study of closed groups of individuals with common interests. In our society, honest and fraudster people are interrelated, and the fraudster take advantage of this circumstance and to commit crimes without being caught. Fraudulent companies take advantage of good faith of honorable companies to conceal fraudulent activities. Terrorists mix with citizens, and drug traffickers often have the appearance of respectable business entrepreneurs, who pay taxes and create jobs, etc. A good example are the so-called *carousel fraud plots*, in which a group of companies cooperates to commit value added tax fraud. This network of companies usually forges relationships with legitimate companies, which are unaware of the fraudulent activity of their partners, suppliers and clients.

The good news is that countries have increasingly improved technologies, capable of saving and monitoring networks of taxpayers, consumers, citizens, businessmen..., and these networks include both the honest and the fraudster subject. However, the above centrality functions do not work in these heterogeneous networks. For example, given a wide collective of companies, some of which have committed carousel fraud, the question is which other companies have a good relation with the companies implicated in fraud? In other words, which companies are suspicious of aiding and abetting the carousel fraud? Centrality measures are unable to identify these suspected companies, since they make computations on the basis of the relations with all nodes, irrespective of whether or not they are fraudsters.

2.1 Complicity Functions

We propose the concept of *complicity* or *suspicion of collaboration* between one actor and another previously marked as toxic (fraudulent), rather than centrality functions. In the following, we assume that the graph G is undirected, and we consider a set of *toxic* nodes in G, $\mathcal{F} \subset V$. Our aim is to study the relation between the nodes in $V - \mathcal{F}$ and the toxic nodes and assign a complicity value to those nodes depending on their relations.

The complicity function must satisfy the following three properties:

1. Nodes directly or indirectly connected to toxic nodes should have higher complicity values than nodes not related to toxic nodes. In fact, nodes without a direct or indirect connection to toxic nodes should be assigned zero complicity.
2. Distance should be penalized, i.e., the complicity of nodes that are closer to toxic nodes is greater.
3. Node connection to toxic nodes should be considered, i.e., the complicity nodes that are connected to a high number of toxic nodes is higher.

The above properties can be stated using the following expression, which represents the complicity of v with respect to the set of toxic nodes $\mathcal{F}$:

$$C(v) = \frac{1}{|\mathcal{F}|} \sum_{u \in \mathcal{F}} L(g(v, u)),$$

where L is a decreasing function and $g(v, u)$ is the geodesic distance between u and v. L can be an exponential or hyperbolic tangent function, as follows:

$$C(v) = \frac{1}{|\mathcal{F}|} \sum_{u \in \mathcal{F}} e^{r(1-g(v,u))} \text{(exponential) or}$$

$$C(v) = \frac{1}{|\mathcal{F}|} \sum_{u \in \mathcal{F}} \frac{1}{2} \left[\frac{e^{r-rg(u,v)} - e^{rg(u,v)-r}}{e^{rg(u,v)-r} + e^{r-rg(u,v)}} + 1 \right] \text{(hyperbolic tangent)},$$

where $g(v, u)$ is the geodesic distance from v to u without passing through any toxic node and r is a constant that models the differences. These complicity functions satisfy the above properties and output values within $[0,1]$.

The reason why we used geodesics that do not traverse the other toxic nodes to compute each summand of the above expression is that the connection with a toxic node could be overestimated if a toxic node is connected to other toxic nodes, which is commonplace. If an actor is connected to a toxic node which is in turn connected to other toxic nodes, then the actor is also connected to the latter, but this would not be an indicator of complicity. However, if the actor is connected repeatedly to the set of toxic nodes, then it is an accomplice.

These functions are suitable when the graph is connected. However, when the graph is formed by two or more disconnected subgraphs (two or more *connected components*), then the distance between two nodes belonging to different subgraphs is ∞, and the complicity of a node with respect to a fraudster in another connected component is zero. Besides, complicity is also zero for a node in the same component as, but at a large distance from, the fraudster. It is critical for the classification model to be able to deal with these two situations; otherwise disconnected plots would not be distinguishable from sets of nodes that are very far apart within the same component. One possibility would be to smooth the decrease by controlling parameter r, but this method does not yield good results in practice since complicity would have to be negligible as of some value of the geodesic distance. In this situation, a finite distance is the same as an infinite distance, which does not solve the problem. To overcome this drawback, we have to substitute a straight line ($y = k > 0$) (small) for the horizontal asymptote at infinity within the hyperbolic tangent model and *put a step in the infinity*, i.e., we control the parameters α and β rather than parameter r in the expression:

$$C(v) = \frac{1}{|\mathcal{F}|} \sum_{u \in \mathcal{F}} \alpha \left[\frac{e^{r-rg(u,v)} - e^{rg(u,v)-r}}{e^{rg(u,v)-r} + e^{r-rg(u,v)}} + \beta \right],$$

such that

$$\lim_{x \to \infty} \alpha \left[\frac{e^{r-rx} - e^{rx-r}}{e^{rx-r} + e^{r-x}} + \beta \right] = \alpha(\beta - 1) = k \text{ and}$$

$$\lim_{x \to -\infty} \alpha \left[\frac{e^{r-rx} - e^{rx-r}}{e^{rx-r} + e^{r-x}} + \beta \right] = \alpha(\beta + 1) = 1, \text{ i.e., } \beta = \frac{1+k}{1-k} \text{ and } \alpha = \frac{1-k}{2}.$$

2.2 Markov Chain-Based Approach

Markov chains are widely used to compute some classic centrality measures. Markov chains can simulate dynamic systems with a set of possible states where it is possible to pass from one state to another with certain probability.

In the above *PageRank algorithm*, given an adjacency matrix $A = (a_{ij})$, where $a_{ij} = 1$ if there is an arc between u_i and u_j and $a_{ij} = 0$ otherwise. If each row is divided by the sum of its elements, then we have a stochastic matrix $T = (t_{ij})$ (where each row is a probability distribution). This matrix is called a *transition matrix* since each element t_{ij} represents the probability of passing from node u_i to node u_j. In fact, matrix $T^n = \left(t_{ij}^{(n)} \right)$ represents the probability of reaching node u_j in a path of length n starting at u_i. This probability will be higher if the number of paths connecting both nodes increase.

The complicity of a toxic node v_j with another node u_i can be computed as:

$$c(v_j, u_i) = \sum_{n=1}^{\infty} \alpha^n t_{ij}^{j(n)},$$

where $\alpha \in (0,1)$ penalizes the distance between the nodes and $t_{ij}^{j(n)}$ are elements of $(T^j)^n$, where T^j is the transition matrix accounting for all nodes except toxic nodes different to v_j.

This function is similar to Katz centrality [12], except that we use the transition rather than the adjacency matrix and it is constrained to the toxic node v_j with respect to the node u_i. In other words, we are interested in the communication channels between toxic nodes and other nodes rather than the relations between all nodes in the graph, whereas Katz centrality sums all these values and outputs a single value for each node using the adjacency matrix.

In practice, it is sufficient to define a limit value, such as the graph *diameter* (length of the longest geodesic in the graph), δ, and compute the powers $T^j, (T^j)2, ..., (T^j)^\delta$ for each $v_j \in \mathcal{F}$. If we take the row in matrix $C^j = \alpha T^j + \alpha^2 (T^j)^2 + ... + \alpha^\delta (T^j)^\delta$ corresponding to the toxic node v_j and remove the element corresponding to that toxic node v_j, then we have a transposed vector denoted by c^j.

Now, let C be the rectangular matrix whose rows are the above transposed vectors c^j for the different toxic nodes, i.e. $\forall v_j \in \mathcal{F}$. Then, the total complicity of each non-toxic node can be computed as the sum of the elements in the column in C associated to that non-toxic node, and the complicity mean by dividing the above amount by the number of toxic nodes:

$$c(u_i) = \frac{1}{|\mathcal{F}|} \sum_{v_j \in \mathcal{F}} c(v_j, u_i) = \frac{1}{|\mathcal{F}|} \sum_{v_j \in \mathcal{F}} \sum_{n=1}^{\delta} \alpha^n t_{ij}^{(n)}.$$

2.3   Detection of Suspects of Complicity

The complicity function is useful for fighting against organized crime, identifying all actors represented by nodes that are highly related to a list of toxic nodes. To do this, we propose Algorithm 1, which accounts for a minimum number $s \in \mathbb{N}$ (for example, $s = 1$) of toxic nodes to which the actor should be related in order to be suspected of complicity. The higher the number is, the more suspect the actor will be. Besides, we consider a percentile $p \in (0, 100)$ as of which complicity is significantly high. To do this, we look at the distribution of the complicity values in the graph, considering the relation between the number of toxic and non-toxic nodes.

Algorithm 1. Detection of partners in crime

Input: (G undirected graph, $\mathcal{F}$ list of toxic nodes, $s \in \mathbb{N}$ minimum number of connected toxic nodes with the suspect node, $p \in (0, 100)$ value for the complicity percentile).

1. For each $w \in V$, compute the number $s(w)$ of elements of $\mathcal{F}$ connected to w without passing through other toxic actors.
2. For each element $v \in \mathcal{F}$:
3.   Remove all nodes $u \in \mathcal{F}$ such that $u \neq v$ from V.
4. $\forall w \in V$ such that $s(w) > s$ calculate $c(w, v) = e^{r(1-g(v,u))}$.
5. For each $w \in V$:
6.   Compute $C(w) = \frac{1}{|\mathcal{F}|} \sum_{v \in \mathcal{F}} c(w, v)$.
7. Compute the percentile of order p from the vector C.
8. Save and rank all nodes with a complicity value above percentile p.

---

3   Detection of Organized Crime Rings

Each toxic node $v_i \in \mathcal{F}$ is associated with a vector $\vec{v_i} = (v_{i1}, ..., v_{ir})$ where $v_{ij} = c(v_i, u_j)$ is the complicity of the toxic node v_i with the actor u_j.

We define the *strength of attraction* between the toxic nodes v_i and v_k by $f_{ik} = \vec{v_i} \cdot \vec{v_k} = \sum_{j=1}^{r} v_{ij} v_{kj}$, which can be normalized as follows: $f_{ik} = \frac{\vec{v_i} \cdot \vec{v_k}}{|\vec{v_i}||\vec{v_k}|}$, representing the cosine of the angle defined by vectors $\vec{v_i}$ and $\vec{v_k}$.

Matrix f is a symmetrical, nonnegatively defined matrix such that the maximum value in the i-th row, $f_{ii} = 1$, is located on the diagonal, i.e., it is a normalized similarity function, in which completely different nodes are in different components, i.e., $f_{ij} = 0$ if and only if $g(u_i, u_j) = \infty$.

It is possible to derive a distance metric in $[0,1]$ from the similarity function S as follows: $d(u, v) = \sqrt{S(u, u) - 2S(u, v) + S(v, v)}$. In our case, $d(u, v) = \sqrt{2 - 2f(u, v)}$. We can use this distance to project all the nodes using multidimensional scaling to represent fraudsters in $[0,1]^2$ with a maximum distance $\sqrt{2}$, which matches the maximum reachable distance by d when $f(u, v) = 0$. If a larger area is required, then a nonnormalized distance could be used, $d' = a \cdot d$.

Multidimensional scaling (MDS) [22] is a visualization algorithm of high-dimensional data arranged in a plane or 3D space. Data are represented by means of points whose distance is proportional to the differences between the data that they represent. The classic example illustrating this procedure is to plot a geographic map in a plane where the only available information is the distance between a set of cities.

Torgerson [22] proves that a matrix with the features of f (symmetrical and nonnegatively defined) can be decomposed as $f = U \Lambda U^t$, where U is the normalized eigenvector matrix and Λ is the eigenvalue diagonal matrix. If we denote $Y = U \Lambda^{1/2}$, then $YY^t = U \Lambda^{1/2} \Lambda^{1/2} U^t = RR^t$.

If two matrices from two different bases have the same matrix of scalar products, then one is the transform of the other by a change of basis. Then, Y is the matrix R in the basis of proper vectors, i.e., each row of Y is the strength vector from a fraudster in the basis of proper vectors.

Now, since $Y = U\Lambda^{1/2}$, if the first p eigenvalues are somewhat larger than the others, we can consider an approximation $\hat{X}$ of the initial fraudster matrix, which can be represented in p dimensions. If $p = 2$, fraudsters can be represented by points in a plane where their distances are proportional to the complicity differences of their related nodes.

Therefore, the percentage of the sum of the first two eigenvalues over the sum of all eigenvalues is a measure of how good the visualization is, i.e., the visualization is more realistic and better represents the differences, the greater the difference is between the sum of the first two eigenvalues and the sum of all eigenvalues.

We proceed to group the nodes according to the distances from the MDS projection on the plane of the set of toxic nodes. A simple and useful algorithm for this purpose is the *density-based spatial clustering of applications with noise* (DBSCAN) [5], which assesses the density of each region of the plane by computing the numbers of points that are within spheres with radius *eps* of each element in the population. Accordingly, the density of a neighborhood is satisfactory if the number of points in this neighborhood is equal to or greater than a prefixed value $MinPt$. The initial parameters of the DBSCAN then are the radius *eps* and the minimum density value $MinPt$. Note that the choice of these values is a critical decision because if *eps* is very small, then the spheres only have one point, whereas if they are too big, all points could belong to the sphere.

In summary, the procedure for ring detection is:

1. Compute the complicity of each node with each toxic node.
2. Compute the strength of attraction f between toxic nodes.
3. Project the set of toxic nodes using multidimensional scaling in a plane in such a way that the distance between nodes are proportional to $\sqrt{2 - 2f}$.
4. Group the points of the projection according to the DBSCAN algorithm.
5. Add each non-toxic node to the group of toxic nodes that maximizes its complicity.

After performing this assignment we can tune the algorithm considering connecting paths connecting partners in crime and toxic nodes.

4   An Illustrative Example

This section illustrates the ring detection procedure with a real example including 835 linked companies, of which eight are fraudulent. Data was provided by the Spanish Tax Agency. These are missing trade companies that are suspicious of belonging to carousel fraud plots (EU VAT plots) for the year 2013. These companies should have paid the full amount of VAT charged on an intra-Community acquisition of goods (a commodity purchase by a EU member country destined for a domestic market) to the national tax authorities. However, the company disappeared after selling the commodity without making this payment.

Figure 1 shows a graph describing the relationships between the 835 companies, highlighting the fraudulent companies in green. These relationships refer to any corporate, family, representation, management, authorization and co-ownership bank accounts for the year 2013.

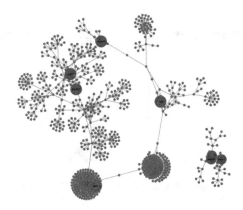

Fig. 1. 835 linked companies (Color figure online)

Our goal is to compute the complicity of companies with the fraudsters, group fraudsters depending on their relationship and complicity with each other and, finally, assign each company to a group of fraudsters accounting for average complicity with fraudsters from each ring following the procedure described in Sect. 3.

Figure 2(a) shows the projection by multidimensional scaling of the fraudster group, whereas Fig. 2(b) shows the resulting cluster after applying the DBSCAN algorithm. Four clusters are identified using different colors. Specifically, companies 582 and 579 (in green) constitute a cluster, whereas companies 17, 34 and 58 (in red) form another cluster. Companies 121 and 460 are in the third cluster. Finally company 240 is itself the fourth cluster.

Next, we build a ring for each cluster including the 827 remaining companies. To do this, we compute the average complicity of each company regarding the fraudster companies in each cluster. Thus, the company will be a member of the ring with which it has the highest average complicity. Each company is located in the selected cluster (ring) at a distance from the centroid of the cluster equal to its average complicity. Figure 2(c) shows the rings for the four clusters. The colors in the respective rings are associated with the percentiles of complicity values. Specifically, companies whose average complicity regarding the fraudsters in the corresponding cluster is under percentile 75 are shaded green, companies above percentile 95 are colored red and yellow is used for companies between the above percentiles.

Finally, Fig. 3 illustrates the original graph including the 835 linked companies in this case highlighting the companies with different colors (red, pink, blue and so on) in the different rings with an average complicity above percentile 95. These are the companies that could be considered as being suspicious of taking part in an organized crime ring.

The Spanish Tax Agency is currently using the proposed approach to rank, on the basis of the above average complicities, companies suspicious of fraud for inspection, since the number of inspectors and their availability are limited. This has led to an increment in the success rate.

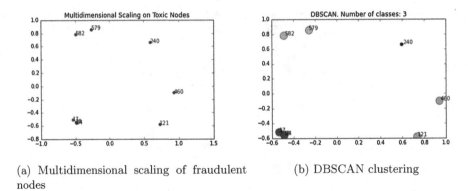

(a) Multidimensional scaling of fraudulent nodes

(b) DBSCAN clustering

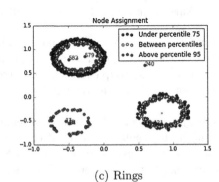

(c) Rings

Fig. 2. Detection of organized crime rings (Color figure online)

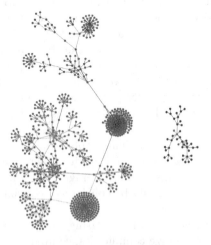

Fig. 3. Resulting organized crime rings (Color figure online)

5   Conclusions

Social network analysis has important implications in the fight against organized crime, business associations with fraudulent purposes or terrorism. We have introduced several functions to measure the degree of complicity between the actors in an heterogenous social network with a set of previously identified *toxic* actors, with the aim of detecting the partners of an organized crime plot. This complicity function naturally induces a similarity function between toxic nodes, the *strength of attraction*, which can be transformed into a distance metric.

Then, toxic nodes can be projected as points in a plane where the distances between points are proportional to the original distances between the toxic nodes (multidimensional scaling algorithm). The DBSCAN algorithm then groups the toxic nodes according to different high-density regions, and nodes with higher complicity are assigned to the different rings accounting for the maximization of average complicity.

The methodology has been illustrated by a real example, including 835 companies, of which eight are fraudulent according to the Spanish Tax Agency. The proposed approach detects organized crime rings and computes, for each ring, the complicity level of its actors. The Spanish Tax Agency is currently using the proposed approach to rank suspicious companies for inspection, since the number of inspectors and their availability are limited. This has led to an increment in the success rate.

We propose as a future research line to conduct a comparative analysis between the proposed method and other group detection methods, and between the complicity function and other centrality functions. Besides, further available information about the considered companies could be incorporated into the analysis and artificial intelligence tools, such as machine learning and statistics, could be used to derive a more robust crime ring detection method.

Acknowledgments. The paper was supported by the project MTM2014-56949-C3-2-R.

References

1. Bavelas, A., Barret, D.: An experimental approach to organizational communication. Personnel **27**, 366–371 (1951)
2. Cohn, B.S., Marriott, M.: Networks and centers in the integration in Indian civilization. J. Soc. Res. **I**(1), 1–9 (1958)
3. Czepiel, J.A.: Word of mouth processes in the diffusion of a major technological innovation. J. Mark. Res. **I**, 1–9 (1974)
4. Eades, P.: A heuristic for graph drawing. Congressus Numeranti **42**(11), 149–160 (1984)
5. Ester, M., Kriegel, H.-P., Sander, J., Xu, X.: A density-based algorithm for discovering clusters in large spatial databases with noise. In: KDD 1996 Proceedings (1996)
6. Faust, K., Wasserman, S.: Blockmodels: interpretation and evaluation. Soc. Netw. **14**(1–9), 5–61 (1992)

7. Freeman, L.C.: Centrality in social networks conceptual clarification. Soc. Netw. **1**, 215–239 (1978)
8. Fruchterman, T.M.J., Reingold, E.M.: Graph drawing by force-directed placement. Softw. Pract. Experience **21**(11), 1129–1164 (1991)
9. Hu, Y.: Efficient and high quality force-directed graph. Math. J. **10**(1), 37–71 (2005)
10. Kamada, T., Kawai, S.: An algorithm for drawing general undirected graphs. Inf. Process. Lett. **31**(1), 7–15 (1989)
11. Karrer, B., Newman, M.E.J.: Stochastic block models and community structure in networks. Phys. Rev. E **83**(1), 016107 (2011)
12. Katz, L.: A new status index derived from sociometric analysis. Psychometrika **18**(1), 39–43 (1953)
13. Kleinberg, J.: Authoritative sources in a hyperlinked environment. J. ACM **46**(5), 604–632 (1999)
14. Langville, A.N., Meyer, C.D.: A survey of eigenvector methods for web information retrieval. SIAM Rev. **47**(1), 135–161 (2005)
15. Lawrence, P., Sergey, B., Rajeev, M., Terry, W.: The pagerank citation ranking: bringing order to the web. Technical report, Stanford University (1998)
16. Leavit, H.J.: Some effects of certain communication pattern on group performance. Massachusetts Institute of Technology (1949)
17. Peixoto, T.P.: Hierarchical block structures and high-resolution model selection in large networks. Phys. Rev. X **4**(1), 011047 (2014)
18. Pitts, F.R.: A graph theoretic approach to historical geography. Prof. Geogr. **17**, 15–20 (1965)
19. Quigley, A., Eades, P.: FADE: graph drawing, clustering, and visual abstraction. In: Marks, J. (ed.) GD 2000. LNCS, vol. 1984, pp. 197–210. Springer, Heidelberg (2001)
20. Schaeffer, S.E.: Graph clustering. Comput. Sci. Rev. **I**, 27–64 (2007)
21. Smith, S.L.: Communication Pattern and the Adaptability of Task-Oriented Groups: An Experimental Study. Group Networks Laboratory, Research Laboratory of Electronics, Massachusetts Institute of Technology, Cambridge (1950)
22. Torgerson, W.S.: Multidimensional scaling: I. Theory and method. Psychometrika **17**(4), 401–419 (1952)

Partial Domain Theories for Privacy

Eva Armengol[1]([✉]) and Vicenç Torra[2]

[1] IIIA - Artificial Intelligence Research Institute,
CSIC - Spanish Council for Scientific Research,
Campus UAB, 08193 Bellaterra, Catalonia, Spain
`eva@iiia.csic.es`
[2] School of Informatics, University of Skövde, Skövde, Sweden
`vtorra@his.se`

Abstract. Generalization and Suppression are two of the most used techniques to achieve k-anonymity. However, the generalization concept is also used in machine learning to obtain domain models useful for the classification task, and the suppression is the way to achieve such generalization. In this paper we want to address the anonymization of data preserving the classification task. What we propose is to use machine learning methods to obtain partial domain theories formed by partial descriptions of classes. Differently than in machine learning, we impose that such descriptions be as specific as possible, i.e., formed by the *maximum* number of attributes. This is achieved by suppressing some values of some records. In our method, we suppress only a particular value of an attribute in only a subset of records, that is, we use local suppression. This avoids one of the problems of global suppression that is the loss of more information than necessary.

Keywords: Machine learning · Lazy learning methods · Partial domain models · k-anonymity · Supression

1 Introduction

Currently, the great amount of data available makes possible their analysis to extract knowledge for a broad range of different purposes. For example, health care data on the population can help in the prevention of certain diseases and improving the quality of life of patients with chronic diseases. For companies, the data about their clients and users, their opinions, their purchasing power, etc. are useful information to improve services as e.g. product recommenders. However, data transmission to data scientists or other experts for their analysis, or free access can violate the privacy of clients and users. In order to avoid the exposure of sensitive information about particular users, it is necessary to anonymize these data prior to their dissemination. There are many algorithms for data anonymisation satisfying privacy requirements. The main goal is to avoid the access to sensitive information about particular individuals whereas the whole data set is still valid for extracting useful information about the population.

© Springer International Publishing Switzerland 2016
V. Torra et al. (Eds.): MDAI 2016, LNAI 9880, pp. 217–226, 2016.
DOI: 10.1007/978-3-319-45656-0_18

There exists several privacy models. One of the most common one is k-anonymization [17,18]. It means that each record in a file is indistinguishable from no fewer than $k-1$ other records. Files compliant with k-anonymity can be constructed by means of generalization, supression and microaggregation. See e.g. Mondrian and Incognito [13,14] as methods to achieve k-anonymity based on generalization, and [5] about using microaggregation for achieving k-anonymity.

Generalization consists on replacing a value by a less specific value. The generalization is usually made using a domain hierarchy taxonomy provided by the domain expert. The main shortcoming of this approach is the need of this taxonomy. First, because it is not always trivial to build a non arbitrary one. The construction of this taxonomy implies to have prior knowledge on the domain. Then, different experts may consider different taxonomies. E.g., we can generalize towns with respect to ZIP codes, counties and bishoprics. Appropriate generalization will depend on the data use, and data protection will be typically different when the different taxonomies do not match. Alternatively, we may consider the construction of an arbitrary hierarchy that minimizes information loss. In this case, the space of possible hierarchies [5] is large and the resulting one may not have a meaningful interpretation.

Suppression refers to removing a certain attribute value and replacing the occurrences of the value with a special one, i.e., *unknown*. Suppression can be seen as a particular type of generalization: the maximum generalization. In other words, when a given value cannot be further generalized it is suppressed. When a single value is suppressed this is known as local suppression. In contrast, we have global suppression as e.g. in Friedman et al. [6] when a certain value of an attribute is suppressed from all the records where it appears. This procedure usually leads to an excessive number of suppressions. Sometimes the value of a specific attribute is related to the value of some other attributes [11]. By taking into account such relation, the value could be maintained in some records.

The generalization concept is also used in machine learning to obtain domain models useful for a classification task. From this point of view, the suppression of one value of a record can be interpreted as a generalization of such record. For instance, given a record $R = (a, b, c, d)$, the record $R_g = (a, ?, c, d)$ is a generalization of R. Inductive learning methods such as decision trees [16] build domain models by generalizing a set of input examples. [12] states that induced domain models are not useful for privacy because those attributes (quasi-identifiers or not) absent from the model can distort some statistics of the anonymized file with respect to the original one. In fact, the idea is that most of times class descriptions are too general from the point of view of privacy. Notice that the aim of anonymising a file is to deal with data that are *almost* as the original ones. Conversely, descriptions forming the induced models are general in order to cover the maximum number of examples, thus they are formed by (too) many unknown values to be considered as similar to original data. In our opinion, this issue can be addressed by imposing a minimum length for the induced descriptions, avoiding in that way the proliferation of unknown values in one record. Fung et al. [8] argue that publishing a classifier instead of the data is useful when users are interested in classification, however, in general, data providers do not know the data use. This means that the classifier is grown to achieve the maximum accuracy but the recipient could be interested in other aspects such as interpretability, recall, etc.

Domain models obtained by inductive learning methods such as decision trees, are global models in the sense that they use all the known examples to build the model. However, it is possible to construct partial domain models by using lazy learning methods. This kind of methods take into account only one input example and try to classify it. As a consequence, they can obtain some kind of description justifying the classification. As we will explain later, this justification can be seen as a partial description of the class. What we propose in this paper is to use such partial descriptions as a way to anonymize a file.

In [1] we have proposed the use of the lazy learning method LID [2] to induce partial domain model theories. Given a problem p, LID proposes a class C for p and gives a justification J of such classification. This justification can be seen as a partial description of the class C since it describes a space of problems that are similar to p and that belong to C. However, the description is partial because there are other elements of C that may not satisfy J. In the experiments performed in [1], we have seen that the descriptions composing the partial domain model are more specific than the ones composing the global model given by inductive learning methods.

Taking into account the previous results, what we propose in this paper is to use LID to obtain a generalized k-anonymous file. Although we found that the descriptions of the partial model are more specific than the ones of the global model, we impose a minimum length for them. In this way we have original records minimally generalized but still satisfying the k-anonymity requirement. We think that this issue is different that what happens using a global classifier that is the subject of the main concern of Fung [8] about using classifiers as anonymized files.

Compared with other methods in the literature, our approach is able to deal with missing values, and we do not need to start with a taxonomy of generalizations. In addition, our method is evaluated satisfactorily with respect to the performance of classifiers built from the protected data set. Note that classifiers are standard tools in machine learning for constructing models of the data.

The structure of the paper is as follows. In Sect. 2 we review the related work. In Sect. 3 we introduce the notation we use. In Sect. 4 we explain the lazy learning method used in the anonymisation experiments. Section 5 we explain the algorithm we propose for k-anonymization in detail. In Sect. 6 we present the experiments we carried out on the Adult data set from the UCI Machine Learning Repository. The paper finishes with some conclusions and lines for future work.

2   Related Work

Approaches similar to the one in this paper are KADET [6] and kACTUS [12]. Both approaches involve decision trees in the anonymisation process. kACTUS is an algorithm oriented to support classification. It wraps a decision tree that induces a domain model from the original file. Then this model is used by kAC-TUS to apply k-anonymity by means of a multi-dimensional suppression method.

KADET [6] is a decision tree induction algorithm that guarantees k-anonymity. In fact, the output of KADET is an anonymous decision tree. kAC-TUS and KADET differ in how they handle the decision tree, so whereas the former grows the tree with an inducer and it is during a sort of pruning phase where the k-anonymity is achieved; the later grows the tree taking into account k-anonymity. Our approach is more similar to kACTUS since we use LID in the usual way to classify a record, and after the classification it is decided whether or not the partial description can be stored.

An example of a lazy construction of a domain model is the *Lazy Decision Trees (LDT)* [7]. Differently from pure inductive techniques, LDT builds a decision tree in a lazy way, i.e. each time that a new problem has to be classified, the system reuses, if possible, the existing tree. Otherwise, a new branch classifying the new problem is added to the tree. Notice that, in fact, the decision tree represents a general model of a domain and LDT builds it in a lazy way. The general procedure of LID is similar to LDT but it does not grows any structure, it only gives a description that can be interpreted as a justification of the proposed classification.

In [3], we proposed C-LID, implemented on top of LID, following a procedure similar to the one of the current approach, although with different interpretation. C-LID was used on the predictive toxicology domain where, most of times LID finishes with a description satisfied by examples of two solution classes. By storing these descriptions we can have evidences of the classification of a new chemical compound based on how many known examples satisfy each description. In the current approach, only the descriptions corresponding to one solution class (in fact, the class of the input example) are stored. Therefore, they perform as true discriminant descriptions of classes like the ones obtained by inductive methods.

Our approach does not need the use of taxonomies to generalize a record. In this sense, it is similar to TDR [8] although this latter work uses an aggressive suppression approach that suppresses a given value in all records where it is present without considering values of other attributes.

3   Preliminaries

Literature on data privacy usually distinguishes three kinds of attributes:

- *Identifiers* are the attributes that unambiguously identify a single individual or entity (f.e., the passport number). They are usually removed or encrypted.
- *Quasi-identifier* attributes that are those that identify an individual with some degree of ambiguity, but when used in combination provide an unambiguous identification of some records. They are usually masked.
- *Confidential* attributes that are those containing sensitive information that are useful for static analysis. They are usually not modified.

In [18] authors prove that removing the identifiers is no enough to protect the identity of an individual. Therefore, the protection must be done on the quasi-identifiers. In our approach, we deal with all attributes as quasi-identifiers. That is, intruders can have access to any attribute. We consider that the only confidential attribute is the class. The confidential attribute is considered a non-quasi-identifier. Currently, our approach is only applicable on categorical attributes. At this moment, to deal with continuous attributes we have to discretize them.

4 Lazy Induction of Descriptions

Lazy Induction of Descriptions (*LID*) is a lazy learning method for classification tasks. LID determines which are the most relevant features of a problem and searches in a case base for cases sharing these relevant features. The problem p is classified when LID finds a set of relevant features shared by a subset of cases all belonging to the same solution class C_i. Then LID classifies the problem as belonging to C_i. We call *similitude term*, D, the description formed by these relevant features and *discriminatory set*, S_D the set of cases satisfying D. In fact, a similitude term D is a generalization of both p and the cases in S_D.

Figure 1 shows the LID algorithm (see a details of LID in [2]). Given a problem p, LID initializes D as a description with no features and the discriminatory set S_D is initialized to the set of cases satisfying D (initially the whole original file). Let D be the current similitude term, when the stopping condition is not satisfied, the next step is to select a feature for specializing D. The specialization of a term D is achieved by adding features to it. The most discriminatory feature is heuristically selected using the López de Mántaras' distance (LM) [15] over the candidate features. Let f_d be the feature assessed as the most discriminatory, in such a situation the specialization of D defines a new similitude term D' by adding to D the feature f_d. After adding f_d to D, the new similitude term $D' = D + f_d$ subsumes a subset of cases in S_D, namely $S_{D'}$. LID is recursively called with $S_{D'}$ and D'. This process continues until either all the cases in $S_{D'}$ belong to the same solution class C_i or D' cannot be further specialized.

Function LID (p, S_D, D, C)
$\quad\quad S_D :=$ Discriminatory-set (D)
$\quad\quad$ if stopping-condition(S_D)
$\quad\quad\quad$ then return $class(S_D)$
$\quad\quad\quad$ else $f_d :=$ Select-feature (p, S_D, C)
$\quad\quad\quad\quad\quad D' :=$ Add-feature(f_d, D)
$\quad\quad\quad\quad\quad S_{D'} :=$ Discriminatory-set (D', S_D)
$\quad\quad\quad\quad\quad$ LID $(S_{D'}, p, D', C)$
$\quad\quad$ end-if
$\quad$ end-function

Fig. 1. The LID algorithm. p is the problem to be solved, D is the similitude term, S_D is the discriminatory set associated to D, C is the set of solution classes, $class(S_D)$ is the class $C_i \in C$ to which all elements in S_D belong.

The similitude term D' is a partial discriminant description of C_i since all the cases satisfying D' belong to C_i (according to one of the stopping conditions of LID). Therefore, the D' is a generalization of knowledge in the sense of inductive learning methods, so similitude terms can be taken as domain rules since they contain the relevant features for classifying a problem. The focus of a term D_j of LID are the features of p. This means that all the input examples that do not share with p some of the features assessed as relevant for the classification of p will not satisfy D_j, although they belong to C_i. The conclusion is that D_j is a description of C_i because classifies correctly objects of that class, however D_j is a *partial description* of the class, because there are cases in C_i that do not satisfy D_j. In other words, descriptions obtained by a lazy learning method represent local descriptions, in contrast to descriptions from inductive methods, since the later describe a space around the problem that has been solved.

5   Using **LID** to k-Anonymize a File

Our approach consists on storing the feature terms built by LID during the classification process to form a domain Fig. 2 shows the procedure we followed to do this. Let us suppose we have an original file where records are of the form $R_i = \langle D_i, C_i \rangle$. Here, D_i is the description of an object formed by a set of pairs attribute value. These attributes are taken as quasi-identifiers. The confidential attribute is the class C_i to which the object belongs. The procedure is based on the leave-one-out method. It takes the description D_i of an object R_i (without the class label C_i) and uses LID to obtain a classification for this description. According to the attribute-value pairs of D_i, LID proposes a classification c_n and also gives a similitude term D_n justifying the classification. When the class c_n proposed by LID is the same that the correct one C_i, the similitude term D_n, as we explained before, can be taken as a partial description of the correct class, and a candidate to be stored. However, previously to keep it, we test if it satisfies two conditions. First, to avoid a high information loss, we check whether D_n has *enough length*. Let us suppose

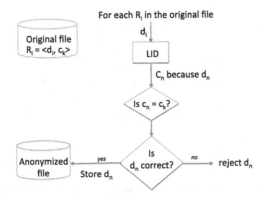

Fig. 2. Scheme of the use of leave one out to generate an anonymized file.

that an object D_i is described by n attributes (again, notice that the class is not considered), we require D_n to have as minimum length $n - long$ attributes otherwise the description is discarded. In our experiments, where domain objects are described by 8 attributes, we consider $long = 2$, and thus, D_n needs a minimum length of 6 attributes. Second, D_n has to be satisfy by at least k original records. This condition is to assure k-anonymity. Therefore, if D_n satisfies the two conditions above, it is stored in the anonymized file, otherwise it is rejected. In order to improve the efficiency of the method, that could be seriously compromised when the file is huge, we remove from the original file all records subsumed by the partial description d_n. In the next section we report some of the experiments we performed using this procedure.

6   Experiments

We performed experiments on the Adult data set from the Machine Learning Repository [4]. This data set is composed of 48842 records (with unknown values) described by 14 attributes. As it was done by Iyengar [10] we considered only eight of these attributes: *age, workclass, education, marital status, occupation, race, sex, native country* and, in addition, the class label *salary*. The attribute *age* is numerical and we discretized it in intervals of 5 (i.e., $[20, 25)$, $[25, 30)$ and so on).

We also considered the labels *low-20* and *high-90* to include those records placed on both sides of the global age range. All the other attributes are categorical. As in [11] we considered the class *salary* as confidential, being all the other attributes quasi-confidential. Commonly, authors [10,11] discard around 3000 records due to unknown values. In our experiments we do not need to do so because the algorithm is able to deal with unknown values. The data set as it is downloaded from the Machine Learning Repository, is already splitted in a training set having 32561 records, and a test set having 16281 records.

In the experiments, we address the classification task. The goal is to classify people with salary up to 50K and down to 50K. We need to fix the input parameters k and $long$, the minimum length of the descriptions. We experimented with $k = 5, 10, 20$ and 30. Concerning $long$, we have set it to 2, that is to say, the maximum number of attributes that can have value *unknown* in the patterns is 2. We want to remark that there are many original records with two or more unknown values, as our goal is to have an anonymized file as similar as possible to the original one, selection of $long$ equal to two seems a reasonable choice.

A simple way to test the equivalence of the original data base and the anonymized one for the classification task, is to induce a domain model (for instance, using a decision tree) for each data base and then evaluate the accuracy of the models on a test set. To construct the models of both the anonymized data base and the original one, we used the J48 inductive learning method, a clone of C4.5 [16] provided by Weka [9]. The accuracy has been evaluated on the test set of the UCI repository. Notice that we cannot use n-fold cross-validation on the anonymized file because what we will test were the generalized descriptions instead of the original data.

Table 1. Accuracy for $k = 5, 10, 20$ and 30 of the models with the original records and the anonymized ones on the training set provided by the UCI ML Repository. We also report the accuracy obtained by the method proposed in [10] (row Iyengar).

File	k	Accuracy	Descriptions
Original with rep.	–	83.12	–
Original without rep.	–	82.16	–
Iyengar	–	82.7/89.5	–
Anonymized	5	82.32	877
Anonymized	10	80.54	115
Anonymized	20	76.38	48
Anonymized	30	76.38	32

Table 1 shows the accuracy of the induced models. We see that the model induced from the original file gives an accuracy around the 83 % and that the anonymized file with $k = 5$ gives an accuracy around 82 %. We also see that high values of k give lower values of accuracy. This is an expected result since as it can already be seen in the Table, high values of k result in a lower number of descriptions forming the model. In fact, when $k = 20$ and $k = 30$ the model is formed only by descriptions for the $\leq 50K$. This explains why both models have exactly the same accuracy (it is the baseline). Our main conclusion is that this approach is only convenient for low values of k. Although they are not completely comparable, Table 1 also reports the accuracy from the method proposed in [10]. Depending on the value of k the accuracy goes from around 82 % for low values of k to 89 % for values of k around 200.

The only preprocessing we applied to the original file was the discretization of the attribute *age*. During a manual analysis of the original file, we have observed that the discretization process produces a file with repeated objects. For instance, let us consider these two records:

$\langle 31, Private, 11th, Divorced, Public, White, Male, USA, LEQ\_50 \rangle$
$\langle 33, Private, 11th, Divorced, Public, White, Male, USA, LEQ\_50 \rangle$

Notice that the only difference between both records is the value of *age*. After discretization both records can collapse into:

$\langle B\_35\_39, Private, 11th, Divorced, Public, White, Male, USA, LEQ\_50 \rangle$

As a consequence, we discovered that there are many records repeated more than k times after preprocessing. This means that these records already represent a k-anonymization as they are. The process followed using LID is not affected by that redundancy since once one valid description is stored, all the records of the original file satisfying it are discarded. Concerning the decision trees, the redundancy influences the accuracy. To avoid overfitting, the J48 algorithm prunes the decision tree. This means that some leaves can have a majority of objects belonging to one class and some others belonging to a different class. Having repeated objects in the original file could lead to leaves with a majority of objects that in fact, are all the same whereas without these repetitions the classification for that leaf could be different. As a consequence of this, we decided to discard all the repeated records and repeat the experiments. The new accuracy is around the 82 % when using the anonymized file with $k = 5$. Figures are given in Table 1. Note that the accuracies obtained for the two cases with the original file and the case of $k = 5$ are rather similar.

7 Conclusions and Future Work

In this paper we introduced a new method for k-anonymization based on using a lazy learning method with leave-one-out. In particular we used the LID method that in addition to classify a domain object is able to give a justification of such classification. Because that justification is composed of the attributes relevant for the classification, we can consider it as a partial description of the class. By imposing that these partial descriptions satisfy both k-anonymity and a minimum length requirement, we can store them and form an anonymized file. We proved that such anonymized file has a similar accuracy with respect to the original file, but only for low values of k. Compared to the original file, the anonymized file drastically reduces its size (from 48842 to 877 for $k = 5$), so we should perform some additional analysis in order to study if it is still useful for tasks different from classification.

We plan to analyze how many values are lost for each attribute and if the resulting file follows the same statistical distributions than the original one. We have also seen that the discretization process has produced repeated records that can be interpreted as k-anonymized descriptions without applying any procedure on them. We want to analyze in depth how discretization processes could be used in privacy. In addition, we should to compare our approach with classical methods of data privacy in order to establish the utility of our method, and also to experiment in other data sets.

Acknowledgments. This research is partially funded by the project RPREF (CSIC Intramural 201650E044) and the grants 2014-SGR-118 from the Generalitat de Catalunya.

References

1. Armengol, E.: Building partial domain theories from explanations. Knowl. Intell. **22**(2), 19–24 (2008)
2. Armengol, E., Plaza, E.: Lazy induction of descriptions for relational case-based learning. In: Flach, P.A., De Raedt, L. (eds.) ECML 2001. LNCS (LNAI), vol. 2167, pp. 13–24. Springer, Heidelberg (2001)
3. Armengol, E., Plaza, E.: Relational case-based reasoning for carcinogenic activity prediction. Artif. Intell. Rev. **20**(1–2), 121–141 (2003)

4. Bache, K., Lichman, M.: UCI machine learning repository (2013)
5. Domingo-Ferrer, J., Torra, V.: Ordinal, continuous and heterogeneous k-anonymity through microaggregation. Data Min. Knowl. Discov. **11**(2), 195–212 (2005)
6. Friedman, A., Wolff, R., Schuster, A.: Providing k-anonymity in data mining. VLDB J. **17**(4), 789–804 (2008)
7. Friedman, J.H.: Lazy decision trees. In: Proceedings of the Thirteenth National Conference on Artificial Intelligence, AAAI 1996, vol. 1, pp. 717–724. AAAI Press (1996)
8. Fung, B.C.M., Wang, K., Yu, P.S.: Anonymizing classification data for privacy preservation. IEEE Trans. Knowl. Data Eng. (TKDE) **19**(5), 711–725 (2007)
9. Hall, M., Frank, E., Holmes, G., Pfahringer, B., Reutemann, P., Witten, I.H.: The WEKA data mining software: an update. SIGKDD Explor. Newsl. **11**(1), 10–18 (2009)
10. Iyengar, V.S.: Transforming data to satisfy privacy constraints. In: Proceedings of the Eighth ACM SIGKDD International Conference on Knowledge Discovery and Data Mining, KDD 2002, pp. 279–288. ACM, New York (2002)
11. Bayardo Jr., R.J., Agrawal, R.: Data privacy through optimal k-anonymization. In: Proceedings of the 21st International Conference on Data Engineering, ICDE 2005, Tokyo, Japan, 5–8 April 2005, pp. 217–228 (2005)
12. Kisilevich, S., Keim, D.A., Rokach, L.: A gis-based decision support system for hotel room rate estimation and temporal price prediction: the hotel brokers' context. Decis. Support Syst. **54**(2), 1119–1133 (2013)
13. LeFevre, K., DeWitt, D.J., Ramakrishnan, R.: Incognito: efficient full-domain k-anonymity. In: Proceedings of the 2005 ACM SIGMOD International Conference on Management of Data, SIGMOD 2005, pp. 49–60. ACM, New York (2005)
14. LeFevre, K., DeWitt, D.J., Ramakrishnan, R.: Mondrian multidimensional k-anonymity. In: Proceedings of the 22nd International Conference on Data Engineering, ICDE 2006, Atlanta, GA, USA, 3–8 April 2006, p. 25 (2006)
15. López de Mántaras, R.: A distance-based attribute selection measure for decision tree induction. Mach. Learn. **6**, 81–92 (1991)
16. Quinlan, J.R.: C4.5: Programs for Machine Learning. Morgan Kaufmann Publishers Inc., San Francisco (1993)
17. Samarati, P.: Protecting respondents' identities in microdata release. IEEE Trans. Knowl. Data Eng. **13**(6), 1010–1027 (2001)
18. Samarati, P., Sweeney, L.: Protecting privacy when disclosing information: k-anonymity and itsenforcement through generalization and suppression. Technical report, SRI (1998)

Privacy-Preserving Cloud-Based Statistical Analyses on Sensitive Categorical Data

Sara Ricci, Josep Domingo-Ferrer[✉], and David Sánchez

UNESCO Chair in Data Privacy, Department of Computer Science and Mathematics,
Universitat Rovira i Virgili,
Av. Països Catalans 26, 43007 Tarragona, Catalonia
{sara.ricci,josep.domingo,david.sanchez}@urv.cat

Abstract. We consider the problem of privacy-preserving cloud-based statistical computation on sensitive categorical data. Specifically, we focus on protocols to obtain the contingency matrix and the sample covariance matrix of the categorical data set. A multi-cloud is used not only to store the sensitive data but also to perform computations on them. However, the multi-cloud is semi-honest, that is, it follows the protocols but is not authorized to learn the sensitive data. Hence, the data must be stored and computed on by the multi-cloud in a privacy-preserving format, which we choose to be vertical splitting among the various clouds. We give a comparison of our proposals, based on the secure scalar product, against a benchmark protocol consisting of downloading plus local computation.

Keywords: Data splitting · Privacy · Categorical data · Cloud computing · Contingency tables · Distance covariance

1 Introduction

Data have become a crucial asset of many enterprises, organizations and public administrations. Collecting and analyzing large amounts of data related to individuals does not only improve research, but it also drives a tremendous business [21]. However, local storage and processing of such *big* data is often unfeasible for the data controllers because of the associated costs (software, hardware, energy, maintenance). The cloud offers a suitable alternative for these data-intensive scenarios, by providing large and highly scalable storage/computation resources at a low cost and with ubiquitous access. However, most controllers holding (potentially) sensitive data are reluctant to embrace the cloud because of security and privacy concerns regarding the cloud service provider (CSP) [3]. On the one hand, CSPs may read, use or even sell the data outsourced by their customers (especially those CSPs that offer their services for free expecting to monetize users' data). On the other hand, CSPs may suffer attacks, accidents or data leakages that may compromise the privacy of the subjects to whom the outsourced data refer.

© Springer International Publishing Switzerland 2016
V. Torra et al. (Eds.): MDAI 2016, LNAI 9880, pp. 227–238, 2016.
DOI: 10.1007/978-3-319-45656-0_19

To allay these issues and win the trust of potential customers in cloud computing, there is a need for secure, efficient and privacy-preserving storage *and processing* methods for the (sensitive) data outsourced to the cloud. This is precisely the main goal of the European project CLARUS [6] in which the current work is framed. CLARUS consists in a proxy located in a domain trusted by the data controller (e.g., a server in her company's intranet or a plug-in in her device) that implements security and privacy-enabling features towards the CSP so that (i) the CSP only receives privacy-protected versions of the controller's (or the controller's users') data, (ii) CLARUS makes the access to such data transparent to the controller's users (by adapting their queries and reconstructing the results retrieved from the cloud) and (iii) it remains possible for the users to leverage the cloud to perform accurate computations on the outsourced data without downloading them.

To do so, CLARUS particularly relies on *data splitting* as a data protection technique: data are partitioned into several fragments, each of which is stored in the *clear* in a cloud provided by a different CSP, see [1,5]. Data splitting is an alternative that is more efficient and functionality-preserving than encryption-based methods (e.g., CipherCloud, PerspecSys, SecureCloud, etc.). In general, even though searchable and homomorphic encryption allow performing some operations on ciphertext [10], computing on encrypted data is extremely limited and costly [15], and it requires careful management of encryption keys. In contrast, the *vertical* data splitting implemented by CLARUS protects privacy (confidential information on an individual is partitioned into fragments that cannot be linked) and allows computation to be performed on clear data.

Yet, computing on split/distributed data is not easy. In fact, [23] acknowledge that mining data from distributed sources remains a challenge and [24] identify correlating the data from the various sources as the main hurdle. In this context, *we assume the CSPs to be semi-honest*: they are not entitled to see the entire data set, but they neither deviate from the protocols nor collude to aggregate the data fragments they hold.

Contribution and Plan of This Paper

In [4], we evaluated several non-cryptographic proposals for statistical computation (basically correlations) on split data, and we enhanced and proposed some protocols adapted to the CLARUS scenario. All these protocols and methods were designed for numerical data. However, many of the (personal) data currently collected from a variety of sources (social networks, surveys, B2C transactions, etc.) are not numerical (for example, user profiles [22], health records [13] or transaction logs [21]).

In this paper, we adapt some of the methods proposed for split numerical data to categorical data. Specifically, we describe two protocols (with and without cryptography, respectively) to compute the contingency table and the sample covariance matrix needed to measure the correlation between two categorical attributes stored in different clouds. We also compare their computational and communication costs against a benchmark consisting of the CLARUS proxy downloading the entire data set and locally computing on the downloaded data set.

The rest of this paper is organized as follows. In Sect. 2, we review some methods to measure the statistical dependence of categorical attributes. Section 3 focuses on the statistical analysis on vertically partitioned data; we give a non-cryptographic protocol and a cryptographic protocol for computing contingency tables and the sample covariance matrix. In Sect. 4, we compare the computational and communication costs of the protocols described in Sect. 3; a benchmark protocol is taken that consists of the CLARUS proxy downloading the entire data set and computing locally on the downloaded data. Finally, Sect. 5 lists some conclusions and future research lines.

2   Statistical Dependence Analyses on Categorical Data

Numerical data (continuous or ordinal) are easy to analyze: correlations, covariances, regressions and classifications can be computed using the standard arithmetic operators. In contrast, analyzing categorical data, consisting of categorical values lacking a total order, is more difficult: they must be either mapped to numbers in some way [8] or they must be processed using methods specifically designed to measure their differences [17], analyze their distributions [18] or estimate their dependence [20]. A well-known, albeit simple, procedure to measure the statistical dependence between two categorical attributes is the χ^2-test of independence [2]. This test uses the contingency tables associated to the categorical attributes as the input for a linear regression analysis. Even if it can measure some degree of statistical dependence, the χ^2-test only considers the similarities between the distributions of categorical labels, but it fails to capture the semantic similarity between the categories themselves.

In [20], a more recent and accurate way to measure the dependence between two categorical attributes is proposed: the *distance covariance/correlation* measure, meant to be an alternative to the standard numerical covariance/correlation. The numerical covariance requires the values of attributes to be totally ordered, and it measures dependence by checking whether greater values of one attribute correspond to greater values of the other attribute, and smaller values to smaller values. This does not work for non-ordinal categorical attributes (i.e., nominal/textual), that lack total order. The *distance covariance* is a viable alternative that quantifies to what extent the two attributes are independently dispersed, where dispersion is measured according to the pairwise distances between all pairs of values of each attribute. Moreover, unlike statistical tests based on contingency tables (i.e., value distributions), pairwise distances can capture the *semantics* inherent to categorical values, which is crucial to properly measure the correlation of non-numerical data [16]. To do so, the pairwise distance can be calculated using similarity/distance measures [17], that quantify how similar are the meanings of the concepts associated to the categorical values, based on the semantic evidences gathered from one or several knowledge sources (e.g., ontologies, corpora).

3   Computation on Vertically Partitioned Data

When storing dynamically changing sensitive data in the cloud, vertical splitting is very convenient: additions/updates are fast, because the other records (those that do not change) do not need to be modified. On the contrary, if data stored in the cloud are masked rather than split, any record addition/update requires re-anonymizing the original data set including the added/updated record and re-uploading the entire re-anonymized data set. Furthermore, if fragments are stored at different CSPs and these do not collude, splitting is more privacy-preserving than masking for dynamic data, because in masking the (single) CSP might infer the value of some original records by comparing the successive anonymized versions of the data set.

In vertical splitting, analyses that involve single attributes (e.g., mean, variance) or attributes within a single fragment are fast and easy to compute: the cloud storing the fragment can compute and send the output of the analysis to the CLARUS proxy. However, statistical dependence analyses may involve attributes stored in different fragments, and thus communication between several clouds. In [4] we focused on computing the sample covariance matrix because many of the statistical dependence analyses on numerical data are based on it. Obtaining the sample covariance matrix in vertical splitting among several clouds can be decomposed into several secure scalar products to be conducted between pairs of clouds. Secure scalar products can be based on cryptography (the protocol in [11] involves homomorphic encryption), or not ([7,12] modify the data before sharing them in such a way that the original data cannot be deduced from the shared data but the final results are preserved).

As discussed in Sect. 2, for categorical data the problem is more complicated. We focus here on the computation of the contingency tables needed for the χ^2-test and of the *distance covariance* matrix needed to measure the statistical dependence between categorical attributes. Section 3.1 reviews the best two computation protocols on split numerical data, one that uses cryptography and another that does not, adapted to the multi-cloud scenario considered in CLARUS [4]. In Sects. 3.2 and 3.3, we modify these protocols to compute contingency tables and the distance covariance matrix for categorical data.

3.1   Secure Scalar Product

Let $\mathbf{x}$ and $\mathbf{y}$ be two n-component vectors, respectively owned by Alice and Bob (who can be two CSPs). The goal is to securely compute the product $\mathbf{x}^T\mathbf{y}$. Shannon wrote in [19]: *"It is shown that perfect secrecy is possible, but requires, if the number of messages is finite, the same number of possible keys."* The privacy of the following protocol relies on the fact that the original vectors $\mathbf{x}$ and $\mathbf{y}$ are not shared at any time; only linear transformations of them are, such that the number of unknowns (randomness) added by the transformations is greater than or equal to the number of private unknowns.

In [9], the protocol is based on what they call a commodity server. Let Alice and Bob be as previously defined and let a third, non-colluding cloud Charlie play the role of the commodity server. In [4], we suggested the following variant:

Protocol 1

i. *Charlie generates two random n-vectors $\mathbf{r_x}$ and $\mathbf{r_y}$ and computes $p = \mathbf{r_x}^T \mathbf{r_y}$ (note that p is a number).*

ii. *Charlie sends $\mathbf{r_x}$ to Alice, $\mathbf{r_y}$ to Bob (or equivalently sends them the seeds for a common random generator). Also, Charlie sends p to CLARUS.*

iii. *Alice computes $\hat{\mathbf{x}} = \mathbf{x} + \mathbf{r_x}$ and sends it to Bob.*

iv. *Bob sends $t = \hat{\mathbf{x}}^T \mathbf{y}$ to CLARUS and sends $\hat{\mathbf{y}} = \mathbf{y} + \mathbf{r_y}$ to Alice.*

v. *Alice computes $s_x = \mathbf{r_x}^T \hat{\mathbf{y}}$ and sends it to CLARUS.*

vi. *CLARUS computes $t - s_x + p = (\mathbf{x} + \mathbf{r_x})^T \mathbf{y} - \mathbf{r_x}^T (\mathbf{y} + \mathbf{r_y}) + \mathbf{r_x}^T \mathbf{r_y} = \mathbf{x}^T \mathbf{y}$.*

Security. Charlie receives nothing from the other clouds. Bob gets n linear equations with n degrees of randomness. Similarly, Alice gets n linear equations with n degrees of randomness. Therefore, neither Alice's vector $\mathbf{x}$ can be computed by Bob, nor Bob's vector $\mathbf{y}$ can be computed by Alice and they are both protected according to the aforementioned Shannons principle.

In [11], the authors propose and justify the security of a cryptographic protocol based on the Paillier homomorphic cryptosystem [14]. The following variant is proposed in [4]:

Protocol 2

Set-up phase:

i. *Alice generates a private and public key pair (s_k, p_k) and sends p_k to Bob.* **Scalar product of Alice's** $\mathbf{x} = (x_1, \ldots, x_n)^T$ **and Bob's** $\mathbf{y} = (y_1, \ldots, y_n)^T$**:**

ii. *Alice generates the ciphertexts $c_i = Enc_{p_k}(x_i; r_i)$, where r_i is a random number in $\mathbb{F}_N$, for every $i = 1, \ldots, n$, and sends them to Bob.*

iii. *Bob computes $\omega = \prod_{i=1}^{n} c_i^{y_i}$.*

iv. *Bob generates a random plaintext s_B, a random number r', sends $\omega' = \omega Enc_{p_k}(-s_B; r')$ to Alice and sends s_B to CLARUS.*

v. *Alice sends $s_A = Dec_{s_k}(\omega') = \mathbf{x}^T \mathbf{y} - s_B$ to CLARUS.*

vi. *CLARUS computes $s_A + s_B = \mathbf{x}^T \mathbf{y}$.*

Protocol 2 works in a finite field $\mathbb{F}_N$, where the order N is the product of two primes p and q of the same length and such that $\gcd(pq, (p-1)(q-1)) = 1$. In case Alice and Bob need to execute this protocol several times, they can reuse public and private keys and thus the set-up step (first step) needs to be executed only once. The complexity of all these operations depends on N: the larger N, the more computationally demanding they are. Since we are computing $\mathbf{x}^T \mathbf{y}$ mod N, if we do not want the result to be modified by the modulus, it must hold that $N > \mathbf{x}^T \mathbf{y}$. Let $M_\mathbf{x} = \max_{x_i \in \mathbf{x}} x_i$ and $M_\mathbf{y} = \max_{y_i \in \mathbf{y}} y_i$. It is sufficient to choose $N > n M_\mathbf{x} M_\mathbf{y}$.

Security. The only modification with respect to the Protocol in [11] is that Alice and Bob do not share their results s_A and s_B, but they send these values to CLARUS. Hence, the security of the protocol is preserved and follows from the security of the Paillier cryptosystem (see [14] for more details).

3.2   Contingency Table Computation

A contingency table (or cross-classification table) is a type of table containing the (multivariate) frequency distributions of the categorical attributes. Let $\mathbf{a}$ and $\mathbf{b}$ denote two categorical attributes, $\mathbf{a}$ with h categories $c_1(\mathbf{a}), \ldots, c_h(\mathbf{a})$ and $\mathbf{b}$ with k categories $c_1(\mathbf{b}), \ldots, c_k(\mathbf{b})$. The contingency table has h rows and k columns displaying the sample frequency counts of the $h \times k$ category combinations.

To obtain the contingency table in vertical splitting among several clouds, one just needs to compute the table cells. Let $(a_1, \ldots, a_n)^T$ and $(b_1, \ldots, b_n)^T$ be the vectors of values from the categorical attributes $\mathbf{a}$ and $\mathbf{b}$, owned by Alice and Bob (who can be two CSPs), respectively. A cell C_{ij} (for every $i = 1, \ldots, h$ and $j = 1, \ldots, k$) is computed by counting the number of records in the original data set containing both the categories $c_i(\mathbf{a})$ and $c_j(\mathbf{b})$. Alice creates a new vector $\mathbf{x} = (x_1, \ldots, x_n)^T$ such that

$$x_l = 1 \text{ if } a_l = c_i(\mathbf{a}), \text{ and } x_l = 0 \text{ otherwise, for } l = 1, \ldots, n. \tag{1}$$

Bob creates $\mathbf{y} = (y_1, \ldots, y_n)^T$ such that

$$y_l = 1 \text{ if } b_l = c_j(\mathbf{b}), \text{ and } y_l = 0 \text{ otherwise, for } l = 1, \ldots, n. \tag{2}$$

The scalar product $\mathbf{x}^T \mathbf{y}$ gives the number C_{ij} of records in the original data set containing both the categories $c_i(\mathbf{a})$ and $c_j(\mathbf{b})$. Hence, Alice and Bob can use Protocol 1 or Protocol 2 to securely compute C_{ij} by just adding two preliminary steps to the scalar product computation part: one step by Alice to generate $\mathbf{x}$ from $(a_1, \ldots, a_n)^T$ using Expression (1), and another step by Bob to generate $\mathbf{y}$ from $(b_1, \ldots, b_n)^T$ using Expression (2).

Security. The only modification with respect to the Protocols 1 and 2 is that Alice and Bob compute $\mathbf{x}$ and $\mathbf{y}$, respectively. These computations are done by the clouds without exchange of information, hence the security of the protocol is preserved.

3.3   Distance Covariance Matrix Computation

As explained in Sect. 2, first of all we need to measure the pairwise (semantic) distance between the categorical values of each attribute; see [17] for a survey of semantic distance measures. Let $\mathbf{x^1} = (x_1^1, \ldots, x_n^1)^T$ and $\mathbf{x^2} = (x_1^2, \ldots, x_n^2)^T$ be vectors of values of two categorical attributes owned by Alice and Bob, respectively. We compute the matrices $X^1 = [x_{ij}^1]_{i,j \leq n}$ and $X^2 = [x_{ij}^2]_{i,j \leq n}$ where $x_{ij}^1 = |x_i^1 - x_j^1|$ and $x_{ij}^2 = |x_i^2 - x_j^2|$ are the semantic distances between two values of the same attribute $\mathbf{x^1}$ and $\mathbf{x^2}$, respectively. We define

$$X_{kl}^1 = x_{kl}^1 - \bar{x}_{k\cdot}^1 - \bar{x}_{\cdot l}^1 + \bar{x}_{\cdot\cdot}^1 \text{ for } k, l = 1, \ldots, n, \tag{3}$$

where

$$\bar{x}_{k\cdot}^1 = \frac{1}{n} \sum_{l=1}^{n} x_{kl}^1, \quad \bar{x}_{\cdot l}^1 = \frac{1}{n} \sum_{k=1}^{n} x_{kl}^1, \quad \bar{x}_{\cdot\cdot}^1 = \frac{1}{n^2} \sum_{k,l=1}^{n} x_{kl}^1.$$

Similarly, we define $X_{kl}^2 = x_{kl}^2 - \bar{x}_{k\cdot}^2 - \bar{x}_{\cdot l}^2 + \bar{x}_{\cdot\cdot}^2$ for $k, l = 1, \ldots, n$.

Definition 1. *The* squared *sample distance covariance is obtained as the arithmetic average of the products* $X_{kl}^1 X_{kl}^2$:

$$d\mathcal{V}_n^2(\mathbf{x^1}, \mathbf{x^2}) = \frac{1}{n^2} \sum_{k,l=1}^{n} X_{kl}^1 X_{kl}^2 \tag{4}$$

and the squared *sample distance variance is obtained as*

$$d\mathcal{V}_n^2(\mathbf{x^1}) = d\mathcal{V}_n^2(\mathbf{x^1}, \mathbf{x^1}) = \frac{1}{n^2} \sum_{k,l=1}^{n} X_{kl}^1 X_{kl}^1.$$

See [20] for details and justification on the above definition. In general, if $\mathbf{X} = (\mathbf{x^1}, \ldots, \mathbf{x^m})$ is a data set with m attributes $\mathbf{x}^j$, $j = 1, \ldots, m$, the distance covariance matrix $\hat{\mathbf{\Sigma}}$ of $\mathbf{X}$ is

$$\hat{\mathbf{\Sigma}} = \begin{pmatrix} d\mathcal{V}_n(\mathbf{x^1}) & d\mathcal{V}_n(\mathbf{x^1}, \mathbf{x^2}) & \cdots & d\mathcal{V}_n(\mathbf{x^1}, \mathbf{x^m}) \\ d\mathcal{V}_n(\mathbf{x^2}, \mathbf{x^1}) & d\mathcal{V}_n(\mathbf{x^2}) & \cdots & d\mathcal{V}_n(\mathbf{x^2}, \mathbf{x^m}) \\ \vdots & \vdots & \ddots & \vdots \\ d\mathcal{V}_n(\mathbf{x^m}, \mathbf{x^1}) & d\mathcal{V}_n(\mathbf{x^m}, \mathbf{x^2}) & \cdots & d\mathcal{V}_n(\mathbf{x^m}) \end{pmatrix}.$$

Note that $d\mathcal{V}_n(\mathbf{x}^i, \mathbf{x}^j)$ is the square root of the number $d\mathcal{V}_n^2(\mathbf{x}^i, \mathbf{x}^j)$, for $i, j = 1, \ldots, m$, and that X^j, X_{kl}^j, $d\mathcal{V}_n(\mathbf{x^j})$, for $j = 1, \ldots, m$, are separately computed by the cloud storing the respective attribute. The most challenging task is therefore calculating the squared sample distance covariance, i.e. Expression (4), which requires performing a secure scalar product of n vectors, each held by two different parties (where "secure" means without any party disclosing her vector to the other party). In fact, calling $\mathbf{X_k^1} = (X_{k1}^1, \ldots, X_{kn}^1)$ and $\mathbf{X_k^2} = (X_{k1}^2, \ldots, X_{kn}^2)$ for $k = 1, \ldots, n$, we can rewrite Expression (4) as

$$d\mathcal{V}_n^2(\mathbf{x^1}, \mathbf{x^2}) = \frac{1}{n^2} \sum_{k=1}^{n} \left(\sum_{l=1}^{n} X_{kl}^1 X_{kl}^2 \right) = \frac{1}{n^2} \sum_{k=1}^{n} (\mathbf{X_k^1})^T \mathbf{X_k^2}, \tag{5}$$

where the n scalar products are $(\mathbf{X_k^1})^T \mathbf{X_k^2}$ for $k = 1, \ldots, n$. Therefore, obtaining the distance covariance matrix in vertical splitting among several clouds can be decomposed into several secure scalar products to be conducted between pairs of clouds. Protocols 1 and 2 are perfectly suited to compute $(\mathbf{X_k^1})^T \mathbf{X_k^2}$ for $k = 1, \ldots, n$ (see Sect. 3.1). The only adaptation needed is to add two preliminary steps: one step by Alice to compute $\mathbf{x} = \mathbf{X_k^1}$ from $\mathbf{x^1}$, an another step by Bob to compute $\mathbf{y} = \mathbf{X_k^2}$ from $\mathbf{x^2}$.

Security. The two preliminary steps added before the secure scalar product are done separately by Alice and Bob, so there is no additional exchange of information between the clouds. Hence the security of Protocols 1 and 2 is preserved.

4   Comparison Among Methods

We compare here the performance of the methods presented in Sects. 3.2 and 3.3 with the following benchmark solution:

Protocol 3

    Set-up phase:
  i. *CLARUS encrypts the original data set* $\mathbf{E} = Enc(\mathbf{X})$.
 ii. *CLARUS sends* $\mathbf{E}$ *to a cloud Alice for storage.*
    Computation phase:
iii. *CLARUS downloads* $\mathbf{E}$ *from Alice, decrypts* $\mathbf{X} = Dec(\mathbf{E})$ *and performs the desired computation.*

Encryption and decryption can be performed using a fast symmetric cryptosystem, such as the Advanced Encryption Standard (AES), which takes time linear in the number of records/vector components n, as well as ciphertexts similar in size to the corresponding plaintexts.

For simplicity, we consider a data set $\mathbf{X}$ with two attributes owned by Alice and Bob, respectively (the generalization to more attributes and clouds is straightforward). We now evaluate the computational cost for Alice, Bob, CLARUS and the total computation under each protocol. Just giving the order of magnitude of the complexity is not accurate enough (e.g. n additions are faster than n multiplications, even if we have $O(n)$ computation in both cases); therefore, we give the complexity in terms of the costliest operation performed in each case. For instance, "read" means reading the vector, "AESdecr" means AES decryption of the vectors and "RNDgen" is the random key generation for the AES encryption. Moreover, operations that do not need to be repeated each time the protocol is executed, e.g. the generation of cryptographic keys in Protocol 2, are separately counted as set-up costs. Assuming that the clouds have unlimited storage, it is reasonable to assume as well that those matrices or vectors need to be generated only once and can be stored for subsequent reuse. In contrast, we do not assume unlimited storage at CLARUS; therefore, we assume the proxy just stores the random seeds and generates the matrices when needed. Also, we have associated the communication cost with the sender and we use a parameter γ to represent the maximum length of the numbers in the vectors and matrices used in the protocols. For the case of Protocol 2, lengths are a function of the size N of the field used by the Paillier cryptosystem: the public key is $3 \log_2 N$ bits long, the secret key is $\log_2 N$ bits long, ciphertexts are $2 \log_2 N$ bits long and plaintexts are $\log_2 N$ bits long. We consider that whenever possible the participants send the seeds of random vectors and matrices, rather than the vectors and matrices themselves.

4.1   Comparison for Contingency Table Computation

In Sect. 3.2, we adapted Protocols 1 and 2 to compute one cell value of the contingency table. We compare here these adaptations with Protocol 3, where, in step (iii), "desired computation" means "count the number of records in $\mathbf{X}$ containing both the categories $c_i(\mathbf{a})$ and $c_j(\mathbf{b})$, where $\mathbf{a}$ and $\mathbf{b}$ are the two attributes in $\mathbf{X}$".

Table 1. Long-term and temporary storage costs of Protocols 1, 2 and 3. n is the number of records/vector components; γ represents the maximum length of the numbers in the vectors and matrices used in the protocols; N is the size of the plaintext field used by Paillier. Note: In protocols not requiring the presence of Bob or Charlie, their costs are indicated with "$-$".

	Storage								
	Long-term				Temporary				
	Alice	Bob	Charlie	CLARUS	Alice	Bob	Charlie	CLARUS
Prot. 1	$(n+1)\gamma$	$(n+1)\gamma$	0	0	$(3n+1)\gamma$	$(3n+1)\gamma$	$(2n+3)\gamma$	3γ
Prot. 2	$n\gamma + 4\log_2 N$	$n\gamma$	$-$	0	$(3n+2)\gamma$	$(2n+4)\gamma$	$-$	3γ
Prot. 3	$2n\gamma$	$-$	$-$	$2n\gamma$	0	$-$	$-$	$(4n+1)\gamma$

Only Protocols 2 and 3 require a set-up phase. When the stored data are updated (that is, records are added, changed or removed), the set-up phase (key generation) of Protocol 2 does not need to be repeated: since $\mathbf{x}$ and $\mathbf{y}$ are binary vectors, we can fix a small order $N > 2$ of the finite field in use (see Sect. 3.1). Protocol 3 requires a set-up phase, that is, the key generation for the AES cryptosystem and the encryption of the private vectors. Compared to Protocol 2, the set-up phase of Protocol 3 needs to be repeated every time that the private vectors are changed. Only Protocols 2 and 3 present a communication cost of the set-up phase, due to the exchange of the public key for the two former protocols and the transmittal of the encrypted private vectors for the latter one.

Table 1 shows the long-term and temporary data storage costs (temporary storage is the one needed only to conduct a certain calculation at some point). In Protocol 1, Alice and Bob need long-term storage for $(a_1, \ldots, a_n)^T$ and $(b_1, \ldots, b_n)^T$, respectively, and for the seed to create the random vector. Charlie needs only temporary storage for $\mathbf{r_x}$, $\mathbf{r_y}$, their seeds and their product p. Alice and Bob also need temporary storage to share $(\mathbf{x}, \mathbf{r_x}, \hat{\mathbf{x}})$ and $(\mathbf{y}, \mathbf{r_y}, \hat{\mathbf{y}})$ and create s_x and t, respectively. In Protocol 2, long-term storage is required by Alice and Bob to store their respective private vectors, as well as to store the public-private key of the Paillier cryptosystem; Alice needs $4\log_2 N$ space for the key pair. The temporary storage depends on the computations as before. All data splitting protocols need the same long-term storage space. On the other hand, only the benchmark Protocol 3 requires CLARUS to store a large amount of data, namely the decrypted data. Note that Protocol 3's long-term storage is $2n\gamma$ as both n-vectors $\mathbf{a}$ and $\mathbf{b}$ are stored.

Table 2. Execution costs for Protocols 1, 2 and 3. n is the number of records/vector components; γ represents the maximum length of the numbers in the vectors and matrices used in the protocols. Charlie only appears in Protocol 1 and Bob does not appear in Protocol 3; we indicate the absence of a cloud with "$-$". The computation cost is presented in terms of the costliest operations performed in each case; the communication cost is the exact amount of transmitted data.

	Computational cost				Communication cost				Crypt
	Alice	Bob	CLARUS	Charlie	Alice	Bob	CLARUS	Charlie	package
Pr.1	n prod.$+ n$ read	n prod. $+ n$ read	2 sum.	$2n$ RND-gen	$(n+1)\gamma$	$(n+1)\gamma$	0	3γ	none
Pr.2	n RNDgen. $+ n$ encr. $+ n$ read	n prod.$+ n$ read	1 sum.	$-$	$(n+1)\gamma$	2γ	0	$-$	Alice
Pr.3	0	$-$	n read $+ n$ AESdecr	$-$	$2n\gamma$	$-$	0	$-$	CLARUS

Table 2 shows the computational and communication costs incurred by the execution of the above mentioned protocols (after set-up). In Protocol 1, the commodity server Charlie computes the scalar product of the two random vectors $(\mathbf{r_x}, \mathbf{r_y})$, stores the result and sends the seeds of $\mathbf{r_x}$ and $\mathbf{r_y}$ to Alice and Bob. Protocols 1 and 2 require reading vectors $(a_1, \ldots, a_n)^T$ and $(b_1, \ldots, b_n)^T$ to compute $\mathbf{x}$ and $\mathbf{y}$. Protocols 1 and 2 have all similar costs for CLARUS, but Protocol 2 needs encryption. Therefore, all in all, Protocol 1 seems to be the most advantageous one. Nevertheless, if a cryptographic package is available, Protocol 2 is suitable and fast.

4.2   Comparison for the Distance Covariance Matrix Computation

Section 3.3 showed that the distance covariance matrix calculation can be decomposed into computing several secure scalar products, each of which is conducted between a pair of clouds. Therefore, we focus on the secure scalar product between two vectors $\mathbf{x} = \mathbf{X}_k^1$ and $\mathbf{y} = \mathbf{X}_k^2$ computed by Alice and Bob, respectively (see Expression (5)). We compare here these adaptations with Protocol 3, where, in Step (iii), "desired computation" means "compute $\mathbf{x}^T\mathbf{y}$".

Note that all the computations to obtain $\mathbf{x}$ and $\mathbf{y}$ are performed by Alice and Bob in all the protocols except for Protocol 3, where they are left to CLARUS. Compared to the protocols in Sect. 4.1, these protocols differ only in the creation of $\mathbf{x}$ and $\mathbf{y}$. Therefore, in Table 1 the long-term storage remains unchanged, but the temporary storage requires a supplement of storage related to Equation (3). In Table 2, showing the execution costs, the sentences "n read" needs to be changed by the complexity of the $\mathbf{x}$ and $\mathbf{y}$ computations: "$(n^2 + 3n)$ sums + semantic distance complexity".

Like in Sect. 4.1, Protocols 1 and 2 present all similar costs for CLARUS, but Protocol 2 needs encryption. Therefore, all in all, Protocol 1 seems to be the most advantageous one. Nevertheless, if a cryptographic package is available, Protocol 2 is suitable and fast.

5   Conclusions and Future Work

We have presented two protocols, a cryptographic one and a non-cryptographic one, that allow statistical computation on protected sensitive categorical data stored in semi-honest clouds. For the sake of flexibility and efficiency, we have considered data splitting as a non-cryptographic method for data protection, rather than the heavier fully homomorphic encryption. We have provided complexity analyses and benchmarking for all proposed protocols, in order to show their computational advantages for the cloud user. In this way, clouds are not only used to store sensitive data, but also to perform computations on them in a privacy-aware manner. This is especially interesting for large sensitive data sets.

In future research, it would be interesting to design protocols for computations other than contingency tables and distance covariance matrices. Also, here we have considered only categorical attributes and in [4] we considered only numerical attributes. Dealing with cloud-stored sensitive data sets containing both types of attributes and using semi-honest clouds to perform computations involving both attribute types would be highly relevant.

Acknowledgments and Disclaimer. Partial support to this work has been received from the European Commission (projects H2020-644024 "CLARUS" and H2020-700540 "CANVAS"), from the Government of Catalonia (ICREA Acadèmia Prize to J. Domingo-Ferrer and grant 2014 SGR 537), and from the Spanish Government (project TIN2014-57364-C2-1-R "SmartGlacis" and TIN 2015-70054-REDC). The authors are with the UNESCO Chair in Data Privacy,but the views in this paper are the authors' own and are not necessarily shared by UNESCO.

References

1. Aggarwal, G., Bawa, M., Ganesan, P., Garcia-Molina, H., Kenthapadi, K., Motwani, R., Srivastava, U., Thomas, D., Xu, Y.: Two can keep a secret: a distributed architecture for secure database services. In: CIDR 2005, pp. 186–199 (2005)
2. Agresti, A., Kateri, M.: Categorical Data Analysis. Springer, Berlin (2011)
3. Armbrust, M., Fox, A., Griffith, R., Joseph, A.D., Katz, R., Konwinski, A., Lee, G., Patterson, D., Rabkin, A., Stoica, I., Zaharia, M.: A view of cloud computing. Commun. ACM **53**(4), 50–58 (2010)
4. Calviño, A., Ricci, S., Domingo-Ferrer, J.: Privacy-preserving distributed statistical computation to a semi-honest multi-cloud. In: IEEE Conference on Communications and Network Security (CNS 2015). IEEE (2015)
5. Ciriani, V., di Vimercati, S.D.C., Foresti, S., Jajodia, S., Paraboschi, S., Samarati, P.: Selective data outsourcing for enforcing privacy. J. Comput. Secur. **19**(3), 531–566 (2011)
6. CLARUS - A Framework for User Centred Privacy and Security in the Cloud, H2020 project (2015–2017). http://www.clarussecure.eu
7. Clifton, C., Kantarcioglu, M., Vaidya, J., Lin, X., Zhu, M.: Tools for privacy preserving distributed data mining. ACM SiGKDD Explor. Newsl. **4**(2), 28–34 (2002)
8. Domingo-Ferrer, J., Sánchez, D., Rufian-Torrell, G.: Anonymization of nominal data based on semantic marginality. Inf. Sci. **242**, 35–48 (2013)

9. Du, W., Han, Y., Chen, S.: Privacy-preserving multivariate statistical analysis: linear regression and classification. In: SDM, SIAM, vol. 4, pp. 222–233 (2004)

10. Dubovitskaya, A., Urovi, V., Vasirani, M., Aberer, K., Schumacher, M.I.: A cloud-based ehealth architecture for privacy preserving data integration. In: Federrath, H., Gollmann, D., Chakravarthy, S.R. (eds.) SEC 2015. IFIP AICT, vol. 455, pp. 585–598. Springer, Heidelberg (2015). doi:10.1007/978-3-319-18467-8_39

11. Goethals, B., Laur, S., Lipmaa, H., Mielikäinen, T.: On private scalar product computation for privacy-preserving data mining. In: Park, C., Chee, S. (eds.) ICISC 2004. LNCS, vol. 3506, pp. 104–120. Springer, Heidelberg (2005)

12. Karr, A., Lin, X., Sanil, A., Reiter, J.: Privacy-preserving analysis of vertically partitioned data using secure matrix products. J. Off. Stat. **25**(1), 125 (2009)

13. Martínez, S., Sánchez, D., Valls, A.: A semantic framework to protect the privacy of electronic health records with non-numerical attributes. J. Biomed. Inform. **46**(2), 294–303 (2013)

14. Paillier, P.: Public-key cryptosystems based on composite degree residuosity classes. In: Stern, J. (ed.) EUROCRYPT 1999. LNCS, vol. 1592, pp. 223–238. Springer, Heidelberg (1999)

15. Ren, K., Wang, C., Wang, Q.: Security challenges for the public cloud. IEEE Internet Comput. **16**(1), 69–73 (2012)

16. Rodríguez-García, M., Batet, M., Sánchez, D.: Semantic noise: privacy-protection of nominal microdata through uncorrelated noise addition. In: 27th IEEE International Conference on Tools with Artificial Intelligence (ICTAI), pp. 1106–1113 (2015)

17. Sánchez, D., Batet, M., Isern, D., Valls, A.: Ontology-based semantic similarity: a new feature-based approach. Expert Syst. Appl. **39**(9), 7718–7728 (2012)

18. Sánchez, D., Batet, M., Martínez, S., Domingo-Ferrer, J.: Semantic variance: an intuitive measure for ontology accuracy evaluation. Eng. Appl. Artif. Intell. **39**, 89–99 (2015)

19. Shannon, C.: Communication theory of secrecy systems. Bell Syst. Tech. J. **28**, 656–715 (1949)

20. Székely, G.J., Rizzo, M.L.: Brownian distance covariance. Ann. Appl. Stat. **3**(4), 1236–1265 (2009)

21. U.S. Federal Trade Commission: Data Brokers, A Call for Transparency and Accountability (2014)

22. Viejo, A., Sánchez, D., Castellà-Roca, J.: Preventing automatic user profiling in Web 2.0 applications. Knowl. Based Syst. **36**, 191–205 (2012)

23. Weiss, G.: Data mining in the real world: experiences, challenges, and recommendations. In: DMIN, pp. 124–130 (2009)

24. Yang, Q., Wu, X.: 10 challenging problems in data mining research. Int. J. Inf. Technol. Decis. Mak. **5**(4), 597–604 (2006)

Machine Learning Combining
with Visualization for Intrusion Detection:
A Survey

Yang Yu, Jun Long, Fang Liu, and Zhiping Cai[(✉)]

College of Computer, National University of Defense Technology,
Changsha 410073, Hunan, China
zpcai@nudt.edu.cn

Abstract. Intrusion detection is facing great challenges as network attacks producing massive volumes of data are increasingly sophisticated and heterogeneous. In order to gain much more accurate and reliable detection results, machine learning and visualization techniques have been respectively applied to intrusion detection. In this paper, we review some important work related to machine learning and visualization techniques for intrusion detection. We present a collaborative analysis architecture for intrusion detection tasks which integrate both machine learning and visualization techniques into intrusion detection. We also discuss some significant issues related to the proposed collaborative analysis architecture.

Keywords: Intrusion detection · Machine learning · Visualization

1 Introduction

Intrusion detection is playing an important role in cybersecurity to prevent lots of malicious attacks and threats. However, intrusion detection faces a variety of huge challenges. First, the massive amounts of network data such as network traffic data are difficult to be handled by using mainstream computing technologies [1]. Besides, large volumes of highly heterogeneous network data compound the analytical difficulties further, which leads to a strong requirement for much more advanced high performance computing technology [2]. Moreover, a wide array of new types of possible attacks like APT (Advanced Persistent Threat) along with the increasing sophistication of diverse network attacks render inefficiency of traditional intrusion detection approaches. In addition, traditional Intrusion Detection Systems (IDSs) [3] which detect anomalies automatically still produce plenty of false positives and false negatives.

Machine learning has received highly concern in both academia and industry. Machine learning in nature is an important branch of artificial intelligence that automatically discovers patterns or knowledge from training data to make predictions or improve system performance [4]. Since machine learning has well adaptive characteristics and mathematical robustness, it has been widely applied in intrusion detection recently and produces many fantastic research results [5]. Machine learning techniques can automatically discover hidden patterns used for detecting network intrusion

© Springer International Publishing Switzerland 2016
V. Torra et al. (Eds.): MDAI 2016, LNAI 9880, pp. 239–249, 2016.
DOI: 10.1007/978-3-319-45656-0_20

behaviors by training historical data. Besides, machine learning is statistically reliable because it relies on methods and ideas from other disciplines like statistics and probability. However, one of the significant problems is that classifiers trained by leveraging machine leaning techniques have poor interpretability. Thus, it is considerably hard for analysts to interpret and trust the detection results.

Fortunately, in combination with visual analysis techniques, many machine leaning approaches can gain better performance and interpretability [6]. In particular, visual perception has the advantages of high interpretability and good pattern recognition capacities. Besides, visualization techniques are also widely leveraged to visualize network security data [7–9]. One problem with visualization techniques is that recognizing a plethora of data can cause the overload problem which could result in unexpected outcomes [10].

While this paper covers a wide spectrum of diverse domains, the topic of this survey paper mainly focuses on solving intrusion detection problems by utilizing machine learning and visualization analysis techniques. The organization of this paper is as follows: Sect. 2 covers some significant machine learning techniques for intrusion detection. Section 3 presents visualization techniques for intrusion detection and machine learning. Section 4 discusses a collaborative analysis architecture and some promising future work on intrusion detection. Finally, Sect. 5 concludes our work presented in this survey paper.

2 Intrusion Detection Using Machine Learning Approaches

Figure 1 illustrates a generic framework for machine learning. In General, a machine learning task incorporates the training phrase and testing phrase. In the training phrase, the historical data are represented as feature vectors by feature extraction or feature selection methods. The human experts then label each feature vector to gain the training samples. Subsequently, the classification model is built through various machine learning algorithms with the training samples as input data. In the testing phase, similar to the way of gaining feature vectors in the training phase, the new data are also transferred into feature vectors first. Then, the classification model is applied on the feature vectors to acquire the ultimate classification results.

Based on the generic framework of machine learning, machine learning-based intrusion detection involves domain-related features in the process of feature extraction and learning algorithms. As features and learning algorithms are two essential components of learning-based intrusion detection, we mainly review some previous important studies related to them.

2.1 Features

Feature extraction or selection which greatly affects algorithm efficiency is regarded as a very crucial part of based-learning intrusion detection. Currently, features utilized in learning-based intrusion detection are classified into three categories, namely classical, sequential and syntactical features [11].

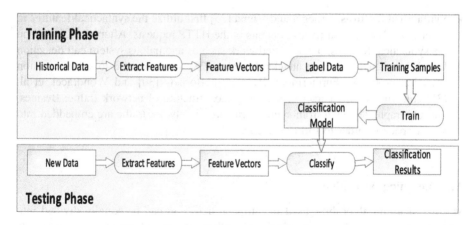

Fig. 1. Generic framework for machine learning

- **Classical features.** Lee and Stolfo [12] totally generalize 41 typical features applied in intrusion classification by employing Data Mining techniques. These typical features are divided into three types, i.e. basic features of the transport layer, content features and statistical features of network traffic. The basic features of the transport layer include protocol type, service type, the number of bytes transferred between source address and destination address, etc. Content features mainly relate to login information and the third type of features incorporate connection times and error messages about SYN. Besides, to test efficiency of different machine learning algorithms used for solving intrusion detection problems, "KDD CUP 1999" [13] utilizes these features to constitute intrusion detection dataset. Ultimately, these features are worldwide regarded as classical features of machine learning-based intrusion detection. The performance of many machine learning algorithms for intrusion detection problems such as anomaly detection [14] and feature selection [15] has been measured on the intrusion detection dataset. While the proposed features considerably facilitate application of machine learning approaches in intrusion detection, they are restricted to attack types at that time and cannot cover all of attack types nowadays.
- **Sequential features.** The majority of network attacks have their specific programming codes, which inevitably results in the discovery of classical sequential patterns existed in the network traffic. Rieck and Laskov et al. [16, 17] elaborate the sequential features and adapt the bag of token technique and q-grams method to handle the application layer payload of network traffic. In fact, sequential features are originally used in Natural Language Processing. Liao and Vemuri [18] treat the symbol strings existed in network traffic as words of documents and apply text categorization approaches based on the vector space model to intrusion detection. Mahoney and Chan [19], Inghamet et al. [20] further cope with intrusion detection problems using text categorization idea. In addition, Kruegel [21], Wang and Stolfo et al. [22] employ q-grams model in Intrusion detection as q-grams model can effectively represent closely related characters existed in the sequence. High-order q-grams are further explored by Rieck, Laskow [23] and Wang et al. [24], where they utilize unsupervised and semi-supervised learning algorithms respectively.

- **Syntactical features.** Kruegel and Vigna [25] first utilize the syntactical features in order to analyze the malicious contents in the HTTP requests. Afterwards, based on the syntactical features, other related work such as anomalous system call detection [26, 27] and SQL injection attacks detection [28] are conducted to solve intrusion detection problems. Furthermore, Pang [29], Borisov [30] and Wondracek et al. [31] construct syntax parser to derive syntax structure of network traffic. Besides, the sub-graphs obtained from syntax structure of network traffic are embedded into vector space in order to form syntactical features [11].

2.2   Learning Algorithms

Machine learning algorithms used in intrusion detection can be equally divided into three kinds of classifiers, i.e. single classifiers, hybrid classifiers and ensemble classifiers [32].

- **Single classifiers.** Generally, a single classifier solely uses one of various machine leaning algorithms to deal with intrusion detection data. Classical algorithms for single classifiers incorporate Naive Bayes classifier [33, 34], Support Vector Machine (SVM) [35–37], Decision Trees [38, 39], Artificial Neural Network [40, 41], Logistic Regression [42] and Nearest Neighbor [43], etc.
- **Hybrid classifiers.** A hybrid classifier integrates more than one kind of machine learning algorithms to enhance efficiency and performance of intrusion detection system. One example for this is that the intermediate results generated from raw data processed by one machine learning algorithm are as the inputs of another algorithm to produce the ultimate results. Specifically, the combination of SVMs and ant colony networks [44] along with integrating decision trees with SVMs [45], is studied recently.
- **Ensemble classifiers.** A variety of weak classifiers aggregated by Adaboost algorithm as well as a series of derivative algorithms constitute an ensemble classifier which has much more better performance than a standalone weak classifier. For instance, Artificial Neural Network and Bayesian Networks classifiers [46], as well as Naive Bayes and Decision Trees classifiers [13] which usually are regards as the weak classifiers can be aggregated into an ensemble classifier.

3   Visualization Techniques for Intrusion Detection and Machine Learning

3.1   Visualization Techniques for Intrusion Detection

Visualization techniques can be helpful in addressing intrusion detection problems because of the powerful cognitive and perceptual capabilities of human beings. Essentially, human perception and thinking capacities are curial components of analysis process. Hence, compared with machine intelligence, visual perception has irreplaceable advantages. Combining intrusion detection with visualization techniques aims at representing numerous cybersecurity data visually to facilitate user interactions so that humans enable to effectively perceive and analyze cyberspace situational awareness. Becker et al. [47] propose using parametric techniques to visualize network data as early as 1995. According to Girardind et al. [48], visualization techniques success in monitoring and analyzing log records.

Currently, most related studies mainly focus on exploring and analyzing network anomalies through various visualization techniques such as heat maps analysis [49], hierarchical edge bundles [50], abstract graphs representation [51], etc. Heat maps analysis can be used to monitor network traffic and allows analysts to clearly observe the network traffic distribution of target hosts from a less complex network topology. Hierarchical edge bundles and abstract graphs representation, which can mitigate the complexity of network nodes and links, are applied in large scale networks.

However, a standalone visualization approach encounters difficulties of monitoring diverse intricate security events. Some researches hence integrate a variety of different visual methods to provide a multilevel security model such as Spiral View [7], NVisionIP [8], VisTracer [9], etc. Specifically, NVisionIP as shown in Fig. 2 [8] integrates three views, namely the galaxy, small multiple, and machine views to monitoring network anomalies.

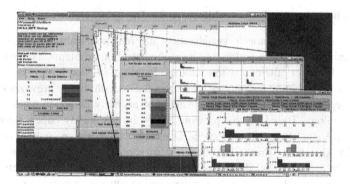

Fig. 2. Three views of NVisionIP [8]

3.2 Visualization Techniques for Machine Learning

In reality, network data are mostly high dimensional and heterogeneous, which leads to both visualization and machine learning facing Big Heterogeneous Data challenges [2]. Actually machine learning and visualization techniques can complement each other in some extent. In general, machine learning techniques leverage effective dimensionality reduction algorithms to achieve visualization of high dimensional network data. Conversely, visualization techniques, which take full advantages of human perception capabilities to quickly handle visual symbols, facilitate the discovery of patterns hidden in a myriad of network data. The studies on the combination of machine learning and visualization have made great process in recent years. Dagstuhl Seminar "Information Visualization, Visual Data Mining and Machine Learning" [52] and "Bridging

Information Visualization with Machine Learning" [10] have been held in 2012 and 2015 respectively, which brought together researchers in information visualization and machine learning to discuss significant challenges and corresponding solutions of integrating the two fields.

Currently, most studies are focusing on classifiers' visualization and visualization interaction of the learning process. While classification models of the majority of machine learning algorithms are difficult to interpret, the interpretability of classification model becomes an important bottleneck in the development of machine learning. Visualization of classifiers emerges as a consequence of the high demands on the interpretable classification model which is helpful for users in discovering and exploring outliers or mislabeled samples, under-fitting or overfitting problems, the spatial modality of diverse classification data [53–55], etc. In addition, visualization interaction of the learning process enables users to interactively adjust parameters of specific algorithms so as to improve learning performance. Related studies cover network alarm triage [56] and interactive optimization which involved in the adjustment of parameters for given performance constrains [6, 57, 58], etc.

4   Discussion and Future Work

4.1   Collaborative Analysis Architecture for Intrusion Detection

Over the past decades, various visualization and machine learning techniques have been respectively applied in addressing intrusion detection problems. To the best of our knowledge, however, there are few studies focusing on combining both machine learning and visualization techniques for intrusion detection. In our previous work [59], we have deeply studied the important features of anomaly detection based on visualization technology. Based on our previous study, in this survey paper, we propose a novel "Brain-Visual Perception-Machine Learning" collaborative analysis architecture for intrusion detection from a holistic view as shown in Fig. 3. Machine Learning automatically analyzes and discovers intrusion patterns hidden in the network data. Meanwhile, as network data, learning process and results are presented by visualization techniques, analysts can interact with a friendly visual interface in real time. In particular, humans play a significant role in this workflow. For instance, after capturing visual information which is then processed by human brain, analysts can tune the parameters of learning algorithms by means of visualization interaction in order to get reliable and interpretable results of intrusion detection.

4.2   Future Work

The following main issues that need to be solved could be promising in the future research:

- **Representation of complex intrusion detection features.** While classical features of intrusion detection are expressed as feature vectors which can be directly utilized by machine learning algorithms, sequential and syntactical features belonging to

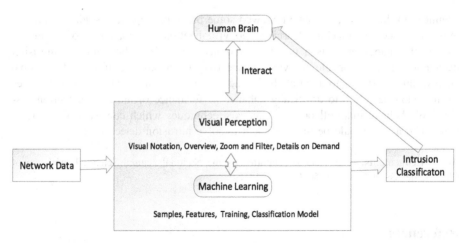

Fig. 3. "Brain-Visual Perception-Machine Learning" collaborative analysis architecture for intrusion detection

complex features are respectively represented by the forms of q-gram and syntax trees [11]. Furthermore, as intrusion features are increasingly sophisticated, the representation of these features requires a new kind of embedding language which can embed intricate structure into the vector space and preserve information of original data as much as possible.

- **Dimensionality reduction problem.** In general, the high dimensional data need to be reduced into low dimensions such as two or three dimensions for visualization purpose by leveraging classical dimensionality reduction algorithms including principal component analysis [60], multidimensional scaling [61], locally linear embedding [62], isomap [63], etc. However, how to combine specific features of intrusion data with those dimensionality reduction algorithms still needs to be further studied.

- **Visualization of parameter space.** Parameter adjustment is crucial to the success of learning process which usually searches for optimal parameters by means of the hill climbing method or heuristic algorithms. Visualizing parameters is helpful in avoiding trapping in local optimum as analysts are involved in parameter adjustment of learning process. Therefore, visualization of parameter space is also an important issue that needs to be addressed.

5   Conclusion

Massive highly heterogeneous, sophisticated and distributed network security data pose great challenges to both academia and industry [64–66]. In addition, applying machine learning and visualization technologies respectively to address intrusion detection issues has been heavily studied over the last decades. In recent years, the combination of machine learning and visualization techniques has attracted many researchers' attention. In this paper, we have reviewed some previous important studies related to intrusion detection, machine learning and visualization. Compared to other survey papers, this paper presents a novel collaborative analysis architecture for intrusion detection from a reasonable and valid perspective. The main idea of this collaborative analysis architecture is to integrate both machine learning and visualization techniques for intrusion detection to drastically enhance performance of intrusion detection systems. We believe that collaborative analysis techniques which consist of three fields mentioned above would be more efficient to solve intrusion detection issues.

Acknowledgements. This work is supported by the National Natural Science Foundation of China under Grant Nos. 61105050, 61379145.

References

1. Ahmed, M., Naser Mahmood, A., Hu, J.: A survey of network anomaly detection techniques. J. Netw. Comput. Appl. **60**, 19–31 (2016)
2. Zuech, R., Khoshgoftaar, T.M., Wald, R.: Intrusion detection and big heterogeneous data: a survey. J. Big Data **2**, 1–41 (2015)
3. Ektefa, M., Memar, S., Sidi, F., Affendey, L.S.: Intrusion detection using data mining techniques. In: International Conference on Information Retrieval & Knowledge Management, pp. 1–14 (2010)
4. Nguyen, H.: Reliable machine learning algorithms for intrusion detection systems. Ph.D. thesis, Faculty of Computer Science and Media Technology Gjøvik University College (2012). http://hdl.handle.net/11250/144371. Accessed August 2015, 2.3. 2, 2.4. 2, 2, 5.3. 3
5. Farah, N., Avishek, M., Muhammad, F., Rahman, A., Rafni, M., Md, D.: Application of machine learning approaches in intrusion detection system: a survey. Int. J. Adv. Res. Artif. Intell. **4**, 9–18 (2015)
6. Kapoor, A., Lee, B., Tan, D., Horvitz, E.: Performance and preferences: interactive refinement of machine learning procedures. In: AAAI Conference on Artificial Intelligence, pp. 113–126 (2015)
7. Bertini, E., Hertzog, P., Lalanne, D.: SpiralView: towards security policies assessment through visual correlation of network resources with evolution of alarms. In: IEEE Symposium on Visual Analytics Science and Technology, 2007, VAST 2007, pp. 139–146. IEEE (2007)
8. Lakkaraju, K., Yurcik, W., Lee, A.J.: NVisionIP: netflow visualizations of system state for security situational awareness. In: Proceedings of the 2004 ACM Workshop on Visualization and Data Mining for Computer Security, pp. 65–72. ACM (2004)
9. Fischer, F., Fuchs, J., Vervier, P.-A., Mansmann, F., Thonnard, O.: Vistracer: a visual analytics tool to investigate routing anomalies in traceroutes. In: Proceedings of the Ninth International Symposium on visualization for Cyber Security, pp. 80–87. ACM (2012)
10. Keim, D.A., Munzner, T., Rossi, F., Verleysen, M., Keim, D.A., Verleysen, M.: Bridging information visualization with machine learning. Dagstuhl Rep. **5**, 1–27 (2015)
11. Rieck, K.: Machine learning for application-layer intrusion detection. In: Fraunhofer Institute FIRST and Berlin Institute of Technology, Berlin, Germany (2009)
12. Lee, W., Stolfo, S.J.: A framework for constructing features and models for intrusion detection systems. ACM Trans. Inf. Syst. Secur. **3**, 227–261 (2000)

13. Kumarshrivas, A., Kumar Dewangan, A.: An ensemble model for classification of attacks with feature selection based on KDD99 and NSL-KDD data set. Int. J. Comput. Appl. **99**, 8–13 (2014)
14. Eskin, E., Arnold, A., Prerau, M., Portnoy, L., Stolfo, S.: A geometric framework for unsupervised anomaly detection. Appl. Data Min. Comput. Sec. **6**, 77–101 (2002)
15. Fan, W., Miller, M., Stolfo, S., Lee, W., Chan, P.: Using artificial anomalies to detect unknown and known network intrusions. In: IEEE International Conference on Data Mining, ICDM, pp. 123–130 (2001)
16. Rieck, K., Laskov, P.: Language models for detection of unknown attacks in network traffic. J. Comput. Virol. **2**, 243–256 (2007)
17. Rieck, K., Laskov, P.: Linear-time computation of similarity measures for sequential data. J. Mach. Learn. Res. **9**, 23–48 (2008)
18. Liao, Y., Vemuri, V.R.: Using text categorization techniques for intrusion detection. In: Proceedings of Usenix Security Symposium, pp. 51–59 (2002)
19. Mahoney, M.V., Chan, P.K.: Learning rules for anomaly detection of hostile network traffic. In: Null, p. 601. IEEE (2003)
20. Ingham, K.L., Inoue, H.: Comparing anomaly detection techniques for HTTP. In: Kruegel, C., Lippmann, R., Clark, A. (eds.) RAID 2007. LNCS, vol. 4637, pp. 42–62. Springer, Heidelberg (2007)
21. Kruegel, C., Valeur, F., Vigna, G., Kemmerer, R.: Stateful intrusion detection for high-speed network's. In: 2002 IEEE Symposium on Security and Privacy, 2002, Proceedings, pp. 285–293 (2002)
22. Wang, K., Stolfo, S.J.: One-class training for masquerade detection. In: IEEE Conference Data Mining Workshop on Data Mining for Computer Security (2003)
23. Rieck, K., Laskov, P.: Detecting unknown network attacks using language models. In: Büschkes, R., Laskov, P. (eds.) DIMVA 2006. LNCS, vol. 4064, pp. 74–90. Springer, Heidelberg (2006)
24. Wang, K., Parekh, J.J., Stolfo, S.J.: Anagram: a content anomaly detector resistant to mimicry attack. In: Zamboni, D., Kruegel, C. (eds.) RAID 2006. LNCS, vol. 4219, pp. 226–248. Springer, Heidelberg (2006)
25. Kruegel, C., Vigna, G.: Anomaly detection of web-based attacks. In: ACM Conference on Computer and Communications Security, pp. 251–261 (2003)
26. Krueger, T., Gehl, C., Rieck, K., Laskov, P.: An architecture for inline anomaly detection. In: 2008 European Conference on Computer Network Defense, pp. 11–18 (2008)
27. Mutz, D., Valeur, F., Vigna, G., Kruegel, C.: Anomalous system call detection. ACM Trans. Inf. Syst. Secur. **9**, 61–93 (2006)
28. Valeur, F., Mutz, D., Vigna, G.: A learning-based approach to the detection of SQL attacks. In: International Conference on Detection of Intrusions & Malware, pp. 123–140 (2005)
29. Pang, R., Paxson, V., Sommer, R., Peterson, L.: binpac: a yacc for writing application protocol parsers. In: ACM SIGCOMM Conference on Internet Measurement, pp. 289–300 (2006)
30. Borisov, N., Brumley, D.J., Wang, H.J., Dunagan, J., Joshi, P., Guo, C.: Generic application-level protocol analyzer and its language. In: Annual Network and Distributed System Security Symposium (2005)
31. Wondracek, G., Comparetti, P.M., Krügel, C., Kirda, E.: Automatic network protocol analysis. In: Proceedings of the 15th Annual Network and Distributed System Security Symposium (NDSS 2008) (2008)
32. Tsai, C.-F., Hsu, Y.-F., Lin, C.-Y., Lin, W.-Y.: Intrusion detection by machine learning: a review. Expert Syst. Appl. **36**, 11994–12000 (2009)

33. Koc, L., Mazzuchi, T.A., Sarkani, S.: A network intrusion detection system based on a Hidden Naïve Bayes multiclass classifier. Expert Syst. Appl. **39**, 13492–13500 (2012)
34. Hou, Y.T., Chang, Y., Chen, T., Laih, C.S., Chen, C.M.: Malicious web content detection by machine learning. Expert Syst. Appl. **37**, 55–60 (2010)
35. Catania, C.A., Bromberg, F., Garino, C.G.: An autonomous labeling approach to support vector machines algorithms for network traffic anomaly detection. Expert Syst. Appl. **39**, 1822–1829 (2012)
36. Kang, I., Jeong, M.K., Kong, D.: A differentiated one-class classification method with applications to intrusion detection. Expert Syst. Appl. **39**, 3899–3905 (2012)
37. Grinblat, G.L., Uzal, L.C., Granitto, P.M.: Abrupt change detection with one-class time-adaptive support vector machines. Expert Syst. Appl. **40**, 7242–7249 (2013)
38. Sahin, Y., Bulkan, S., Duman, E.: A cost-sensitive decision tree approach for fraud detection. Expert Syst. Appl. **40**, 5916–5923 (2013)
39. Wu, S.Y., Yen, E.: Data mining-based intrusion detectors. Expert Syst. Appl. **36**, 5605–5612 (2009)
40. Devaraju, S.: Detection of accuracy for intrusion detection system using neural network classifier. In: International Conference on Information, Systems and Computing-ICISC, pp. 1028–1041 (2013)
41. Wu, H.C., Huang, S.H.S.: Neural networks-based detection of stepping-stone intrusion. Expert Syst. Appl. **37**, 1431–1437 (2010)
42. Min, S.M., Sohn, S.Y., Ju, Y.H.: Random effects logistic regression model for anomaly detection. Pharmacol. Biochem. Behav. **37**, 7162–7166 (2010)
43. Davanzo, G., Medvet, E., Bartoli, A.: Anomaly detection techniques for a web defacement monitoring service. Expert Syst. Appl. **38**, 12521–12530 (2011)
44. Feng, W., Zhang, Q., Hu, G., Huang, J.X.: Mining network data for intrusion detection through combining SVMs with ant colony networks. Future Gener. Comput. Syst. **37**, 127–140 (2014)
45. Ranjan, R., Sahoo, G.: A new clustering approach for anomaly intrusion detection. Eprint Arxiv **4**, 29–38 (2014)
46. Farid, D.M., Zhang, L., Hossain, A., Rahman, C.M., Strachan, R., Sexton, G., Dahal, K.: An adaptive ensemble classifier for mining concept drifting data streams. Expert Syst. Appl. **40**, 5895–5906 (2013)
47. Becker, R.A., Eick, S.G., Wilks, A.R.: Visualizing network data. IEEE Trans. Visual. Comput. Graph. **1**, 16–28 (1995)
48. Girardin, L., Brodbeck, D.: A visual approach for monitoring logs. In: LISA, pp. 299–308 (2001)
49. Zhao, Y., Liang, X., Fan, X., Wang, Y., Yang, M., Zhou, F.: MVSec: multi-perspective and deductive visual analytics on heterogeneous network security data. J. Visual. **17**, 181–196 (2014)
50. Fischer, F., Mansmann, F., Keim, D.A., Pietzko, S., Waldvogel, M.: Large-scale network monitoring for visual analysis of attacks. In: Goodall, J.R., Conti, G., Ma, K.-L. (eds.) VizSec 2008. LNCS, vol. 5210, pp. 111–118. Springer, Heidelberg (2008)
51. Tsigkas, O., Thonnard, O., Tzovaras, D.: Visual spam campaigns analysis using abstract graphs representation. In: Proceedings of the Ninth International Symposium on Visualization for Cyber Security, pp. 64–71. ACM (2012)
52. Keim, D.A., Rossi, F., Seidl, T., Verleysen, M., Wrobel, S., Seidl, T.: Information visualization, visual data mining and machine learning. Dagstuhl Rep. **2**, 58–83 (2012)
53. Schulz, A., Gisbrecht, A., Bunte, K., Hammer, B.: How to visualize a classifier. In: New Challenges in Neural Computation, pp. 73–83 (2012)

54. Schulz, A., Gisbrecht, A., Hammer, B.: Using discriminative dimensionality reduction to visualize classifiers. Neural Process. Lett. **42**, 27–54 (2014)
55. Gisbrecht, A., Schulz, A., Hammer, B.: Discriminative dimensionality reduction for the visualization of classifiers. In: Fred, A., De Marsico, M. (eds.) ICPRAM 2013. AISC, vol. 318, pp. 39–56. Springer, Heidelberg (2015)
56. Amershi, S., Lee, B., Kapoor, A., Mahajan, R., Christian, B.: CueT: human-guided fast and accurate network alarm triage. In: Proceedings of the SIGCHI Conference on Human Factors in Computing Systems, pp. 157–166. ACM (2011)
57. Kapoor, A., Lee, B., Tan, D., Horvitz, E.: Interactive optimization for steering machine classification. In: SIGCHI Conference on Human Factors in Computing Systems, pp. 1343–1352 (2010)
58. Amershi, S., Chickering, M., Drucker, S.M., Lee, B., Simard, P., Suh, J.: ModelTracker: redesigning performance analysis tools for machine learning. In: ACM Conference on Human Factors in Computing Systems, pp. 337–346 (2015)
59. Zhao, Q., Long, J., Fang, F., Cai, Z.: The important features of anomaly detection based on visualization technology. In: Proceedings of the 12th International Conference on Modeling Decisions for Artificial Intelligence (MDAI 2015), Skovde, Sweden (2015)
60. Abdi, H., Williams, L.J.: Principal component analysis. Wiley Interdisc. Rev. Comput. Stat. **2**, 433–459 (2010)
61. Kruskal, J.B.: Multidimensional scaling by optimizing goodness of fit to a nonmetric hypothesis. Brain Res. **1142**, 159–168 (2007)
62. Saul, L.K., Roweis, S.T.: An Introduction to Locally Linear Embedding. Report at AT&T Labs – Research (2000)
63. Choi, H., Choi, S.: Robust kernel Isomap. Pattern Recogn. **40**, 853–862 (2010)
64. Cai, Z., Wang, Z., Zheng, K., Cao, J.: A distributed TCAM coprocessor architecture for integrated longest prefix matching, policy filtering, and content filtering. IEEE Trans. Comput. **62**(3), 417–427 (2013)
65. Chen, J., Yin, J., Liu, Y., Cai, Z., Li, M.: Detecting distributed denial of service attack based on address correlation value. J. Comput. Res. Dev. **46**(8), 1334–1340 (2009)
66. Liu, F., Dai, K., Wang, Z., Cai, Z.: Research on the technology of quantitative security evaluation based on fuzzy number arithmetic operation. Fuzzy Syst. Math. **18**(4), 51–54 (2004)

Sampling and Merging for Graph Anonymization

Julián Salas[✉]

Department of Computer Engineering and Mathematics,
Universitat Rovira i Virgili, Tarragona, Spain
julian.salas@urv.cat

Abstract. We propose a method for network anonymization that consists on sampling a subset of vertices and merging its neighborhoods in the network. In such a way, by publishing the merged graph of the network together with the sampled vertices and their locally anonymized neighborhoods, we obtain a complete anonymized picture of the network. We prove that the anonymization of the merged graph incurs in lower information loss, hence, it has more utility than the direct anonymization of the graph. It also yields an improvement on the quality of the anonymization of the local neighbors of a given subset of vertices.

Keywords: Degree sequence · Graph anonymization · Graph sampling · Network privacy · k-anonymity

1 Introduction

Networks can be used to represent very diverse objects, such as organizations, neural or metabolic networks, distribution networks, or social networks. Their vertices and edges may have very different properties which may be used for representing different things, such as individuals, objects, acquaintances, locations, incomes, geographical proximity, among many others.

Online social networks have become a part of the daily lifes of most people. With their increasing use and pervasiveness, researchers and enterprises have found an oportunity to analyze them and extract information that may be valuable for the benefit of our society.

In order to extract that knowledge from networks they must be published for their study. But this, in time, yields the possibility of revealing personal characteristics of the members of the network, with the risk of reidentifying the individuals behind the nodes and revealing their private attributes. This implies the need for modification of characteristics that may lead to reidentification or to revealing attributes that may be considered private, i.e., they must be anonymized.

In this work we propose a method of sampling and merging the local neighborhoods of given vertices of the graph. In such a way, the set of neighborhoods can be anonymized separately from the rest of the graph. In [17], was shown that anonymizing such set of non-overlapping neighborhoods improved the overall

© Springer International Publishing Switzerland 2016
V. Torra et al. (Eds.): MDAI 2016, LNAI 9880, pp. 250–260, 2016.
DOI: 10.1007/978-3-319-45656-0_21

information loss in information rich graphs. However, if only the local neighbors of the sampled vertices are published as in [17], then the global properties of the original graph (e.g., connectivity, shortest paths, average path length) are completely lost.

Therefore, in order to keep the global information of the original network, a solution that we propose in this paper is the method of sampling and merging local neighborhoods. Afterwards, publishing the anonymized local neighborhoods together with the merged graph. In this way a local and a global picture of the graph are obtained, by following this procedure different anonymization techniques may be used simultaneously for the local graphs and for the merged graph. In this paper we focus the study only on the merged graph, we provide k-degree subset anonymity for such graph and show that the information loss is improved with respect to the k-degree anonymization of the original network.

2 Related Work

It was found by Backstrom et al. [3] that simple anonymization of a social network (only replacing names or identifiers by a code name) may allow an attacker to learn relations between targeted pairs of nodes and recover the original and private names from the structure of the simply anonymized copy of a social network.

With the aim of properly anonymizing the networks, there is always a tradeoff between the utility and protection. Several measures have been considered for social networks, most of them come from statistical disclosure control, such is the case of k-anonymity [19,23]. The concept of k anonymity has several different definitions for graphs depending on the assumption of the attacker's knowledge, e.g., k-degree anonymity [15], k-neighborhood anonymity [28], in general all of them can be resumed as k-Candidate anonymity [11] or $k - \mathcal{P}$-anonymity [6], i.e., for a given a structural property $\mathcal{P}$ and a vertex in the graph G there are at least $k - 1$ other vertices with the same property $\mathcal{P}$.

Note that the most restrictive of all the structural properties, in the sense that it implies all the others, is when $\mathcal{P}$ is the neighborhood, cf. [22]. On the other hand, all these definitions imply k-degree anonymity, therefore the minimum number of edge modifications needed to obtain k-degree anonymity is a lower bound for all the other properties.

The standard approach to anonymization is to take into account all the vertices of the network, and to anonymize all with the same parameter, e.g., k-anonymity, t-closeness, and even differential privacy considers the entire set of vertices. However, as [7] points out, altering the structure of one vertex necessitates altering the structure of another simultaneously. Thus, they define a relaxation of k-degree anonymity that they call k-degree-based subset anonymization problem (k-D-SAP) in which there is only a target subset $X \subseteq V$, to be anonymized. Formally, the problem is to find a graph $G' = (V, E \cup E')$, such that X is k-degree-anonymous in G' and the number of new edges added, $|E'|$, is minimized. This, has been noted since the work of Liu and Terzi [15] that is very restrictive, therefore a relaxation of this condition may be considered in which edge modifications are allowed and not only edge additions. This is done in such a way that the number of edges that the original and the anonymized graph have in common is as large as possible.

Graph anonymization strategies are characterized in [29] as clustering based or graph modification approaches. The modification based approaches anonymize a graph by inserting or deleting edges or vertices, that can be done in a greedy or a randomized way. Hay et al. [11] study a randomization method by changing the original graph with m edge deletions and m insertions chosen uniformly at random, and calculate the protection it provides.

Liu and Terzi proposed the model of k degree anoymity in [15], they obtained a k-degree anonymous graph in two steps, first generating a k-anonymous sequence by a dynamic programming method to minimize the degree anonymization cost L_1 and then constructing the graph with the given sequence. We use the approach of [12] to calculate the k-anonymous sequence that minimizes the L_2 distance to the original sequence, and then obtain the graph by following the algorithm sketched in [20].

Some sufficient conditions for degree sequences to be graphic and for applications to k-degree anonymization can be found in [20,21], these conditions (namely P-Stability) may also be used for edge randomization while preserving the degree sequences.

The clustering approaches for graph anonymization, consider clustering edges (e.g., [27]) or vertices (e.g., [4,5,11]). Following the clustering approach, in [24] the authors define an anonymized social network as $\mathcal{AG} = (\mathcal{AN}, \mathcal{AE})$ where $\mathcal{AN} = \{C_1, \ldots, C_v\}$ are the nodes of the graph $\mathcal{AG}$, and C_i is a node representing the centroid of c_i a given cluster of nodes in G, with the additional information of how many nodes and edges are in c_i (inter-cluster generalization pair). And $\mathcal{AE}$ is the set of edges between nodes in $\mathcal{AN}$ such that each edge $(C_i, C_j) \in \mathcal{AE}$ has the additional information of how many edges has one end-vertex in cluster c_i and the other in cluster c_j (intra-cluster generalization value).

Hence, all nodes from a cluster c are collapsed into a super-node C, all the edges also are collapsed into super-edges, and the information of how many edges and nodes have been collapsed is published. In order to satisfy k-anonymity all clusters must have at least k nodes.

Note that satisfying this k-anonymity definition implies the strictest definition of k-anonymity for graphs, they are protected against any structural property, yet, this also means that the information loss increases (with respect to less restrictive definitions of k-anonymity).

One of the difficulties of evaluating the utility of a graph that is k-anonymized by a clustering technique, is that if the original graph has n nodes and all the nodes in each cluster are collapsed into a supernode, then the anonymized graph has at most n/k nodes. Thus, it is difficult to retain the distance properties of the original graph (such as the diameter, or average path length). Therefore, for the evaluation of the utility of the anonymized graphs in [24], the authors "de-anonymize" the graph to compare it with the original, they re-generate a graph with n nodes by randomly generating graphs with the given number of nodes and edges for each c_i, and attach them randomly to a set of nodes c_j with a set of edges that is pre-specified as the inter-cluster generalization value for the edge $(C_i, C_j) \in \mathcal{AE}$.

Our proposed approach to anonymization is in between clustering and k-degree anonymization. First we group the nodes in the neighborhoods of a given subset of vertices and merge them altogether, then we modify the graph obtained by this procedure to anonymize it. The set of anonymized neighborhoods may also be published to give a more accurate local and global picture of the graph, cf. [17]. We will explain our method after introducing some definitions and notations in the following section.

3   Definitions

We denote the *rectilinear* and *Euclidean* distances respectively by L_1 and L_2. Given two vectors $v_1 = (v_{11}, v_{12}, \ldots v_{1n})$ and $v_2 = (v_{21}, v_{22}, \ldots v_{2n})$, we define $L_1(v_1, v_2) = \sum_{i=1}^{n} |v_{1i} - v_{2i}|$ and $L_2(v_1, v_2) = \sum_{i=1}^{n} (v_{1i} - v_{2i})^2$.

We denote a graph by $G = (V, E)$ where $V = V(G)$ denotes the set of vertices and $E = E(G)$ the set of edges of G.

A graph is *simple* if it does not contains loops nor multiple edges.

The *component* of a vertex is the set of all vertices that can be reached from paths starting from the given vertex. A graph is *connected* if it has only one component.

We consider undirected, simple and connected graphs. The connectivity restriction is due to the fact that diameter, average path length and centrality measures are meaningful for connected graphs.

The *neighborhood* $N(v)$ of a vertex v is the set of all adjacent vertices to v.

The *degree* is the number of edges connected to a vertex, in the case of simple graphs it is the same as the number of vertices adjacent to a vertex.

Given a graph G and a sampled subset of nodes S, we denote the *merged graph* by G_S, that is defined next.

4   Sampling and Merging for Graph Anonymization

Given a graph G, we sample a dense subset of nodes $S = \{s_1, \ldots, s_t\}$, such that their corresponding neighborhoods $\{N_1, \ldots, N_t\}$ do not intersect, and merge all the nodes in N_i onto a single node s_i to obtain the merged graph G_S.

Formally, G_S is the graph where $v \in V(G_S)$ if either $v \in S$ or $v \in R = V(G) \setminus \bigcup_{i=1}^{t} N_i$. And, $(u_i, u_j) \in E(G_S)$ if $u_i, u_j \in R$ and $(u_i, u_j) \in E(G)$; or $u_i \in R$, $u_j = s_j \in S$ and there is a vertex $w \in V(G)$ such that $(u_i, w), (w, s_j) \in E(G)$; or $u_i = s_i \in S$, $u_j = s_j \in S$ and there are two vertices x and w such that $(u_i, x), (x, w), (w, s_j) \in E(G)$ (Figs. 1 and 2).

Note that the assumption that the local subgraphs $\{N_1, \ldots, N_t\}$ do not intersect may be omitted. However, by taking the subgraphs $\{N_1, \ldots, N_t\}$ to be disjoint, they can be anonymized without the most notable difficulty of graph anonymization, that is, altering the neighborhood of a vertex also modifies other vertices neighborhoods.

Since it is possible to define a categorical and structural distance between the neighborhoods $\{N_1, \ldots, N_t\}$ corresponding to nodes $\{s_1, \ldots, s_t\}$ and the modifications done to each neighborhood do not alter the others, then the "average neighborhood graph" can be defined and the furthest graphs from the average can be calculated. Hence, an approach to anonymization such as MDAV [8,9] can now be applied to such graphs. In [17] the method used for anonymizing the neighborhoods $\{N_1, \ldots, N_t\}$ is a graph matching algorithm cf. [16]. The method of sampling dense subgraphs (i.e., such that any vertex in the graph is at distance at most 2 from a given seed node) was tested for synthetic information rich graphs and it improves the overall information loss measures compared with the cases when overlapping neighbors are allowed, cf. [17].

Further properties are that it simplifies the networks (by merging the given neighborhoods) and reduces the number of vertices in the merged graph, retaining a correct global picture of the original graph, hence, it may increase the efficiency of the experiments to be realized on the published graphs.

Fig. 1. Karate network G

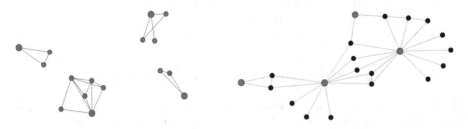

Fig. 2. Subgraphs $\{N_1, \ldots, N_t\}$ and merged network G_S

In [14], different sampling techniques are defined, these are in three different categories, deletion, contraction and exploration methods. These techniques are applied to a given graph until a given number of nodes or edges is reached. Our sampling technique is similar to the edge contractions, but has the particularity that all the edges incident to a given vertex are contracted at a time.

A more general definition for the method of sampling and merging could allow for arbitrarily (or algorithmically) choosing the nodes which are to be merged.

4.1 Experimental Evaluation

In our experiments, we compare two algorithms for k-degree anonymity, in fact, we use the same algorithm but on two assumptions, the one of [15], and ours from [20]. We do not use a dynamic program as Liu and Terzi, but from [12] we know we obtain optimal partitions.

We use the L_2 metric for anonymization instead of the L_1 distance considered in [15], that is, by taking the the mean value of the degree in a group of nodes (for k-degree anonymization) instead of the median, we obtain a k-degree anonymous graph which has more edges in common with the original graph. However, both measures remain close, presumably because the mean and the median of the subsets in our datasets are similar.

We compare a k-degree anonymity algorithm for the original and the merged graph. Our approach to subset anonymization is the following: We obtain the best approximation (considering a distance such as L_2) for the subset of nodes X by taking only the degree sequence of the vertices in X, denoted as d_X, and following the microaggregation procedure from [12] applied to d_X.

We evaluate our techniques on real datasets coming from different kinds of networks, that are Karate [26], PolBooks [13], Celegans [25], PolBlogs [1] and Netscience [18], see Table 1. We consider the graph to be the main connected component as we explained in the introduction. The global metrics considered are explained in next section.

Table 1. Graph metrics

	Karate	PolBooks	Celegans	PolBlogs	Netscience
$|V|$	34	105	297	1222	1589
$|E|$	78	441	2148	16714	2742
$Diam$	5	7	5	8	17
APL	2.40820	3.07875	2.45532	2.73753	5.82324
C_C	0.29819	0.17195	0.37331	0.29432	0.00029
C_B	0.40556	0.12044	0.29947	0.09666	0.02229
C_D	0.37611	0.15962	0.40384	0.26507	0.01924
γ	2.12582	2.62186	3.34269	3.66713	3.60673

4.2 Information Loss Measures

The *diameter* is the maximum of the distances among all pairs of vertices in the graph.

We will denote the *average path length* as APL and is the average length of the shortest path between all pairs of reachable nodes.

Next, we write the definitions of the degree, betweenness, and closeness centrality measures for a network, cf. [10].

The degree centrality C_D of a node v is the number of edges adjacent to the node (degree) normalized to the interval $[0, 1]$. Thus, $C_D(v) = \frac{deg(v}{n-1}$.

The *degree centrality* of a graph G is defined as:
$$C_D(G) = \frac{\sum_{i=1}^{n} |C_D(v^*) - C_D(v_i)|}{n-2},$$ where v^* is the node that has the maximum degree from all nodes in G.

The betweenness centrality of a node v is the normalized sum of the number of shortest paths between any pair of nodes going through that node, divided by the number of shortest paths between any pair of nodes. i.e., $C_B(v) = \frac{2 \sum_{s \neq v \neq t} \frac{\sigma_{st}(v)}{\sigma_{st}}}{(n-1)(n-2)}$, where σ_{st} is the number of shortest paths from s to t and $\sigma_{st}(v)$ is the number of shortest paths from s to t that go through v.

Thus, the *betweeness centrality* of a graph G is defined as:
$$C_B(G) = \frac{\sum_{i=1}^{n} |C_B(v^*) - C_B(v_i)|}{n-1},$$ where v^*, in this case, is the node that has the maximum betweeness centrality in G.

The closeness centrality of a node v is defined as the inverse of the average of shortest paths lengths between the node v and all other nodes from G, normalized to $[0, 1]$. i.e., $C_C(G) = \frac{n-1}{\sum_{i=1}^{n} d(v_i, v)}$, where $d(v, w)$ is the length of the shortest path from v to w, or equivalently, the distance from v to w.

Therefore, the *closeness centrality* of a graph G is defined as: $C_C(G) = \frac{\sum_{i=1}^{n} |C_C(v^*) - C_C(v_i)|}{(n-1)(n-2)/(2n-3)}$, where v^*, in this case, is the node that has the maximum closeness centrality in G.

We fitted a *power law* (cf. [2]) for the original and merged graphs, the value of γ denotes the corresponding exponent for the power law.

As additional measure we consider the proportion of edges of the intersection of the graph G and the anonymized graph G_k with respect to the total number of edges in G and in G_k

4.3 Experimental Results

First we present the metrics for the merged graphs in Table 2.

Next, we apply k-degree anonymization technique to the original graph, and on the other hand, sample/merging followed by k-degree anonymization. We prove that by using the sampling/merging technique the utility is improved. In particular, we apply k-degree anonymity and then compare how many edges in proportion have been modified by both methods.

We denote the method to minimize the degree anonymization cost L_1 as *Median* the method that minimizes the L_2 distance to the original sequence as *Mean*, and the method of sampling and merging followed by k-degree anonymization as *Merged*. See Figs. 3, 4, 5 and 6.

The value "% precision" is the value $m(\tilde{G})$ expressed as a percentage of $m(G)$, where G denotes either the original or the merged graph, $\tilde{G}$ is the k-degree anonymized graph and $m = APL, C_B, C_D, C_C$.

Table 2. Merged graphs metrics

	Karate$_S$	PolBooks$_S$	Celegans$_S$	PolBlogs$_S$	Netsciences$_S$
$\|S\|$	3	9	11	103	452
$\|V\|$	26	53	238	1004	805
$\|E\|$	53	207	1516	13693	874
% of $\|V(G)\|$	76	51	80	82	50
% of $\|E(G)\|$	67	47	70	81	31
$Diam$	4	4	4	5	10
APL	2.19692	2.25181	2.24090	2.41158	4.70828
C_C	0.33353	0.37289	0.50269	0.45409	0.00069
C_B	0.41056	0.22225	0.35184	0.20729	0.03557
C_D	0.43692	0.29209	0.52853	0.45436	0.04581
γ	2.08879	3.00645	2.86769	2.89354	3.20339

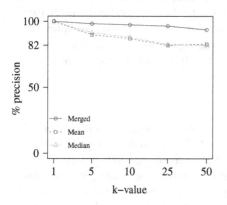

Fig. 3. APL for PolBooks

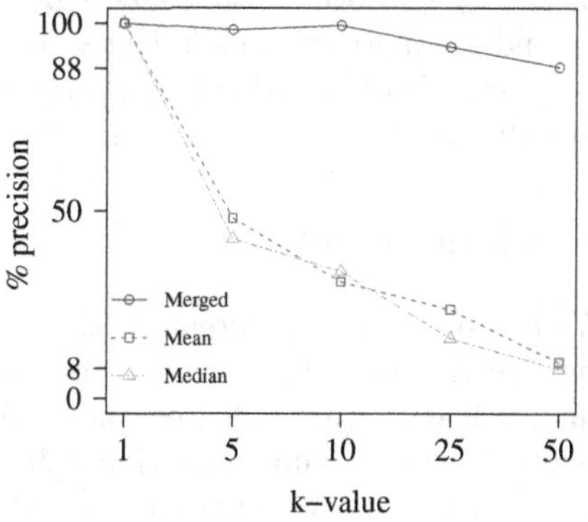

Fig. 4. C_B for Celegans

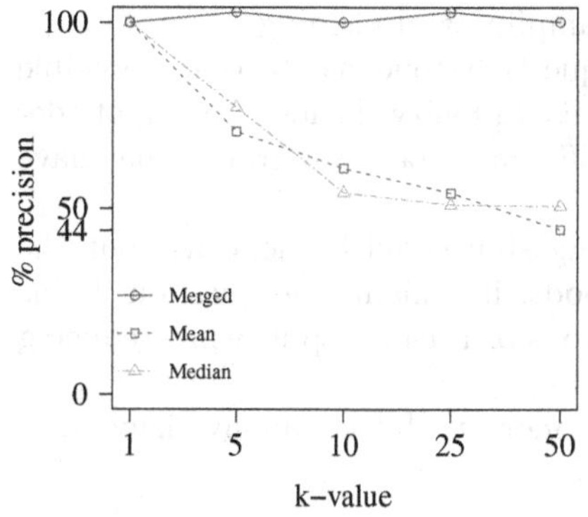

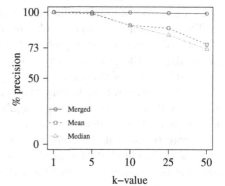

Fig. 5. C_D for networkscience

Fig. 6. C_C for PolBlogs

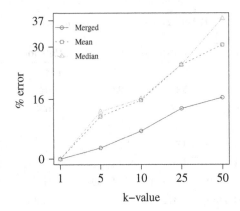

Fig. 7. Error percentage for all measures and datasets

Finally, in Fig. 7 we present a graph considering the proportion of the errors (% error) of both methods k-degree anonymity for the original graph (with L_1 and L_2 metrics) and k-degree anonymity for the graph after sampling and merging. We obtained this measure, by averaging among all different datasets, all different measures, and taking the difference from the original graphs.

5   Conclusions

We compared two different methods for k-degree anonymity, namely optimizing the distance L_1 and L_2, also implemented a version of k-degree subset anonymization (for k-degree anonymizing the merged graphs) and proposed a new method for anonymization consisting of sampling and merging neighborhoods in graphs. We showed that this approach may be used to provide different anonymization guarantees for different subsets of nodes on the same graph. We conducted an empirical evaluation of our algorithm on well-known network datasets and obtained a considerable improvement in the information loss measures comparing a graph before and after sampling and merging.

There remain many open problems and questions, one may be that a possible way of evaluating the quality of the method could follow the ideas of [24] of "de-anonymizing" the graphs to evaluate the differences between graphs that have a similar amount of vertices.

Another observation is that the sampling step could be done without the requirement of non-overlapping neighborhoods, it remains the question if this may decrease the quality of the anonymizations or it can keep it, while reducing even more the size of the original graphs.

It is probable that this method will be more useful for anonymizing very large datasets, since it reduces their size.

Acknowledgements. Support by Spanish MCYT under project SmartGlacis TIN2014-57364-C2-1-R is acknowledged.

References

1. Adamic, L.A., Glance, N.: The political blogosphere and the 2004 US election. In: Proceedings of the WWW-2005 Workshop on the Weblogging Ecosystem (2005)
2. Barabasi, A.L., Albert, R.: Emergence of scaling in random networks. Science **286**, 509–512 (1999)
3. Backstrom, L., Dwork, C., Kleinberg, J.: Where art thou R3579X? Anonymized social networks, hidden patterns, and structural steganography. In: Proceedings of 16th International World Wide Web Conference (2007)
4. Campan, A., Truta, T.M.: A clustering approach for data and structural anonymity in social networks. In: Proceedings of the 2nd ACM SIGKDD International Workshop on Privacy, Security, and Trust in KDD (PinKDD 2008), in conjunction with KDD 2008, Las Vegas, Nevada, USA (2008)
5. Campan, A., Truta, T.M.: Data and structural k-anonymity in social networks. In: Bonchi, F., Ferrari, E., Jiang, W., Malin, B. (eds.) PinKDD 2008. LNCS, vol. 5456, pp. 33–54. Springer, Heidelberg (2009)
6. Chester, S., Kapron, B.M., Ramesh, G., Srivastava, G., Thomo, A., Venkatesh, S.: Why Waldo befriended the dummy? k-Anonymization of social networks with pseudo-nodes. Soc. Netw. Anal. Min. **3**(3), 381–399 (2013)
7. Chester, S., Kapron, B., Srivastava, G., Venkatesh, S.: Complexity of social network anonymization. Soc. Netw. Anal. Min. **3**(2), 151–166 (2013)
8. Domingo-Ferrer, J., Mateo-Sanz, J.M.: Practical data-oriented microaggregation for statistical disclosure control. IEEE Trans. Knowl. Data Eng. **14**(1), 189–201 (2002)
9. Domingo-Ferrer, J., Torra, V.: Ordinal, continuous and heterogeneous k-anonymity through microaggregation. Data Min. Knowl. Discov. **11**(2), 195–212 (2005)
10. Freeman, L.C.: Centrality in social networks: conceptual clarification. Soc. Netw. **1**(3), 215–239 (1979)
11. Hay, M., Miklau, G., Jensen, D., Towsley, D.: Resisting structural identification in anonymized social networks. In: Proceedings of the 34th International Conference on Very Large Databases (VLDB 2008). ACM (2008)
12. Hansen, S.L., Mukherjee, S.: A polynomial algorithm for optimal univariate microaggregation. IEEE Trans. Knowl. Data Eng. **15**(4), 1043–1044 (2003)
13. Krebs, V.: (unpublished). http://www.orgnet.com/
14. Krishnamurthy, V., Faloutsos, M., Chrobak, M., Cui, J., Lao, L., Percus, A.: Sampling large internet topologies for simulation purposes. Comput. Netw. **51**(15), 4284–4302 (2007)
15. Liu, K., Terzi, E.: Towards identity anonymization on graphs. In: Proceedings of the ACM SIGMOD International Conference on Management of Data, pp. 93–106 (2008)
16. Nettleton, D.F., Dries, A.: Local neighbourhood sub-graph matching method, European Patent application number: 13382308.8 (Priority 30/7/2013). PCT application number: PCT/ES2014/065505 (Priority 18 July 2014)
17. Nettleton, D.F., Salas, J.: A data driven anonymization system for information rich online social network graphs. Expert Syst. Appl. **55**, 87–105 (2016)
18. Newman, M.E.J.: Finding community structure in networks using the eigenvectors of matrices. Preprint Physics/0605087 (2006)
19. Samarati, P.: Protecting respondents identities in microdata release. IEEE Trans. Knowl. Data Eng. **13**(6), 1010–1027 (2001)

20. Salas, J., Torra, V.: Graphic sequences, distances and k-degree anonymity. Disc. Appl. Math. **188**, 25–31 (2015)
21. Salas, J., Torra, V.: Improving the characterization of P-stability for applications in network privacy. Disc. Appl. Math. **206**, 109–114 (2016)
22. Stokes, K., Torra, V.: Reidentification and k-anonymity: a model for disclosure risk in graphs. Soft Comput. **16**(10), 1657–1670 (2012)
23. Sweeney, L.: k-anonymity: a model for protecting privacy. Int. J. Uncertainty Fuzziness Knowl.-Based Syst. **10**(5), 557–570 (2002)
24. Truta, T.M., Campan, A., Ralescu, A.L.: Preservation of structural properties in anonymized social networks. In: CollaborateCom, pp. 619–627 (2012)
25. Watts, D.J., Strogatz, S.H.: Collective dynamics of 'small-world' networks. Nature **393**, 440–442 (1998)
26. Zachary, W.W.: An information flow model for conflict and fission in small groups. J. Anthropol. Res. **33**(4), 452–473 (1977)
27. Zheleva, E., Getoor, L.: Preserving the privacy of sensitive relationships in graph data. In: Bonchi, F., Malin, B., Saygın, Y. (eds.) PInKDD 2007. LNCS, vol. 4890, pp. 153–171. Springer, Heidelberg (2008)
28. Zhou, B., Pei, J.: Preserving privacy in social networks against neighborhood attacks. In: ICDE (2008)
29. Zhou, B., Pei, J., Luk, W.S.: A brief survey on anonymization techniques for privacy preserving publishing of social network data. ACM SIGKDD Explor. Newsl. **10**(2), 12–22 (2008)

Data Mining and Applications

Diabetic Retinopathy Risk Estimation Using Fuzzy Rules on Electronic Health Record Data

Emran Saleh[1], Aida Valls[1(✉)], Antonio Moreno[1], Pedro Romero-Aroca[2],
Sofia de la Riva-Fernandez[2], and Ramon Sagarra-Alamo[2]

[1] Departament d'Enginyeria Informàtica i Matemàtiques,
Universitat Rovira i Virgili, Tarragona, Spain
emran.saleh@estudiants.urv.cat, {aida.valls,antonio.moreno}@urv.cat
[2] Ophthalmic Service, University Hospital Sant Joan de Reus,
Institut d'Investigació Sanitària Pere Virgili (IISPV),
Universitat Rovira i Virgili, Tarragona, Spain
pedro.romero@urv.cat, delariva.sofia@gmail.com, rsagarraalamo@gmail.com

Abstract. Diabetic retinopathy is an ocular disease that involves an important healthcare spending and is the most serious cause of secondary blindness. Precocious and precautionary detection through a yearly screening of the eye fundus is difficult to make because of the large number of diabetic patients. This paper presents a novel clinical decision support system, based on fuzzy rules, that calculates the risk of developing diabetic retinopathy. The system has been trained and validated on a dataset of patients from Sant Joan de Reus University Hospital. The system achieves levels of sensitivity and specificity above 80%, which is in practice the minimum threshold required for the validity of clinical tests.

Keywords: Diabetic retinopathy · Fuzzy expert systems · Induction of fuzzy decision trees

1 Introduction

Diabetes Mellitus has become a deep-seated disease with a high spread, as it is suffered by 9 % of adults around the world [1]. Moreover, it is estimated that 46 % of diabetic patients are not even diagnosed [8]. The spread of diabetes has been steadily growing in the last decades. In Spain, for example, the National Health Surveys detected that diabetes has increased from 4.1 % of the population in 1993 to 6.4 % in 2009, and it is expected to grow to 11.1 % by 2030 [19].

Diabetic retinopathy (DR) is one of the main complications of diabetes, being a common cause of blindness for this kind of patients. As diabetes prevalence grows, it does also the number of people suffering DR, being a main concern for health care centres. Early screening may be done by means of non-mydriatic fundus cameras [17]. Regular screening of diabetic patients may decrease the economic impact of the therapy and minimize the development of blindness. However, because of the large number of diabetic patients, it is too resource-consuming and costly to make a preventive screening to all of them.

© Springer International Publishing Switzerland 2016
V. Torra et al. (Eds.): MDAI 2016, LNAI 9880, pp. 263–274, 2016.
DOI: 10.1007/978-3-319-45656-0_22

Doctors in the Ophthalmology Department of Sant Joan de Reus University Hospital are screening around 15000 patients for diabetic retinopathy yearly. They confirmed an incidence of about 8 % to 9 % of patients in 2012–2015 [18]. Although it is increasing, this low proportion indicates that many patients could be safely screened only every 2 or 3 years, so that the screening resources could be focused on the part of the diabetic population with more risk to develop diabetic retinopathy [5,16].

In this scenario, the goal of this work is to build a *Clinical Decision Support System* (CDSS) to help clinicians to estimate the risk of developing diabetic retinopathy of new patients. A good prediction system would have three important advantages: first, it would save costs to health care centres because they would not need to make such a large number of screening tests and the workload of ophthalmologist services would also be reduced; second, it would save time to many patients that do not need a yearly screening test; and third, it would permit to detect early stages of DR because those patients with more probability of developing this disease could be screened with more frequency.

This new CDSS consists on a set of binary classification rules, that assess whether a new patient has a high risk of developing DR or not. It is well known that medical diagnosis has to deal with imprecision and uncertainty [21] because Medicine is not a matter of precise numerical values. Doctors usually work with linguistic assertions based on ranges of values. Although most of the indicators have some established intervals corresponding to good/bad states, the limits of these intervals are fuzzy because they may depend on other factors of each patient. As a result, classification algorithms working with crisp data usually do not give a high accuracy. For this reason, this study proposes a *fuzzy rule-based system* (FRBS).

Fuzzy rule based systems are a good choice for dealing with medical data for several reasons: (1) they represent the domain knowledge with a linguistic model that can be easily interpreted and understood by doctors, (2) they naturally deal with uncertainty and imprecision, (3) FRBS give also the degree of fulfillment of the classification output, which is an interesting value for the doctor, and (4) FRBS usually give a higher classification accuracy than crisp decision trees. In classification problems for diagnosis, doctors are interested not only on the classification result but also on how the system derived the answer [7]. In contrast, neural networks and linear programming models [15] usually get a high classification accuracy but the decision process is a black box.

The rest of the paper is organized as follows. Section 2 presents related work on induction of fuzzy decision trees. In Sect. 3 we introduce the algorithm of induction of fuzzy decision trees used in this paper. Section 4 explains the dataset, variables and fuzzification procedure. Section 5 presents the experimental results. Finally, Sect. 6 shows the conclusions and future work.

2   Related Works

In the literature there are diverse methods for the induction of *fuzzy decision trees* (FDT). The classic ones, from the 1990s, are an extension of the ID3 method for crisp data. Two main approaches can be found to identify the best attribute at each step of the construction of the tree: the ones based on *information theory* [22] and the ones based on *classification ambiguity* measures [25].

There are different ways of calculating the *fuzzy entropy* in the information theory model. Umanol et al. [22] initiated this approach with a fuzzy version of the crisp entropy. Later, Levashenko and Zaitseva [13] defined a fuzzy conditional entropy between attributes, which is based on joint information, to choose an attribute for tree expansion. Some other authors have focused on the Hartley function as a more general measure of uncertainty (which coincides with Shannon's entropy measure in the case of a uniform probability distribution). For example, Li and Jiang [14] used a weight function (cut-standard) to solve the influence of different fuzzy levels with Hartley measures. In [11] the authors applied a generalized Hartley metric model with a fuzzy consciousness function to consider the non-linearity in membership functions.

Yuan and Shaw proposed the minimization of the *classification ambiguity* as a criterion to build the tree, instead of *entropy* ([25]). The new children of a node must reduce the parent's ambiguity to continue growing the tree or the process is stopped. The classification ambiguity is a measure of non-specificity of a possibilistic measure. In [24] Xiao proposed another classification ambiguity function based on the probability distribution of class on data. Using a similar approach, Wang and Yeung [23] proposed a selection criterion based on the degree of influence of each attribute in a good classification. In this case a set of weighted rules was obtained, and the reasoning method consisted on a weighted average of similarity.

Some recent works combine different techniques, such as the construction of a fuzzy rough tree in [2]. Fuzzy rough set theory is used for choosing the nodes and partitioning branches when building the tree. Attributes are evaluated with a fuzzy dependency function. Chang, Fan and Dzan presented a hybrid model that merged three techniques ([6]). First, a case-based method built a weighted distance metric that was used in a clustering algorithm. After that, a fuzzy entropy measure was applied for node selection. Finally, a genetic algorithm was used to increase the accuracy of the decision tree by detecting the best number of terms to consider for each attribute.

Fuzzy decision trees have been applied in many fields, including health care. In the medical area, fuzzy rules have been applied to some well-known diseases [20]. For example, in [9] fuzzy trees were used in the diagnosis of breast cancer, diabetes and liver malfunction. However, as far as we know there is no previous work on the use of decision trees in the diagnosis of diabetic retinopathy.

Wang and Yeung [23] made a comparative study between the entropy-based induction algorithm of Umanol et al. [22] and the classification ambiguity induction algorithm of Yuan and Shaw [25]. The comparative study covered the attribute selection criteria, the complexity of the methods, the reasoning accuracy, the reasoning technique and the tree comprehensibility. This study concluded that the performance is quite similar in both cases. The number of rules is slightly larger in the method of Yuan and Shaw, but the accuracy of the prediction both in the training and testing sets is quite similar (slightly worse in Yuan and Shaw for some datasets). After this analysis, we chose the fuzzy tree induction algorithm proposed by Yuan and Shaw [25] because it takes a possibilistic approach (i.e. a patient can belong to a certain degree to both classes) rather than a probabilistic one.

3   Methodology

To construct the CDSS we have a labeled training dataset of patients from the Sant Joan de Reus University Hospital. The data will be explained in Sect. 4. The obtained set of rules is used to classify a different test dataset. We propose a modification of the classic Mamdani inference procedure in subsect. 3.2 in order to detect undecidable cases. The algorithm for the induction of a fuzzy decision tree is described in this section.

3.1   Preliminaries

Let us consider the universe of discourse $U = \{u_1, u_2, ..., u_m\}$, where u_i is an object described by a collection of attributes $A = \{a_1, ..., a_n\}$. In this case U denotes the set of users (patients).

Each attribute $a \in A$ takes values on a linguistic fuzzy partition [3] $T = \{t_1, ..., t_s\}$ with membership functions $\mu_{t_i} \in \mu_T$. These membership functions can be understood as a possibility distribution.

The U-uncertainty (or non-specificity measure) of a possibility distribution π on the set $X = \{x_1, x_2, ..., x_d\}$ is defined in [25] as:

$$g(\pi) = \sum_{i=1}^{d} (\pi_i^* - \pi_{i+1}^*) \ln i \tag{1}$$

where $\pi^* = \{\pi_1^*, \pi_2^*, ..., \pi_d^*\}$ is a permutation of $\pi = \{\pi(x_1), \pi(x_2), ..., \pi(x_d)\}$ such that $\pi_i^* \geq \pi_{i+1}^*$, for $i = 1, ..., d$, and $\pi_{d+1}^* = 0$.

3.2   Fuzzy Tree Induction

The induction algorithm proposed in [25] is an extension of the classic ID3 method for crisp data. It incorporates two parameters to manage the uncertainty:

– The *significance level* (α) is used to filter evidence that is not relevant enough. If the membership degree of a fuzzy evidence E is lower than the level α, it is not used in the rule induction process.

$$\mu_{E_\alpha}(u_i) = \begin{cases} \mu_E(u_i) & if \ \mu_E(u_i) \geq \alpha \\ 0 & if \ \mu_E(u_i) < \alpha \end{cases}$$

– The *truth level threshold* (β) fixes the minimum truth of the conclusion given by a rule. Thus, it controls the growth of the decision tree. Lower values of β may lead to smaller trees but with a lower classification accuracy.

The main steps of the induction process of a fuzzy decision tree are the following:

1. Select the best attribute for the root node v: the one with the **smallest ambiguity**.
2. Create a new branch for each of the values of the attribute v for which we have examples with support at least α.
3. Calculate the **truth level of classification** of the objects within a branch into each class.
4. If the truth level of classification is above β for at least one of the classes C_i, terminate the branch with a leaf with label C_i, corresponding to the class with the highest truth level.
5. If the truth level is smaller than β for all classes, check if an additional attribute will further reduce the classification ambiguity.
6. If it does, select the attribute with the **smallest classification ambiguity with the accumulated evidence** as a new decision node from the branch. Repeat from step 2 until no further growth is possible.
7. If it doesn't, terminate the branch as a leaf with a label corresponding to the class with the highest truth level.

The three measures shown in bold in the previous algorithm control the construction of the tree. They are explained in the following paragraphs. Some of these measures are based on the concept of Fuzzy Evidence, which is a fuzzy set defined on the linguistic values taken by one or more attributes (*i.e.* a condition given by one branch of the decision tree).

Ambiguity of an Attribute $a \in A$: Considering that attribute a takes values on a linguistic fuzzy partition $T = \{t_1, ..., t_s\}$ with membership functions $\mu_{t_i} \in \mu_T$, its ambiguity is calculated as

$$Ambiguity(a) = \frac{1}{m} \sum_{i=1}^{m} g(\pi_T(u_i)) \tag{2}$$

where π_T is the normalized possibility distribution of μ_T on U:

$$\pi_{t_r}(u_i) = \mu_{t_r}(u_i)/max_{1 \leq j \leq s}\{\mu_{t_j}(u_i)\} \tag{3}$$

Truth Level of Classification: Having a set of classes $C = \{C_1, ..., C_p\}$, the truth level of classification indicates the possibility of classifying an object u_i into a class $C_k \in C$ given the fuzzy evidence E

$$Truth(C_k|E) = S(E, C_k)/max_{1 \leq j \leq p}\{S(E, C_j)\} \tag{4}$$

where S is the subsethood of the fuzzy set X on the fuzzy set Y

$$S(X,Y) = \frac{M(X \cap Y)}{M(X)} = \frac{\sum_{i=1}^{m} min(\mu_X(u_i), \mu_Y(u_i))}{\sum_{i=1}^{m} \mu_X(u_i)} \tag{5}$$

and $M(X)$ is the cardinality or sigma count of the fuzzy set X. For the case considered in this paper, the classes C are crisp, so μ_{C_k} will be just 0 or 1.

The truth level can be understood as a possibility distribution on the set U. As before, $\pi(C|E)$ is the corresponding normalisation, which is used to define the next concept.

Classification Ambiguity: Having a fuzzy partition $P = \{E_1, ..., E_k\}$ on fuzzy evidence F, the classification ambiguity, denoted by $G(P|F)$, is calculated as

$$G(P|F) = \sum_{i=1}^{k} W(E_i|F)g(\pi(C|E_i \cap F)) \qquad (6)$$

where $W(E_i|F)$ is the weight which represents the relative size of the subset $(E_i \cap F)$ with respect to F (i.e. $W(E_i|F) = M(E_i \cap F)/\sum_{i=1}^{k} M(E_i \cap F)$).

3.3   Classification Using the Fuzzy Rules

The Mamdani inference procedure is used for the binary classification in the classes 0 (no DR) and 1 (suffering DR). An additional step is added at the end of the classic procedure:

1. Calculate the satisfaction degree of a rule $\mu_R(u)$ using the t-norm minimum.
2. Calculate the membership to the conclusion class as the product between the satisfaction degree $\mu_R(u)$ and the degree of support of the rule.
3. Aggregate all the memberships for the same class, given by different rules, using the t-conorm maximum.
4. Compare the membership degrees for class 0 and class 1:

$$Class(u) = \begin{cases} \text{``Unknown''} & , \quad if |\mu_{c0}(u) - \mu_{c1}(u)| < \delta \\ argmax_{C_k \in \{c0, c1\}}(\mu_{C_k}(u)) , & otherwise \end{cases} \qquad (7)$$

The difference threshold (δ) is used to check if an object belongs to the two different classes with a similar membership degree. If the degree of membership is not significantly different, the object is classified as "Unknown". With this additional step we can detect cases where conflicting inference is made, because some variables support the classification to class 0 while others support class 1. The identification of these cases is of extreme importance in this medical application, because this corresponds to an atypical patient that needs to be manually assessed by the doctors as the system is not able to determine the correct class.

4   The Data

The method presented in the previous section has been applied to data stored in the Electronic Health Records (EHR) of patients treated at Sant Joan de Reus University Hospital. This hospital serves an area of Catalonia with a population of 247,174 inhabitants, having 17,792 patients with Diabetes Mellitus [18]. Various units of non-mydriatic cameras are used to screen these diabetic patients.

Since 2007 several analytical, metabolic and demographic data have been systematically collected and stored in the Electronic Health Records of the different units. An statistical analysis for the 8-year period from 2007 to 2014 was made in order to determine the changes in the incidence of diabetic retinopathy [18]. It is observed that incidence was stable between 2007 and 2011 (around 8.1 %) but since 2011 it has continuously increased until almost 9 %. This study also analysed which are the main risk factors for developing DR. Out of the results of this previous work, a set of 8 attributes have been taken for the construction of the fuzzy rule based system. Most of the attributes are numerical (e.g. age, body mass index) but there are also some categorical attributes (e.g. sex, medical treatment). We only present some of these attributes for confidentiality reasons related to the research project development.

These data are used to build and test the classification model which helps the doctor to decide whether or not the patient has a high risk of developing DR on the basis of the selected attributes. The rules also associate a membership degree to the conclusion that indicates the confidence on the class assignment.

4.1 Data Fuzzyfication

The first step was the definition of the fuzzy sets of the linguistic terms for each numerical attribute. The meaning of the intervals is of great importance to understand the classification rules that will be obtained. Thus, the membership functions have been defined according to the standard intervals of some indicators (such as Body Mass Index BMI) or according to the medical knowledge and the findings of the statistical analysis [18], such as the age division (Fig. 1).

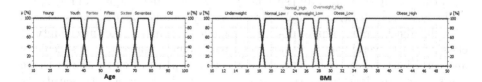

Fig. 1. Definition of linguistic labels for age and body mass index.

4.2 Preprocessing for Class Imbalance

In medical applications it is common that the low incidence of some diseases generates imbalanced datasets. In this case, the hospital gave us the EHR of 2323 diabetic patients, who were already labeled regarding the diabetic retinopathy disease (579 patients with DR (class 1) and 1744 patients not suffering from DR (class 0)). The dataset was divided in two parts, one for training with 1212 examples (871 from class 0 and 341 from class 1) and another for testing with 1111 patients (873 from class 0 and 238 from class 1).

In this dataset only 25 % of the patients belong to class 1. Although the incidence of the disease in diabetic patients is much greater than the one in the population (which is below 9 %) it still represents a high imbalance in favour of class 0. This situation may cause some problems to the learning algorithm because, as shown previously, it is based on proportions between both classes.

To avoid imbalanced data, a bunch of solutions exist at the algorithmic and data levels [12]. In our case, for the training datasets, we have made a random oversampling to balance the class distribution by replicating class 1 examples until they become equal to the number of examples of class 0. Thus, finally the training dataset had 871 patients of each class.

5   Experimental Results

In this section we study the influence of the parameters α, β and δ on the quality of the classification rules. The usual classification measures applied in the medical field are used for the evaluation: specificity and sensitivity.

For values of β below 0.5 the resulting tree was not useful because all the rules predicted class 0 in the all cases (the truth level of classification is too low and the rules cannot find the differential features of the patients with RD). Figure 2 shows the results with $\beta = 0.6$ and $\beta = 0.7$. Higher values of β generated very large and complex trees without improving either the sensitivity or the specificity.

Figures 2(a), (b) and (c) show the influence of δ on the sensitivity and speci-ficity, for values of α between 0 and 1. With $\alpha \leq 0.5$ the results are quite stable (for $\beta = 0.6$ they do not change), but when the significance level is increased, the sensitivity improvesand the specificity decreases.The figures show that increasing δ increases the sensitivity and specificity for both values of β. The best balance between sensitivity and specificity is found when $\beta = 0.7$, $\alpha = 0.2$ and $\delta = 0.30$, in which sensitivity = 82.11 and specificity = 82.38 (see black square in Fig. 2(c)).

In fuzzy classification models one object can be classified in many classes with different membership values. In our model, δ removes the uncertainty of classifying to several classes. When the δ parameter is changed, the specificity and sensitivity results behave in the same way depending on α. However, we can see that the increase of the δ value brings an increase of both sensitivity and specificity. The best value for δ is 0.30. Results do not improve with higher values of δ.

Fixing $\delta = 0.3$ we can study the number of unclassified patients (Fig. 2 (d)) and the number of generated rules (Fig. 2 (e)). Increasing β usually increases the number of unclassified patients. Figure 2 (d) shows that the difference between the number of unclassified patients for $\beta = 0.6$ and $\beta = 0.7$ is quite stable for values of α lower than 0.8 (between 10 and 15 patients). Increasing α (up to 0.7) increases the number of unclassified patients, but this number decreases when α is 0.8 or 0.9.

β plays an important role in the construction of the fuzzy decision tree, because it is used to decide if the current node is a leaf or not. Only the nodes with a *truth level of classification* higher than β become leaves of the tree. Thus, when β is high, the number and length of the rules increase, but it is still a manageable number for a doctor. Figure 2(e) shows the effect of α on the number

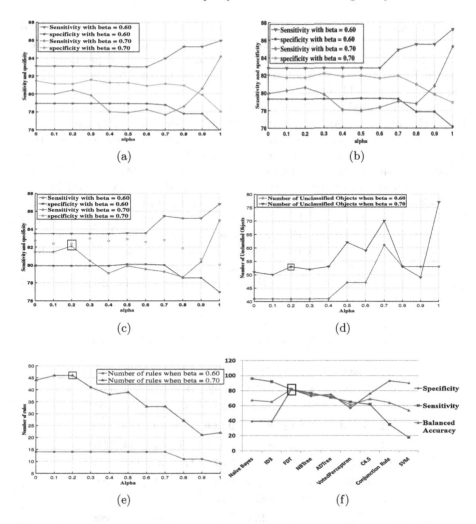

Fig. 2. Sensitivity and specificity for $\delta = 0.10$ (a), $\delta = 0.20$ (b), $\delta = 0.30$ (c), number of unclassified objects (d), number of rules (e), comparison with other learning algorithms (f)

of rules generated by the algorithm with $\beta = 0.7$ and $\beta = 0.6$. When α increases the system uses less information and the number of rules is smaller. We can see that the number of rules is always around 14 when $\beta = 0.6$, but it is higher for $\beta = 0.7$ (between 20 and 46 rules).

Table 1 shows some more measures with $\delta = 0.3$ and $\beta = 0.7$ because this is the one with better results, highlighted by the black square in Fig. 2. Column (NPV) shows the negative predictive value measure, TP is the number of true positives, FN are the false negatives, FP are the false positives and TN is the number of true negatives. In Clinical Decision Support Systems a False Negative is more dangerous than a False Positive, because an ill person is being classified as healthy and is not being properly treated. Thus, it would be desirable to have a low number of FNs. For this reason we want both sensitivity and specificity to be high. In Table 1 the combination that gives the best balance between sensitivity and specificity is highlighted.

The method used in this work (FDT) gives better results than some well-known classification algorithms on the current dataset. Figure 2(f) shows a comparison between the results of some algorithms based on decision trees (ID3, NBTree, ADTree, C4.5), rule-based algorithms (Conjunction Rule), and other classification algorithms like Support Vector Machines (SVM),Bayes (Naive Bayes) and Functions (VotedPerceptron) [10]. Balanced accuracy, sensitivity and specificity are used for the comparison as they are good performance measures for imbalanced datasets [4]. Balanced accuracy is the mean of specificity and sensitivity. Naive Bayes and ID3 give high sensitivity but they have a low specificity and a low balanced accuracy, whereas on the other side Conjunction Rule and SVM give a good specificity but a bad sensitivity. The method in this work gives the best combination of sensitivity and specificity, and the highest balanced accuracy as well.

Table 1. Classification results with $\beta = 0.70$ and $\delta = 0.30$

α	Precision	Sensit	Spec	NPV	TP	FN	FP	TN
0	55.38	81.45	82.72	94.42	180	41	145	694
0.1	54.88	81.45	82.38	94.41	180	41	148	692
0.2	**54.74**	**82.11**	**82.38**	**94.66**	**179**	**39**	**148**	**692**
0.3	55.31	80.45	82.96	94.18	177	43	143	696
0.4	54.55	79.09	82.70	93.78	174	46	145	693
0.5	55.21	79.91	82.89	93.99	175	44	142	688
0.6	54.69	79.55	82.57	93.85	175	45	145	687
0.7	54.78	79.26	82.77	93.81	172	45	142	682
0.8	53.23	78.64	81.86	93.59	173	47	152	686
0.9	52.63	80.36	80.67	93.89	180	44	162	676
1	52.46	84.98	80.02	95.36	181	32	164	657

6   Conclusion and Future Work

Developing clinical decision support systems for diabetic retinopathy may improve the diagnosis and avoid unnecessary screenings on patients, reducing the workload of ophthalmologist services, focusing the use of resources on the patients that really need them, and saving the time of doctors and patients.

The fuzzy decision tree induction method presented in the paper (based on minimizing the classification ambiguity) provides quite good results on the tests done with data coming from the Sant Joan de Reus Hospital. It provides better results than other non-fuzzy approaches, with a specificity and sensitivity above 80 % (as required for clinical systems). Therefore, it seems that fuzzy sets are a suitable way of dealing with the ambiguity of the data stored in the EHR. Moreover, obtaining linguistic rules is appreciated by doctors, because they can easily understand the knowledge model and accept it as a valid estimation tool.

The next step will be the validation of the model with a larger dataset obtained from different medical centres in the region. Although the studied patients are representative of the area of Tarragona, it is important to know if the model keeps the same level of prediction accuracy when used with people living in other provinces. As future work, we are planning to extend the model in order to predict also the severity of the diabetic retinopathy disease according to the data available in the EHR. More complex classification models, such as fuzzy random forests, are also currently being investigated.

Acknowledgements. This study was funded by the Spanish research projects PI12/01535 and PI15-/01150 (Instituto de Salud Carlos III) and the URV grants 2014PFR-URV-B2-60 and 2015PFR-URV-B2-60.

References

1. World health organisation: global status report of non communicable diseases 2014. WHO Library Cataloguing-in-Publication Data (ISBN: 978-92-4-156485-4) (2014)
2. An, S., Hu, Q.: Fuzzy rough decision trees. In: Yao, J.T., Yang, Y., Słowiński, R., Greco, S., Li, H., Mitra, S., Polkowski, L. (eds.) RSCTC 2012. LNCS, vol. 7413, pp. 397–404. Springer, Heidelberg (2012)
3. Bodjanova, S.: Fuzzy Sets and Fuzzy Partitions, pp. 55–60. Springer, Heidelberg (1993)
4. Brodersen, K.H., Ong, C.S., Stephan, K.E., Buhmann, J.M.: The balanced accuracy and its posterior distribution. In: 2010 20th international conference on Pattern recognition (ICPR), pp. 3121–3124. IEEE (2010)
5. Chalk, D., Pitt, M., Vaidya, B., Stein, K.: Can the retinal screening interval be safely increased to 2 years for type 2 diabetic patients without retinopathy? Diabetes Care **35**(8), 1663–1668 (2012)
6. Chang, P.C., Fan, C.Y., Dzan, W.Y.: A CBR-based fuzzy decision tree approach for database classification. Expert Syst. Appl. **37**(1), 214–225 (2010)
7. Fan, C.Y., Chang, P.C., Lin, J.J., Hsieh, J.: A hybrid model combining case-based reasoning and fuzzy decision tree for medical data classification. Appl. Soft Comput. **11**(1), 632–644 (2011)
8. Federation, I.D.: IDF Diabetes Atlas 6th (edn.) (ISBN: 2-930229-85-3) (2013)
9. Gadaras, I., Mikhailov, L.: An interpretable fuzzy rule-based classification methodology for medical diagnosis. Artif. Intell. Med. **47**(1), 25–41 (2009)
10. Witten, H.I., Frank, E.: Data Mining: Practical Machine Learning Tools and Techniques. The Morgan Kaufmann Series in Data Management Systems. Morgan Kaufmann, Burlington (2013)

11. Jin, C., Li, F., Li, Y.: A generalized fuzzy ID3 algorithm using generalized information entropy. Knowl.-Based Syst. **64**, 13–21 (2014)
12. Kotsiantis, S., Kanellopoulos, D., Pintelas, P., et al.: Handling imbalanced datasets: a review. GESTS Int. Trans. Comput. Sci. Eng. **30**(1), 25–36 (2006)
13. Levashenko, V.G., Zaitseva, E.N.: Usage of new information estimations for induction of fuzzy decision trees. In: Yin, H., Allinson, N.M., Freeman, R., Keane, J.A., Hubbard, S. (eds.) IDEAL 2002. LNCS, vol. 2412, pp. 493–499. Springer, Heidelberg (2002)
14. Li, F., Jiang, D.: Fuzzy ID3 algorithm based on generating hartley measure. In: Gong, Z., Luo, X., Chen, J., Lei, J., Wang, F.L. (eds.) WISM 2011, Part II. LNCS, vol. 6988, pp. 188–195. Springer, Heidelberg (2011)
15. Mangasarian, K.: Neural network training via linear programming. Adv. Optim. Parallel Comput., 56–67 (1992)
16. Olafsdottir, E., Stefansson, E.: Biennial eye screening in patients with diabetes without retinopathy: 10-year experience. Br. J. Ophthalmol. **91**(12), 1599–1601 (2007)
17. Romero Aroca, P., Reyes Torres, J., Sagarra Alamo, R., Basora Gallisa, J., Fernández-Balart, J., Pareja Ríos, A., Baget-Bernaldiz, M.: Resultados de la implantación de la cámara no midriática sobre la población diabética. Salud (i) cienc. **19**(3), 214–219 (2012)
18. Romero-Aroca, P., de la Riva-Fernandez, S., Valls-Mateu, A., Sagarra-Alamo, R., Moreno-Ribas, A., Soler, N.: Changes observed in diabetic retinopathy: eight-year follow-up of a spanish population. Br. J. Ophthalmol. (2016 in press)
19. Shaw, J.E., Sicree, R.A., Zimmet, P.Z.: Global estimates of the prevalence of diabetes for 2010 and 2030. Diabetes Res. Clin. Pract. **87**(1), 4–14 (2010)
20. Sikchi, S.S., Sikchi, S., Ali, M.: Fuzzy expert systems (FES) for medical diagnosis. Int. J. Comput. Appl. 63(11) (2013)
21. Szolovits, P., et al.: Uncertainty and decisions in medical informatics. Methods Inf. Med. **34**(1), 111–121 (1995)
22. Umano, M., Okamoto, H., Hatono, I., Tamura, H., Kawachi, F., Umedzu, S., Kinoshita, J.: Fuzzy decision trees by fuzzy ID3 algorithm and its application to diagnosis systems. In: Fuzzy Systems, 1994. In: Proceedings of the Third IEEE Conference on IEEE World Congress on Computational Intelligence, pp. 2113–2118. IEEE (1994)
23. Wang, X., Yeung, D.S., Tsang, E.C.C.: A comparative study on heuristic algorithms for generating fuzzy decision trees. IEEE Trans. Syst. Man Cybern. Part B: Cybern. **31**(2), 215–226 (2001)
24. Xiao, T., Huang, D.M., Zhou, X., Zhang, N.: Inducting fuzzy decision tree based on discrete attributes through uncertainty reduction. Applied Mechanics & Materials (2014)
25. Yuan, Y., Shaw, M.J.: Induction of fuzzy decision trees. Fuzzy Sets Syst. **69**(2), 125–139 (1995)

Optimizing Robot Movements in Flexible Job Shop Environment by Metaheuristics Based on Clustered Holonic Multiagent Model

Houssem Eddine Nouri[1](✉), Olfa Belkahla Driss[2],
and Khaled Ghédira[1]

[1] Institut Supérieur de Gestion de Tunis, Université de Tunis,
41, Avenue de la Liberté, Cité Bouchoucha, Bardo, Tunis, Tunisia
houssemeddine.nouri@gmail.com,
khaled.ghedira@isg.rnu.tn
[2] École Supérieure de Commerce de Tunis,
Université de la Manouba, Tunis, Tunisia
olfa.belkahla@isg.rnu.tn

Abstract. In systems based robotic cells, the control of some elements such as transport robot has some difficulties when planning operations dynamically. The Flexible Job Shop scheduling Problem with Transportation times and Many Robots (FJSPT-MR) is a generalization of the classical Job Shop scheduling Problem (JSP) where a set of jobs have to be transported between them by several transport robots. This paper proposes hybrid metaheuristics based on clustered holonic multiagent model for the FJSPT-MR. Computational results are presented using a set of literature benchmark instances. New upper bounds are found, showing the effectiveness of the presented approach.

Keywords: Scheduling · Robots · Flexible job shop · Genetic algorithm · Tabu search · Holonic multiagent

1 Introduction

The Flexible Job Shop scheduling Problem with Transportation times and Many Robots (FJSPT-MR) is a generalization of the classical Job Shop scheduling Problem (JSP) where a set of jobs have to be processed on a set of alternative machines and additionally have to be transported between them by several transport robots. In the FJSPT-MR, we have to consider two NP-hard problems simultaneously: the flexible job-shop scheduling problem [5] and the robot routing problem, which is similar to the pickup and delivery problem [10].

For the literature of the Flexible Job Shop scheduling Problem with Transportation times and Many Robots, most of the researchers have considered the machine and robot scheduling as two independent problems. Therefore, only few researchers have emphasized the importance of simultaneous scheduling of jobs and several robots. Bilge and Ulusoy [2] proposed an iterative heuristic based on the decomposition of the master problem into two sub-problems, allowing a simultaneous resolution of this

© Springer International Publishing Switzerland 2016
V. Torra et al. (Eds.): MDAI 2016, LNAI 9880, pp. 275–288, 2016.
DOI: 10.1007/978-3-319-45656-0_23

scheduling problem with time windows. A local search algorithm is proposed by Hurink and Knust [6] for the job shop scheduling problem with a single robot, where they supposed that the robot movements can be considered as a generalization of the travelling salesman problem with time windows, and additional precedence constraints must be respected. The used local search is based on a neighborhood structure inspired from [11] to make the search process more effective. Abdelmaguid et al. [1] addressed the problem of simultaneous scheduling of machines and identical robots in flexible manufacturing systems, by developing a hybrid approach composed by a genetic algorithm and a heuristic. The genetic algorithm is used for the jobs scheduling problem and the robot assignment is made by the heuristic algorithm. Deroussi and Norre [4] considered the flexible Job shop scheduling problem with transport robots, where each operation can be realized by a subset of machines and adding the transport movement after each machine operation. To solve this problem, an iterative local search algorithm is proposed based on classical exchange, insertion and perturbation moves. Then a simulated annealing schema is used for the acceptance criterion. A hybrid metaheuristic approach is proposed by Zhang et al. [12] for the flexible Job Shop problem with transport constraints and bounded processing times. This hybrid approach is composed by a genetic algorithm to solve the assignment problem of operations to machines, and then a tabu search procedure is used to find new improved scheduling solutions. Lacomme et al. [8] solved the machines and robots simultaneous scheduling problem in flexible manufacturing systems, by adapting a memetic algorithm using a genetic coding containing two parts: a resource selection part for machine operations and a sequencing part for transport operations.

In this paper, we propose a hybridization of two metaheuristics based on clustered holonic multiagent model for the flexible job shop scheduling problem with transportation robots. This new approach follows two principal hierarchical steps, where a genetic algorithm is applied by a scheduler agent for a global exploration of the search space. Then, a tabu search technique is used by a set of cluster agents to guide the research in promising regions. Numerical tests were made to evaluate the performance of our approach using the flexible data set of [4] and completed by comparisons with other approaches.

The rest of the paper is organized as follows. In Sect. 2, we define the formulation of the FJSPT-MR with its objective function and a simple problem instance. Then, in Sect. 3, we detail the proposed hybrid approach with its holonic multiagent levels. The experimental and comparison results are provided in Sect. 4. Finally, Sect. 5 ends the paper with a conclusion.

2 Problem Formulation

There is a set of n jobs $J = \{J_1, \ldots, J_n\}$ to be processed without preemption on a set $M = \{M_0, M_1, \ldots, M_m\}$ of $m + 1$ machines (M_0 represents the load/unload or LU station from which jobs enter and leave the system). Each job J_i is formed by a sequence of ni operations $\{O_{i,1}, O_{i,2}, \ldots, O_{i,ni}\}$ to be performed successively according to the given sequence. For each operation $O_{i,j}$, there is a machine $\mu_{i,j} \in \{M_0, \ldots, M_m\}$

and a processing time $p_{i,j}$ associated with it. In addition, each job $J_i(J_1, \ldots, J_n)$ is composed by $ni - 1$ transport operations $\{T_{i,1}, T_{i,2}, \ldots, T_{i,ni-1}\}$ to be made by a set of r robots $R = \{R_1, \ldots, R_r\}$ from one machine to another. In fact, for each transport operation $T_{i,j}$ there is two types of movements: travel transport operation and empty transport operation.

Firstly, travel transport operation $t^{Rh}\mu_{i,j}, \mu_{i,j+1}$ must be considered for robot $R_h \in \{R_1, \ldots, R_r\}$ when an operation $O_{i,j}$ is processed on machine $\mu_{i,j}$ and operation $O_{i,j+1}$ is processed on machine $\mu_{i,j+1}$. These transportation times are job-independent and robot-dependent. Each transportation operation is assumed to be processed by only one transport robot R_h which can handle at most one job at one time. For convenience, $t^{Rh}\mu_{i,j}, \mu_{i,j+1}$ is used to denote both a transportation operation and a transportation time.

Secondly, empty transport operation $t'^{Rh}_{i,j}$ have to be considered while the robot R_h moves from machine M_i to machine M_j without carrying a job. So, it is possible to assume, for each robot R_h, that $t'^{Rh}_{i,i} = 0$ and $t^{Rh}_{i,j} \geq t'^{Rh}_{i,j}$. All data $p_{i,j}, t^{Rh}\mu_{i,j}, \mu_{i,j+1}, t'^{Rh}\mu_{i,j}, \mu_{i,j+1}$ are assumed to be nonnegative integers.

The objective is to determine a feasible schedule which minimizes the makespan $Cmax = Max_{j=1,n}\{C_j\}$ where C_j denotes the completion time of the last operation $O_{i,ni}$ of job J_i including the processing times of machine operations and transport operations.

3 Hybrid Metaheuristics Based on Clustered Holonic Multiagent Model

In this work, we propose a hybrid metaheuristic approach based on clustering processing two general steps: a first step of global exploration using a genetic algorithm to find promising areas in the search space and a clustering operator allowing to regroup them in a set of clusters. In the second step, a tabu search algorithm is applied to find the best individual solution for each cluster. The global process of the proposed approach is implemented in two hierarchical holonic levels adopted by a recursive multiagent model, named a hybrid Genetic Algorithm with Tabu Search based on clustered HolonicMultiagent model (GATS + HM), see Fig. 1. The first holonic level is composed by a Scheduler Agent which is the Master/Super-agent, preparing the best promising regions of the search space, and the second holonic level containing a set of Cluster Agents which are the Workers/Sub-agents, guiding the search to the global optimum solution of the problem. Each holonic level of this model is responsible to process a step of the hybrid metaheuristic approach and to cooperate between them to attain the global solution of the problem.

3.1 Non Oriented Disjunctive Graph

In this work, we chose to use the disjunctive graph of [6] but with new extension for the job shop problem with transportation times and many robots. Hurink and Knust [6] presented in their disjunctive graph all conflicts for scheduling a set of machines and one robot in a job shop environment. So, we have to improve this graph only by

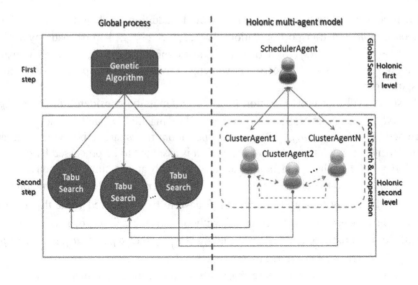

Fig. 1. Hybrid metaheuristics based clustered holonic multiagent model

integrating the assignment of many robots to transport operations. To explain this graph, a sample problem of three jobs and five machines with their transportation times for each robot R_h is presented in Table 1.

Table 1. One instance of flexible job shop problem with two robots

		Processing times for each job Ji						Transportation times for each robot Rh				
		M1	M2	M3	M4	M5		M1	M2	M3	M4	M5
J1	O_{11}	2	9	4	5	1	M1	0	1	2	3	4
	O_{12}	-	6	-	4	-	M2	1	0	1	2	3
J2	O_{21}	1	-	5	-	6	M3	2	1	0	1	2
	O_{22}	3	8	6	-	-	M4	3	2	1	0	1
	O_{23}	-	5	9	3	9	M5	4	3	2	1	0
J3	O_{31}	-	6	6	-	-						
	O_{32}	3	-	-	5	4						

The disjunctive graph $G = (V_m \cup V_t, C \cup D_m \cup D_r)$, see Fig. 2, is composed by: a set of vertices V_m containing all machine operations, a set of vertices V_t is the set of transport operations obtained by an assignment of a robot to each transport operation, and two dummy nodes 0 and $*$. Also, this graph consists of: a set of conjunctions C representing precedence constraints $O_{i,k} \rightarrow t^{Rh}\mu_{i,k}, \mu_{i,k+1} \rightarrow O_{i,k+1}$, undirected disjunctions for machines D_m, and undirected disjunctions for transport robots D_r. For each job J_i, $ni - 1$ transport operations $t^{Rh}\mu_{i,k}, \mu_{i,k+1}$ are introduced including precedence $O_{i,k} \rightarrow t^{Rh} \mu_{i,k}, \mu_{i,k+1} \rightarrow O_{i,k+1}$. Each robot R_h may be considered as an additional "machine" which can process all these transport operations. The arcs from machine node to transport node are weighted with the machine operation durations. Edges between machine operations represent disjunctions for machine operations which have to be processed on the same machine and cannot use it simultaneously.

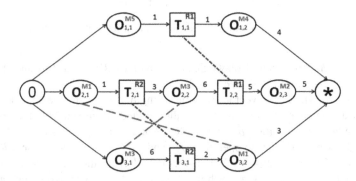

Fig. 2. Non oriented disjunctive graph

As for the classical job shop, the conjunctions C model the execution order of operations within each job J_i. In addition to the classical set of undirected machine disjunctions D_m, it is necessary to consider the set of undirected robot disjunctions D_r. To solve the scheduling problem it is necessary to turn all undirected arcs in $D_m \cup D_r$ into directed ones, and to assign one robot R_h to each transport operation, where the final graph becomes an oriented disjunctive graph.

3.2 Scheduler Agent

The Scheduler Agent (SA) is responsible to process the first step of the hybrid algorithm by using a genetic algorithm called NGA (Neighborhood-based Genetic Algorithm) to identify areas with high average fitness in the search space during a fixed number of iterations *MaxIter*. Then, a clustering operator is integrated to divide the best identified areas by the NGA in the search space to different parts where each part is a cluster $CL_i \in CL$ the set of clusters, where $CL = \{CL_1, CL_2, \ldots, CL_N\}$. According to the number of clusters N obtained after the integration of the clustering operator, the SA creates N Cluster Agents (CAs) preparing the passage to the next step of the global algorithm. After that, the SA remains in a waiting state until the reception of the best solutions found by the CAs for each cluster CL_i. Finally, it finishes the process by displaying the final solution of the problem.

Individual's Solution Presentation Based Oriented Disjunctive Graph. The Flexible Job Shop scheduling Problem with Transportation times and Many Robots is composed by two sub-problems: firstly the machines and robots selection, secondly the operations scheduling problem, that is why the chromosome representation is encoded

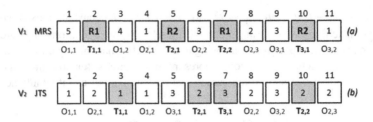

Fig. 3. The chromosome representation of a scheduling solution

in two parts: Machines and Robots Selection part (MRS), and Job and Transport operation Sequence part (JTS), see Fig. 3.

The first part MRS is a vector V_1 with a length L equal to the total number of operations and where each index represents the selected machine or robot to process an operation indicated at position p, see Fig. 3(a). For example $p = 3$ and $p = 7$, $V_1(3)$ is the selected machine M_4 for the operation $O_{1,2}$ and $V_1(7)$ is the selected robot R_1 for the operation $T_{2,2}$. The second part JTS is a vector V_2 having the same length of V_1 and where each index represents a machine operation $O_{i,j}$ or a transport operation $T_{i,j}$ according to the predefined operations for each job, see Fig. 3(b). For example this operation sequence 1-2-1-1-3-2-3-2-3-2-2 can be translated to: $(O_{1,1}, M_5) \rightarrow (O_{2,1}, M_1) \rightarrow (T_{1,1}, R_1) \rightarrow (O_{1,2}, M_4) \rightarrow (O_{3,1}, M_3) \rightarrow (T_{2,1}, R_2) \rightarrow (T_{3,1}, R_2) \rightarrow (O_{2,2}, M_3) \rightarrow (O_{3,2}, M_1) \rightarrow (T_{2,2}, R_1) \rightarrow (O_{2,3}, M_2)$. In addition, for each job J_i $(J_1, .., J_i)$ $ni - 1$ transport operations are generated $T_{1,1}, T_{2,1}, T_{2,2}$ and $T_{3,1}$, and scheduled following the presented solution in vector JTS, allowing to fix the final path to be considered by each robot R_h during the shop process.

To model an oriented disjunctive graph we should consider some rules. Let the example in Fig. 4, if the edge is oriented in the direction $O_{i,k} \rightarrow O'_{j,k}$ it gets the weight $p_{i,k}$, else it takes $p'_{j,k}$ in the inverse case. If an arc is added from $T_{i,k}$ to $T_{j,k'}$, its gets the weight $t^{Rh}O_{i,k}, O_{i,k+1} + t'^{Rh}O_{i,k+1}, O'_{j,k}$ and $t^{Rh}O'_{j,k}, O'_{j,k+1} + t'^{Rh}O'_{j,k+1}, O_{i,k}$ if it is oriented in the other direction.

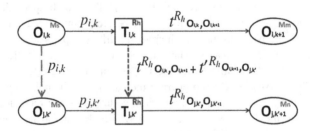

Fig. 4. Example of oriented machine and robot disjunctions

Thus, basing on [6] we can define a fixed machine selection S_m called directed Machine Disjunctions and a fixed transport selection S_r called directed Transport

Disjunctions, with their precedence relations C called operation Conjunctions. So, a fully oriented disjunctive graph can be obtained using $\hat{S} = C \cup S_m \cup S_r$, which is called a complete selection. In fact, the selections of the two sets of disjunctions S_m and S_r with their set of conjunctions C are based on the two proposed vectors MRS and JTS, where MRS allows to present the selected machines to process job operations and the selected robots to process transport operations. JTS presents the execution order of the job and transport operations in their selected machines and robots allowing to fix the final Machine and Transport Disjunctions $S_m \cup S_r$ with their set of Conjunctions C representing the precedence relations. The union $C \cup S_m \cup S_r = \hat{S}$ fully describes a solution if the resulting oriented disjunctive graph $G = (V_m, V_t, \hat{S})$ is acyclic. A feasible schedule can be constructed by longest path calculation which permits to obtain the earliest starting time of both machine and transport operations and fully defines a semi-active schedule with the *Cmax* given by the length of the longest path from node 0 to *, see Fig. 5.

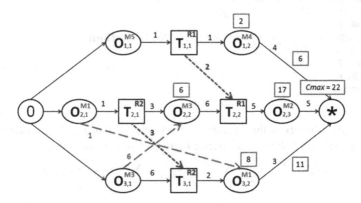

Fig. 5. Oriented disjunctive graph

Noting that the chromosome fitness is calculated by *Fitness(i)* which is the fitness function of each chromosome i and *Cmax(i)* is its makespan value, where $i \in \{1, \ldots, P\}$ and P is the total population size, see Eq. (1).

$$Fitness(i) = \frac{1}{Cmax(i)} \tag{1}$$

Population Initialization. The initial population is generated randomly basing on a neighborhood parameter inspired from [3], calculated by verifying the difference between two chromosomes in terms of the placement of each machine operation $O_{i,j}$ on its assigned machine $\mu_{i,j}$ and the placement of each transport operation $T_{i,j}$ on its assigned robot R_h in the Machines and Robots Selection V_1 (MRS). Also, we have to verify the execution order of all the shop operations $O_{i,j}$ and $T_{i,j}$ in the Job and Transport operation Sequence V_2 (JTS). Let $Chrom_1(MRS_1, JTS_1)$ and $Chrom_2(MRS_2, JTS_2)$ two chromosomes of two different scheduling solutions, $M(O_{i,j})$ the number of alternative machines for each machine operation $O_{i,j}$, $R(T_{i,j})$ the number of alternative robots for each transport operation $T_{i,j}$, L is the total number of operations of all jobs and $Dist$ is the dissimilarity distance. The distance is calculated firstly by measuring the difference between the Machines and Robots Selection vectors MRS_1 and MRS_2 which is in order of $O(n)$, then by verifying the execution order difference of the Job and Transport operation Sequence vectors JTS_1 and JTS_2 which is in order of $O(1)$, we give here how to proceed:

```
Begin
 Dist    0, k   1
 For k from 1 to L Do
    If Chrom₁(MRS₁(k)) ≠ Chrom₂(MRS₂(k)) and
 IsMachineOperation(MRS₁(k)) Then

       Dist    Dist + M(O₁,ⱼ)
    Else If Chrom₁(MRS₁(k)) ≠ Chrom₂(MRS₂(k)) and
 IsTransportOperation(MRS₁(k)) Then
       Dist    Dist + R(T₁,ⱼ)
    End If
    If Chrom₁(JTS₁(k)) ≠ Chrom₂(JTS₂(k)) Then
       Dist    Dist + 1
    End If
 End For
 Return Dist
End.
```

Noting that $Distmax$ is the maximal dissimilarity distance and it is calculated by Eq. (2), representing 100 % of difference between two chromosomes.

$$Distmax = \sum_{i=1}^{n} \sum_{i,1}^{i,ni} M(O_{i,j}) + \sum_{i,1}^{i,ni-1} R(T_{i,j}) + L \qquad (2)$$

Selection Operator. The selection operator is based on the fitness function and the neighborhood parameter, where we propose a new Fitness-Neighborhood Selection Operator (FNSO) allowing to add the dissimilarity distance parameter to the fitness function to select the best parents for the crossover step. The FNSO chooses in each iteration two parent individuals until engaging all the population to create the next generation. The first parent takes successively in each case a solution i, where $i \in \{1,. .., P\}$ and P is the total population size. The second parent obtains its solution j randomly by the roulette wheel selection method based on the two Fitness and Neighborhood parameters relative to the selected first parent, where $j \in \{1,. .., P\} \setminus \{i\}$ in the P population and where $j \neq i$. In fact, to use this random method, we should calculate the Fitness-Neighborhood total FN for the population, see Eq. (3), the selection probability sp_k for each individual I_k, see Eq. (4), and the cumulative probability cp_k, see Eq. (5). After that, a random number r will be generated from the uniform range $[0,1]$. If $r \leq cp_1$ then the second parent takes the first individual I_1, else it gets the kth individual $I_k \in \{I_2,. .., I_P\} \setminus \{I_i\}$ and where $cp_{k-1} < r \leq cp_k$. For Eqs. (3), (4) and (5), $k = \{1, 2,. .., P\} \setminus \{i\}$.

- The Fitness-Neighborhood total FN for the population:

$$FN = \sum_{k=1}^{P} [1/(Cmax[k] \times Neighborhood[i][k])] \tag{3}$$

- The selection probability sp_k for each individual I_k:

$$sp_k = \frac{1/(Cmax[k] \times Neighborhood[i][k])}{FN} \tag{4}$$

- The cumulative probability cp_k for each individual I_k:

$$cp_k = \sum_{h=1}^{k} sp_h \tag{5}$$

Crossover Operator. The crossover operator is applied with two different techniques successively for the parent's chromosome vectors MRS and JTS. Firstly, the MRS crossover operator generates in each case a mixed vector between two parent vectors $Parent_1$-MRS_1 and $Parent_2$-MRS_2, allowing to obtain two new children, $Child_1$-MRS'_1 and $Child_2$-MRS'_2. This uniform crossover is based on two assignment cases, if the generated number r is less than 0.5, the first child $Child_1$ gets the current machine value of $Parent_1$ and the second child $Child_2$ takes the current machine value of $Parent_2$. Else, the two children change their assignment direction, first child $Child_1$ to $Parent_2$ and the second child $Child_2$ to $Parent_1$. Secondly, the JTS crossover operator is an improved precedence preserving order-based on crossover (iPOX), inspired from [9], and is adapted for the parent operation vector JTS. This iPOX operator is applied following four steps, a first step is selecting two parent operation vectors (JTS_1 and JTS_2) and generating randomly two job sub-sets Js_1/Js_2 from all jobs. A second step is allowing to copy any element in JTS_1/JTS_2 that belong to Js_1/Js_2 into child individual JTS'_1/JTS'_2 and retain them in the same position. Then the third step deletes the elements that are already in the sub-set Js_1/Js_2 from JTS_1/JTS_2. Finally, fill orderly the empty positions in JTS'_1/JTS'_2 with the reminder elements of JTS_2/JTS_1 in the fourth step.

Mutation Operator. The mutation operator is integrated to promote the children generation diversity. Firstly, the MRS mutation operator uses a random selection of a transport operation index from the vector MRS. Then, it replaces the current number in the selected index by another belonging to the alternative set of machines (if machine operation $O_{i,j}$) or the set of robots (if transport operation $T_{i,j}$). Secondly, the JTS mutation operator selects randomly two indexes $index_1$ and $index_2$ from the vector JTS. Next, it changes the position of the job number in the $index_1$ to the second $index_2$ and inversely.

Replacement Operator. The replacement operator has an important role to prepare the remaining surviving population to be considered for the next iterations. This operator replaces in each case a parent by one of its children which has the best fitness in its current family.

Clustering Operator. By finishing the maximum iteration number *MaxIter* of the genetic algorithm, the Scheduler Agent applies a clustering operator using the hierarchical clustering algorithm of [7] to divide the final population into N Clusters to be treated by the Cluster Agents in the second step of the global process. The clustering operator is based on the neighbourhood parameter which is the dissimilarity distance between individuals. The clustering operator starts by assigning each individual *Indiv* (i) to a cluster CL_i, so if we have P individuals, we have P clusters containing just one individual in each of them. For each case, we fixe an individual *Indiv(i)* and we verify successively for each next individual *Indiv(j)* from the remaining population (where i and $j \in \{1,..., P\}$, $i \neq j$) if the dissimilarity distance *Dist* between *Indiv(i)* and *Indiv* (j) is less than or equal to a fixed threshold *Distfix* (representing a percentage of difference $X\%$ relatively to *Distmax*, see Eq. (6)) and where *Cluster(Indiv(i))* $\neq$ *Cluster (Indiv(j))*. If it is the case, *Merge(Cluster(Indiv(i)),Cluster(Indiv(j)))*, else continue the search for new combination with the remaining individuals. The stopping condition is by browsing all the population individuals, where we obtained at the end N Clusters.

$$Distfix = Distmax \times X\% \tag{6}$$

3.3   Cluster Agents

Each Cluster Agent CA_i is responsible to apply successively to each cluster CL_i a Tabu Search algorithm to guide the research in promising regions of the search space. Let E the elite solution of a cluster CL_i, $E' \in N(E)$ is a neighbor of the elite solution E, GL_i is the Global List of each CA_i to receive new found elite solutions by the remaining CAs, each CL_i plays the role of the tabu list with a dynamic length and *Cmax* is the makespan of the obtained solution. So, the search process of this local search starts from an elite solution E using the move and insert method of [11], where each Cluster Agent CA_i changes the position of a machine operation $O_{i,j}$ from a machine M_m to another machine M_n belonging to the alternative set of machines and the position of a transport operation $T_{i,j}$ from a robot R_k to another robot R_h belonging to the alternative set of robots in the vector MRS. In addition, modifies the execution order of an operation from an index i to another index k in the vector JTS, searching to generate new scheduling combination $E' \in N(E)$. After that, verifying if the makespan value of this new generated solution *Cmax(E')* dominates *Cmax(E)*, and if it is the case CA_i saves E' in its tabu list (which is CL_i) and sends it to all the other CAs agents to be placed in their Global Lists *GLs(E',CA_i)*, to ensure that it will not be used again by them as a search point. Else continues the neighborhood search from the current solution E. The stopping condition is by attaining the maximum allowed number of neighbors for a solution E without improvement. We give here how to proceed:

```
Begin
  E   Elite(CLᵢ)
  While N(E) ≠ ∅ Do
    Ej   Move-and-insert(E) | Ej ∈ N(E) | E' ∉ CLᵢ
    If Cmax(Ej) < Cmax(E) and Ej ∉ GLᵢ Then
      E   Ej
      CLᵢ   Ej
      Send-to-all(Ej, CAᵢ)
    End If
  End While
  Return E
End.
```

By finishing this local search step, the CAs agents terminate the process by sending their last best solutions to the SA agent, which takes the best one for the FJSPT-MR.

4 Experimental Results

4.1 Experimental Setup

The proposed GATS + HM is implemented in java language on a 2.10 GHz Intel Core 2 Duo processor and 3 Gb of RAM memory, using the *eclipse* IDE to code the approach and the multiagent platform *Jade* to create the holonic multiagent model. To evaluate its efficiency, numerical tests are made based on the data instances of Deroussi and Norre [4]. The used parameter settings for our algorithm are adjusted experimentally and presented as follow: the crossover probability = 1.0, the mutation probability = 0.5, the maximum number of iterations = 1000 and the population size = 200. The computational results are presented by three metrics in Table 2, such as the best makespan, CPU time of our GATS + HM in minutes and the gap between our approach and the best results in the literature of the FJSPT-MR, which is calculated by Eq. (7). The *Mko* is the makespan obtained by Our approach and *Mkc* is the makespan of one of the chosen algorithms for Comparisons.

$$Gap = [(Mko - Mkc)/Mkc] \times 100\,\% \tag{7}$$

4.2 Experimental Comparisons

To show the efficiency of our GATS + HM approach, we compare its obtained results from the previously cited data set with other well known algorithms in the literature of the FJSPTMR. The chosen algorithms are: The shifting bottleneck (SBN) and the tabu search procedure (Tabu) of [13] which are standard heuristic and metaheuristic methods. The hybrid genetic algorithm-tabu search procedure (GATS) of [12] and the combined genetic algorithm-tabu search-shifting bottleneck (GTSB) of [13] which are two recent hybrid metaheuristic approaches.

Table 2. Results of [4] data instances

Instance	SBN	GapSBN	Tabu	GapTabu	GTSB	GapGTSB	GATS	GapGATS	GATS + HM	CPU time
fjspt1	156	−29.49	160	−31.25	146	−24.66	144	−23.61	**110**	0.51
fjspt2	124	−26.61	128	−28.91	118	−22.88	118	−22.88	**91**	0.30
fjspt3	140	−25.71	162	−35.80	124	−16.13	124	−16.13	**104**	0.32
fjspt4	132	−32.58	126	−29.37	124	−28.23	124	−28.23	**89**	0.41
fjspt5	96	−15.63	100	−19.00	94	−13.83	94	−13.83	**81**	0.14
fjspt6	148	−21.62	152	−23.68	144	−19.44	144	−19.44	**116**	0.50
fjspt7	132	−36.36	132	−36.36	122	−31.15	124	−32.26	**84**	1.00
fjspt8	191	−24.08	188	−22.87	181	−19.89	180	−19.44	**145**	0.58
fjspt9	154	−22.08	162	−25.93	146	−17.81	150	−20.00	**120**	0.35

From results in Table 2, it can be seen that our approach GATS + HM is the best one which solves all the Deroussi and Norre [4] instances, where we attain nine new upper bounds. In fact, our approach shows its superiority to the first two methods SBN and Tabu in all instances, with a same maximum gap of −36.36 % for the fjspt7 instance. For the comparison with the second two hybrid metaheurstic methods GTSB and GATS, our approach GATS + HM outperforms them in all the nine instances, with a maximum gap of −31.15 % for the GTSB and −32.26 % for the GATS by solving the fjspt7 instance.

5 Conclusion

In this paper, we present a new metaheuristic hybridization approach based on clustered holonic multiagent model, called GATS + HM, for the flexible job shop scheduling problem with transportation times and many robots. To measure its performance, numerical tests are made and where new upper bounds are found showing the effectiveness of the presented approach. In the future work, we will search to treat other extensions of the FJSPT-MR, such as by considering the constraints of a non-unit transport capacity for the moving robots, where the problem becomes a flexible job shop scheduling problem with moving robots and non-unit transport capacity. So, we will make improvements to our approach to adapt it to this new transformation and study its effects on the makespan.

References

1. Abdelmaguid, T.F., Nassef, A.O., Kamal, B.A., Hassan, M.F.: A hybrid GA/heuristic approach to the simultaneous scheduling of machines and automated guided vehicles. Int. J. Prod. Res. 42(2), 267–281 (2004)
2. Bilge, U., Ulusoy, G.: A time window approach to simultaneous scheduling of machines and material handling system in an FMS. Oper. Res. 43(6), 1058–1070 (1995)
3. Bozejko, W., Uchronski, M., Wodecki, M.: The new golf neighborhood for the flexible job shop problem. In: The International Conference on Computational Science, pp. 289–296 (2010)
4. Deroussi, L., Norre, S.: Simultaneous scheduling of machines and vehicles for the flexible job shop problem. In: International Conference on Metaheuristics and Nature Inspired Computing, Djerba, Tunisia, pp. 1–2 (2010)
5. Garey, M.R., Johnson, D.S., Sethi, R.: The complexity of flow shop and job shop scheduling. Math. Oper. Res. 1(2), 117–129 (1976)
6. Hurink, J., Knust, S.: Tabu search algorithms for job-shop problems with a single transport robot. Eur. J. Oper. Res. 162(1), 99–111 (2005)
7. Johnson, S.C.: Hierarchical clustering schemes. Psychometrika 32(3), 241–254 (1967)
8. Lacomme, P., Larabi, M., Tchernev, N.: Job-shop based framework for simultaneous scheduling of machines and automated guided vehicles. Int. J. Prod. Econ. 143(1), 24–34 (2013)

9. Lee, K., Yamakawa, T., Lee, K.M.: A genetic algorithm for general machine scheduling problems. In: The Second IEEE International Conference on Knowledge-Based Intelligent Electronic Systems, pp. 60–66 (1998)
10. Lenstra, J.K., Kan, A.H.G.R.: Complexity of vehicle routing and scheduling problems. Networks 11(2), 221–227 (1981)
11. Mastrolilli, M., Gambardella, L.: Effective neighbourhood functions for the flexible job shop problem. J. Sched. 3(1), 3–20 (2000)
12. Zhang, Q., Manier, H., Manier, M.A.: A genetic algorithm with tabu search procedure for flexible job shop scheduling with transportation constraints and bounded processing times. Comput. Oper. Res. 39(7), 1713–1723 (2012)
13. Zhang, Q., Manier, H., Manier, M.A.: Metaheuristics for Job Shop Scheduling with Transportation. Wiley, New York (2013). pp. 465–493. ISBN 9781118731598

A Novel Android Application Design Based on Fuzzy Ontologies to Carry Out Local Based Group Decision Making Processes

J.A. Morente Molinera[1,5]($\boxtimes$), R. Wikström[2,3], C. Carlsson[2,3],
F.J. Cabrerizo[5], I.J. Pérez[4], and E. Herrera-Viedma[5]

[1] Department of Engineering, Universidad Internacional de la Rioja (UNIR),
Logroño, La Rioja, Spain
[2] Laboratory of Industrial Management, Abo Akademi University, Abo, Finland
{robin.wikstrom,christer.carlsson}@abo.fi
[3] Institute for Advanced Management Systems Research,
Abo Akademi University, Abo, Finland
[4] Department of Computer Engineering, University of Cádiz, Cádiz, Spain
ignaciojavier.perez@uca.es
[5] Department of Computer Science and Artificial Intelligence,
University of Granada, Granada, Spain
{jamoren,cabrerizo,viedma}@decsai.ugr.es

Abstract. The appearance of Web 2.0 and mobile technologies, the increase of users participating on the Internet and the high amount of available information have created the necessity of designing tools capable of making the most out of this environment. In this paper, the design of an Android application that is capable of aiding some experts in carrying out a group decision making process in Web 2.0 and mobile environments is presented. For this purpose, Fuzzy Ontologies are used in order to deal with the high amount of information available for the users. Thanks to the way that they deal with the information, they are used in order to retrieve a small set of alternatives that the users can utilize in order to carry out group decision making processes with a feasible set of valid alternatives.

Keywords: Group decision making · Fuzzy ontologies · Decision support system

1 Introduction

In its recent days, Internet was a static platform where a small minority of experts uploaded all the information that the Internet users could download. Since a small set of experts controlled all the information, the data was proved valid and trustful. Nevertheless, the quantity of the information was low and users did not have the chance to provide their own information and experience to the users net.

Nowadays, situation has completely changed. The appearance of Web 2.0 [1] technologies have allowed all the users from the Internet to share and consume

© Springer International Publishing Switzerland 2016
V. Torra et al. (Eds.): MDAI 2016, LNAI 9880, pp. 289–300, 2016.
DOI: 10.1007/978-3-319-45656-0_24

information at the same time. This paradigm change has provoke an increase in the quantity of information available on the Internet. Nevertheless, since anyone is able to become an information provider, validity and veracity of the information can no longer be taken for granted. Furthermore, there is a need of tools that are capable of sorting and organizing the information in a way that users can benefit from it.

Along with the Web 2.0 technologies revolution, mobile phones have been transformed from calling devices to small and portable computers. This has provoked mobile phones, recently called smartphones, to become the item that users want to use as assistant devices in their everyday life. Thanks to 3G/4G technologies, smartphones have acquired the ability to connect and make use of all the information that is stored on the Internet. Thanks to Web 2.0 and mobile technologies combination, users can benefit from any kind of information stored on the Internet, at any time, independently of their location.

There is no doubt that Web 2.0 and mobile technologies have drastically transformed the way that users used to retrieve information and communicate among themselves. Therefore, there is a need for traditional computational fields to adapt and benefit from this new paradigm.

One of these traditional fields is group decision making [2]. Traditionally, group decision making processes were carried out by a set of experts that reunite in a single room and debate over a small set of alternatives. Nowadays, because of the paradigm change, users need to carry out group decision making processes about an unspecified high amount set of alternatives. Also, thanks to mobile phones, experts do not want to reunite in an specific place any more. They are willing to carry out decision making processes independently of where they are located communicating among themselves using their mobile phones.

In order to achieve these goals, a novel Android decision support system is presented in this paper. This system uses fuzzy ontologies [3] in order to reduce the high amount of available alternatives into a feasible set. Also, experts use their mobile phones in order to communicate and carry out the group decision making process. Mobile phones features, like the GPS, are used in order to help the fuzzy ontology to reduce the available set of alternatives. Thanks to this novel developed method, experts can carry out group decision making processes at any time independently of where they are located. In order to increase the comprehensibility of the presented method, we will work with a concrete example. In the proposed situation, experts are reunited in a restaurant and they need to select a specific wine from a set containing a high amount of possibilities. Fuzzy ontologies will help to reduce the available amount of alternatives into a feasible one. Finally, a group decision making process is carried out in order for the experts to select an specific wine from the list.

In Sect. 2, basis of all the tools needed to understand the designed Android decision support system are exposed. In Sect. 3, the designed decision making support system is presented. In Sect. 4, advantages and drawbacks of the developed method are highlighted. Finally, some conclusions are pointed out.

2 Preliminaries

In this section, basic concepts needed to understand the presented design are exposed. In Sect. 2.1, basis of group decision making methods are exposed. In Sect. 2.2, basis of fuzzy ontologies are described.

2.1 Group Decision Making Methods

A common Group Decision Making method problem can be formally defined as follows:

Let $E = \{e_1, \ldots, e_n\}$ be a set of experts and $X = \{x_1, \ldots, x_m\}$ a set of alternatives that the experts must discuss about. A group decision making problem consists in sorting X, or choosing the best alternative x_b, using the preferences values P^k, $\forall k \in [1, n]$, provided by the experts.

In order to carry out this process, the following steps are usually followed:

1. **Providing Preferences:** Experts provide their preferences about the alternatives. In the proposed method, linguistic label sets can be used. This way, people can express themselves in a more comfortable way using words instead of numbers. Also, preference relations are the chosen representation mean due to the fact that they allow experts to carry out pairwise comparisons. Although preference relations are not the unique possible choice, we have selected them since they are capable of carrying out pairwise comparisons among the alternatives. This way, experts can provide their preferences in a comfortable and user-friendly way. A linguistic label set is provided for each pair of alternatives.
2. **Aggregation Step:** After all the experts provide their preferences, they are aggregated into a single collective preference relation. For this task, OWA operator [4, 5] or the mean operator can be used. When working with linguistic information, it is possible to use specific operators such as the LOWA [6] one, that is capable of aggregating linguistic information without having to deal with any numeric conversion.
3. **Selection Step:** Once that the collective preference matrix has been calculated, selection operators are used in order to generate the ranking or select the most preferred alternative. For this purpose, guided dominance degree and guided non dominance degree are used [7, 8].
4. **Consensus Calculation Step:** When carrying out a group decision making process, it is very important to try to bring opinions closer and make all the experts to converge to an unique solution. It is desirable for experts to debate, express their points of view and select a solution together. For this purpose, several group decision making rounds can be performed until they reach a specific level of agreement. This level of agreement can be calculated using consensus measures [9]. Therefore, if, after providing the preferences, the level of agreement is low, experts are asked to carry out more debate and resend their preferences. On the contrary, if the consensus level is high, final decision results are calculated.

A scheme of this process can be seen in Fig. 1.

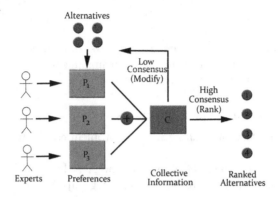

Fig. 1. Group decision making scheme.

2.2   Fuzzy Ontologies

Fuzzy ontologies are constructions whose main purpose consists in representing data in a sorted way using an organized structure. Using their representation capabilities, it is possible to retrieve specific pieces of data from them using queries that search elements with certain characteristics. A fuzzy ontology can be formally expressed as follows:

Definition 1. A fuzzy ontology [10,11] is a quintuple $O_F = \{I, C, R, F, A\}$ where I represents a set of individuals, C is a set of concepts, R is a the set of relations, F is a set of fuzzy relations and A is the set of axioms. A regular fuzzy ontology scheme is showed in Fig. 2.

The purpose of each of the required sets are specified below [3,12,13]:

- **Individuals:** Represent the entities that are stored in the fuzzy ontology. These entities are usually formed by the set of elements that conform the environment that we are describing.
- **Concepts:** They are the perceptions used to describe the individuals. Thanks to the concepts, it is possible to carry out thorough descriptions of the individuals that conform the ontology.
- **Relations:** Their main purpose is to apply concepts to individuals. This way, relations are in charge of determining which concepts are fulfilled by each of the individuals. Individuals can be related among themselves or with concepts. Relations among concepts are not usual. Information about relations should be obtained from a trustful source of information.
- **Axioms:** They establish rules that must always be fulfilled by the defined fuzzy ontology.

A graphical representation of a Fuzzy Ontology is shown in Fig. 2.

In order to increase the comprehension of the article, an example is given below.

Imagine that we want to design an ontology describing all the smartphones that are available at the market at a certain time. One of the possible ways

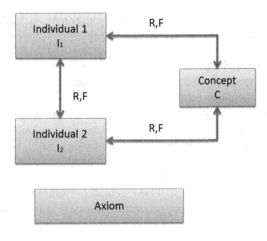

Fig. 2. Fuzzy ontology scheme.

of representing the information is as follows. First, a list of the smartphones and features that need to appear in the ontology are selected. Each smartphone will be represented as an individual and each feature is modeled as a concept. Relations link each individual with each concept at a certain degree. A linguistic label set of 5 elements $B = \{very\_low, low, medium, high, very\_high\}$ can be used for the representation of every fuzzy relation. Smartphones features taken into account in the design of the ontology can be the following ones:

- **Screen Size:** Represents the size of the screen. There are people that prefer small screen size smartphones because they are easy to transport and there are others that prefer bigger size ones.
- **Screen Resolution:** It refers to the definition of the screen. Some buyers can prefer a high resolution smartphone in order to watch videos and others probably do not care and prefer a normal or low resolution in order to decrease the price.
- **Processor Speed:** Smartphone processor performance.
- **Components Quality:** It refers to the reliability of the smartphones components.
- **Brand:** It allows the buyer to search for a specific brand. This concept is not fuzzy.
- **Capacity:** Smartphone capacity. Depending on the use there can be different preferences.
- **Price:** This will allow the search of smartphones that have specific price ranges.
- **Weight:** Smartphone weight. A light smartphone is more comfortable to transport that a heavy one but is likely to be more expensive.
- **Battery:** Type of battery that the smartphone has inside.

It is possible to extract information from fuzzy ontologies using queries. Users specify the characteristics that they want the individuals to have and the most similar entries can be retrieved from the fuzzy ontology. In order to carry out this process, the following steps are followed [14]:

1. **Query Providing Step:** The expert specifies the query that he/she wants to perform. The queries are formed by the specific data that the expert wants the retrieved elements to fulfil. That is, a set of concepts and relation values. Once that the query is performed, data is sent to the ontology reasoner.
2. **Ontology Searching Step:** The ontology reasoner calculates the similarity of each of the individuals relations of the ontology with the provided query. The most similar individuals are selected and ordered according to their ranking value. The rest of them are discarded.
3. **Results Presentation Step:** The calculated ranking is presented to the user. He/she can work with the provided information as needed.

3   Android Platform Design

The designed Android platform follows a client-server model in order to carry out the necessary operations. Thanks to this, heavy computational operations can be held in the server while light ones can be performed on the experts smartphones. In this paper, for the sake of comprehensibility, we work with an specific example. A set of experts are attending a dinner in a restaurant and they need to select a wine from a set of 600 alternatives. Information about the alcohol and acidity level, country of origin and price is known for each of them. The designed Fuzzy Ontology is created as follows:

- **Individuals:** Each individual of the fuzzy ontology refers to each of the wines.
- **Concepts:** Each concept is assigned to one of the properties (alcohol and acidity levels, price, ...).
- **Relations:** Each of the individuals is related to all of the concepts of the fuzzy ontology. Moreover, individuals are not related among them in this case.

For example, the individual *Zenato_Veneto_Rosso* is a wine which is related to all of the fuzzy ontology concepts. The relation values for each of them are specified as follows:

- **Price:** Its numerical value is 9.99. A linguistic label set of five labels is used for its representation. s_1 is the linguistic label that represents the relation.
- **Country:** The country of origin is Italy. Since this is a crisp relation there is no need of using the fuzzy sets computational environment.
- **Alcohol:** Its numerical value is 13.5. A linguistic label set with a granularity value of 3 is used for representing this relation. The linguistic label that represents the relation is s_1.
- **Acidity:** Its numerical value is 5.1. Same representation as the alcohol is used. The linguistic label s_0 is used in this case.

The process that the designed system follows in order to carry out the presented group decision making problem is exposed below:

1. **Location Request:** First, location of the experts is calculated using the smartphones GPS technologies. This information will be used in order to reduce the initial alternatives set and ease computations.
2. **Parameters Providing Step:** Experts provide the parameters that will help the ontology reasoner to generate a reduced set of alternatives. They are described below:
   (a) *Context*: It refers to the dinner scenario and purpose. Depending on the context, the list of adequate wines might change. For example, for a formal meeting, a cheap wine might not be suitable. Experts are asked to select among three options: candle, friends and formal.
   (b) *Food*: It indicates the type of food that the experts are going to consume. It is well-known that depending on the food, some wines can be more suitable than others. For instance, red wines are more suitable for meat dishes than white ones. Experts are asked to select among five different options: game, fish, grilled food, chicken and shellfish.
   (c) *Number of people*: The number of people that are participating in the meal. This parameter is used in order for the system to known the number of experts participating in the group decision making process.
   (d) *Number of wines*: Minimum number of wines that the experts want to discuss about. Minimum value is set to four. The purpose of this parameter is to let experts decide the size of the alternatives set that they want to discuss about.
3. **Ontology Reasoning Process:** In order to generate the ranking, four different criteria are followed. This way, experts can analyse results that comes from different points of view. The four followed criteria are described below:
   - *Most famous wine*: This criteria selects the most popular wine in the place where the experts are located. This criteria provides a chance for the experts to taste a famous local wine. Typically, this criteria leads to the most popular wine of the location or the one that is most consumed there.
   - *Lowest price wine*: This criteria selects the cheapest option available at the chosen place. This criteria is suitable for people that is not fond on wines and want to go for the economic option.
   - *Most voted wine*: This criteria selects the most chosen wine in an specific location. Previous group decision making results are stored in a database in order to find out how many times each wine is selected in each location. If there is no wine selected at an specific location, this criterion is not taken into account.
   - *Best wines according to the context and food*: This criteria takes into account the context and the food of the meeting. It selects the best wines according to the options selected by the experts in the previous step.

Using these four criteria and the parameters specified in the previous step, a list of wines and the criteria used for selection is showed to the experts.

4. **Group Decision Making Process:** A group decision making process like the one described in Subsect. 2.1 is carried out. Experts must discuss and decide which wine they want to order among the presented ones. The approach follows classical linguistic Group Decision Making methods [2,15,16].

5. **Updating Wine Information:** After choosing the wine that they want to consume, the number of times that they wine has been selected is updated and stored in the server database. This way, it is possible to keep track of how many times a specific wine has been chosen in an specific location. This information will be used in posteriors group decision making processes. A server applet running in the server is in charge of carrying out this task.

Android screenshots can be seen in Figs. 3 and 4. Information about locations, available wines and the number of times that a wine has been selected by the experts is stored in a separate database. The database its quite simple, two tables and one relationship is enough to store the required information:

- **Wine Table:** The main purpose of the wine table is to store the names of all the wines that conform the ontology.
- **Location Table:** All the locations available in the system are stored here. Thanks to the wine-location separability, it is possible to add wines and locations in a comfortable way at any time. We believe that scalability is very important in this system since the fuzzy ontology should be updated continuously in order to keep the information up to date.
- **Wine_Location Relationship:** Wines and locations are related in order to determine which wines are available in each location. A wine can belong to many locations and, in each location, several wines can be found. The number of times that a wine is chosen in each location is also stored here.

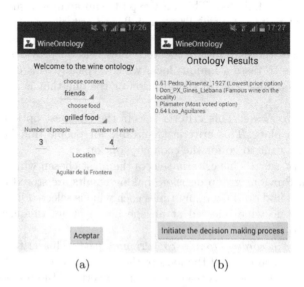

(a)   (b)

Fig. 3. Android application: search information screenshot and wine ontology results screenshot.

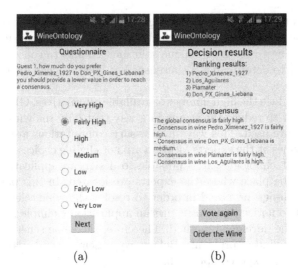

Fig. 4. Android application: questionnaire screenshot and temporary results decision screenshot.

All the incoming requests from the mobile devices are handled by the server. The server is in charge of creating a queue with all the petitions and attend them. It connects with the fuzzy ontology and is also capable of retrieving information from the wine-location database. Every time that a set of experts carry out a group decision making process, the server retrieves the wines available at their location and resolve the fuzzy ontology query carrying out the similarity comparisons on the available wines. Results obtained are sent back to the mobile device that has performed the request.

4   Discussion

In this paper, fuzzy ontologies and group decision making methods have been combined in other to take advantage of the recent novelties that Internet and mobile phones have brought to us. The main purpose of our method is to allow experts to use their mobile phones in order to carry out group decisions making processes including a high amount of alternatives.

Thanks to fuzzy ontologies, it is possible to deal with imprecise and linguistic knowledge. Wine world is a clear example of a field where almost all the information associated to the wines is imprecise and based on experts opinions. In this kind of environments is where fuzzy ontologies benefits are more clear. Linguistic modelling also allow experts to communicate with the system in a more comfortable way using words instead of numbers.

Our developed method provides advice about which wine should experts choose for the dinner. Using wine information and ontology reasoning techniques, experts can benefit from a high amount of information in order to make a right choice. Thanks to group decision making methods and consensus measures, the decision can be carried out in a efficient, organized and consensual way using debates and trying to reach an unique solution. Thanks to consensus measures [9,17], it is easy to determine is a consensus has been reached or if more debate is needed.

In order to reduce the high number of available alternatives, GPS technologies are used in order to reduce the search in a way that only the wines available at a certain place are used. This way, unnecessary computations are avoided since not available wines at the experts location are not eligible choices. Consequently, thanks to GPS technologies, it is possible to design an application capable of adapt itself to the place where the experts are located. In this paper, we have used the novel design proposed in order to resolve the wine selection problem. It is important to notice that this is just an application example in order to test the validity of the developed method. Therefore, the same scheme can be used for solving other kind of problems. Below, we present several example that can be solved using the exposed approach:

- It would be possible to use the proposed approach in order to provide information about loans from banks located in the place where experts are living. The different loans and their characteristics can be stored in the fuzzy ontology in order to help experts to choose the one that is more suitable to them.
- It would also be useful to store in the fuzzy ontology information about apartments available for rent or sale in an specific location. This way, the expert can make use of the information in order to find his/her most suitable apartment.
- It would be possible to use the proposed system in order to help experts to choose the place that they want to go on holidays.
- The presented system is also suitable for storing companies generated knowledge. This way, experts can use our system in order to carry out critical decisions about the company management.
- The presented approach is also useful in order to help investors to choose the places where they can invest money in a way that they can obtain the highest benefit.

It is also possible to carry out several upgrades to the proposed method. Some of them are described below:

- It will be desirable to allow experts to select several wines for the dinner, one for each ordered dish.
- In the presented method, the number of available wines varies depending on the location of the user. If GPS coordinates are precise enough, it would also be possible to select the wines available at the restaurant where the experts are attending the meal.

5   Conclusions

In this paper, a novel android decision support system is developed. The designed system uses fuzzy ontologies in order to sort the information in an organized way and to provide the experts with a feasible set of alternatives that they can use to carry out the group decision making process. Experts can express their preferences in a linguistic way and they can benefit from consensus measures in order to carry out a thorough debate before calculating the final results.

Thanks to smartphones, experts can make use of the fuzzy ontology information at any place. Thanks to GPS technologies, it is possible to reduce the number of available alternatives in order to focus only on those ones available at the place where experts are located.

Acknowledgments. This paper has been developed with the financing of FEDER funds in TIN2013-40658-P and Andalusian Excellence Project TIC-5991.

References

1. Fuchs, C., Boersma, K., Albrechtslund, A., Sandoval, M.: Internet and surveillance: the challenges of Web 2.0 and social media, **16** (2013)
2. Kacprzyk, J.: Group decision making with a fuzzy linguistic majority. Fuzzy Sets Syst. **18**(2), 105–118 (1986)
3. Maedche, A.: Ontology Learning for the Semantic Web, vol. 665. Springer Science and Business Media, Heidelberg (2012)
4. Yager, R.R.: On ordered weighted averaging aggregation operators in multicriteria decisionmaking. IEEE Trans. Syst. Man Cybern. **18**(1), 183–190 (1988)
5. Yager, R.R.: Quantifier guided aggregation using OWA operators. Int. J. Intell. Syst. **11**(1), 49–73 (1996)
6. Herrera, F., Herrera-Viedma, E.: Aggregation operators for linguistic weighted information. IEEE Trans. Syst. Man Cybern. Part A: Syst. Hum. **27**(5), 646–656 (1997)
7. Chiclana, F., Herrera, F., Herrera-Viedma, E.: Integrating three representation models in fuzzy multipurpose decision making based on fuzzy preference relations. Fuzzy Sets Syst. **97**(1), 33–48 (1998)
8. Herrera, F., Herrera-Viedma, E., Verdegay, J.L.: A sequential selection process in group decision making with a linguistic assessment approach. Inf. Sci. **85**(4), 223–239 (1995)
9. Alonso, S., Pérez, I.J., Cabrerizo, F.J., Herrera-Viedma, E.: A linguistic consensus model for web 2.0 communities. Appl. Soft Comput. **13**(1), 149–157 (2013)
10. Calegari, S., Ciucci, D.: Fuzzy ontology, fuzzy description logics and fuzzy-OWL. In: Masulli, F., Mitra, S., Pasi, G. (eds.) WILF 2007. LNCS (LNAI), vol. 4578, pp. 118–126. Springer, Heidelberg (2007)
11. Baader, F.: The Description Logic Handbook: Theory, Implementation and Applications. Cambridge University Press, Cambridge (2003)
12. Lukasiewicz, T., Straccia, U.: Managing uncertainty and vagueness in description logics for the semantic web. Web Semant.: Sci. Serv. Agents World Wide Web **6**(4), 291–308 (2008)
13. Bobillo, F.: Managing Vagueness in Ontologies. University of Granada, Granada (2008)
14. Carlsson, C., Mezei, J., Brunelli, M.: Fuzzy ontology used for knowledge mobilization. Int. J. Intell. Syst. **28**, 52–71 (2013)

15. Morente-Molinera, J.A., Al-hmouz, R., Morfeq, A., Balamash, A.S., Herrera-Viedma, E.: A decision support system for decision making in changeable and multi-granular fuzzy linguistic contexts. J. Multiple-Valued Logic Soft Comput. **26**(3–5), 485–514 (2016)
16. Shen, F., Xu, J., Xu, Z.: An outranking sorting method for multi-criteria group decision making using intuitionistic fuzzy sets. Inf. Sci. **334**, 338–353 (2016)
17. Cabrerizo, F.J., Chiclana, F., Al-Hmouz, R., Morfeq, A., Balamash, A.S., Herrera-Viedma, E.: Fuzzy decision making and consensus: challenges. J. Intell. Fuzzy Syst. **29**(3), 1109–1118 (2015)

Automatic Detection and Segmentation of Brain Tumor Using Random Forest Approach

Zoltán Kapás[1], László Lefkovits[1], and László Szilágyi[1,2,3(✉)]

[1] Sapientia University of Transylvania, Tîrgu-Mureş, Romania
lalo@ms.sapientia.ro
[2] Department of Control Engineering and Information Technology,
Budapest University of Technology and Economics, Budapest, Hungary
[3] University of Canterbury, Christchurch, New Zealand

Abstract. Early detection is the key of success in the treatment of tumors. Establishing methods that can identify the presence and position of tumors in their early stage is a current great challenge in medical imaging. This study proposes a machine learning solution based on binary decision trees and random forest technique, aiming at the detection and accurate segmentation of brain tumors from multispectral volumetric MRI records. The training and testing of the proposed method uses twelve selected volumes from the BRATS 2012/13 database. Image volumes were preprocessed to extend the feature set with local information of each voxel. Intending to enhance the segmentation accuracy, each detected tumor pixel is validated or discarded according to a criterion based on neighborhood information. A detailed preliminary investigation is carried out in order to identify and enhance the capabilities of random forests trained with information originating from single image records. The achieved accuracy is generally characterized by a Dice score up to 0.9. Recommendation are formulated for the future development of a complex, random forest based tumor detection and segmentation system.

Keywords: Decision tree · Random forest · Machine learning · Image segmentation

1 Introduction

The early detection of brain tumors is utmost important as it can save human lives. The accurate segmentation of brain tumors is also essential, as it can assist the medical staff in the planning of treatment and intervention. The manual segmentation of tumors requires plenty of time even for a well-trained expert. A fully automated segmentation and quantitative analysis of tumors is thus a highly beneficial service. However, it is also a very challenging one, because of the high variety of anatomical structures and low contrast of current imaging

Research supported in part by Marie Curie Actions, IRSES Project no. FP7-318943.

V. Torra et al. (Eds.): MDAI 2016, LNAI 9880, pp. 301–312, 2016.
DOI: 10.1007/978-3-319-45656-0_25

techniques which make the difference between normal regions and the tumor hardly recognizable for the human eye [1]. Recent solutions, usually assisted by the use of prior information, employ various image processing and pattern recognition methodologies like: combination of multi-atlas based segmentation with non-parametric intensity analysis [2], AdaBoost classifier [3], level sets [4], active contour model [5], graph cut distribution matching [6], diffusion and perfusion metrics [7], 3D blob detection [8], support vector machine [9], concatenated random forests [10,11], and fuzzy c-means clustering [12].

The main goal of our research work is to build a reliable procedure for brain tumor detection from multimodal MRI records, based on supervised machine learning techniques, using the MICCAI BRATS data set that contains several dozens of image volumes together with ground truth provided by human experts. In this paper we propose a solution based on binary decision trees and random forest, and present preliminary results together with a recommendations towards a complex and reliable brain tumor detection system.

2   Materials and Methods

The main goal of this study is to establish a machine learning solution to detect and localize tumors in MRI volumes. This paper presents preliminary results obtained using the random forest technique. The algorithm is trained to separate three tissue types, which are labeled as tumor, edema, and negative. The primary focus is on establishing the presence or absence of the tumor, while the accurate segmentation is a secondary goal.

2.1   BRATS Data Sets

Brain tumor image data used in this work were obtained from the MICCAI 2012 Challenge on Multimodal Brain Tumor Segmentation [13]. The challenge database contains fully anonymized images originating from the following institutions: ETH Zürich, University of Bern, University of Debrecen, and University of Utah. The image database consists of multi-contrast MR scans of 30 glioma patient, out of which 20 have been acquired from high-grade (anaplastic astrocytomas and glioblastoma multiform tumors) and 10 from low-grade (histological diagnosis: astrocytomas or oligoastrocytomas) glioma patients. For each patient, multimodal (T1, T2, FLAIR, and post-Gadolinium T1) MR images are available. All volumes were linearly co-registered to the T1 contrast image, skull stripped, and interpolated to 1 mm isotropic resolution. Each records contains approximately 1.5 millions of feature vectors. All images are stored as signed 16-bit integers, but only positives values are used. Each image set has a truth image which contains the expert annotations for "active tumor" and "edema". Each voxel in a volume is represented by a four-dimensional feature vector:

$$\boldsymbol{x} = [x^{(\text{T1})}, x^{(\text{T2})}, x^{(\text{T1C})}, x^{(\text{FLAIR})}]^T. \tag{1}$$

These are the observed features of each pixel. Since these four values do not incorporate any information regarding the location or the neighborhood of the pixel, there is a strong need to extend the feature vector with further, computed features.

2.2 Data Preprocessing

Preprocessing steps in our application have three main goals.

1. Histogram normalization. Whether we like it or not, absolute intensity values in magnetic resonance imaging say nothing about the observed tissue. Intensities are relative and frequently contaminated with intensity inhomogeneity [14]. Treating the latter problem stays outside the scope of this study, as the MICCAI BRATS data set is free from intensity inhomogeneity. However, the histogram of each volume needs to be mapped on a uniform scale. In this order, all intensity values underwent a linear transformation $x \rightarrow \alpha x + \beta$, where parameters α and β were established separately for each feature such a way that the middle fifty percent of the data fell between 600 and 800 after the transformation. Further on, we set up a minimum and maximum limit for intensity values at 200 and 1200, respectively. Intensities situated beyond the limit were replaced by the corresponding limit value.
2. Computed features. Since the observed data vectors bear no information on the position of the pixel, we included eight more features into the feature vector. For each of the four channels, two locally averaged intensities were computed within 10-element and 26-element neighborhood. The former contained eight direct neighbors within the slice and the two closest ones from neighbor slices. The latter contained all neighbors of the given pixel situated within a $3 \times 3 \times 3$-sized cube. Pixels having no valid neighbors in the specific neighborhood inherited the own intensity value of the pixel itself in the given channel.
3. Missing Data. Some pixels have zero valued observed features standing for a missing value. These pixels were not included in the main data processing. However, all existing features were used at the computation of averaged features, so pixels with missing values may have contributed to their neighbors, before being discarded.

2.3 Decision Tree

Binary decision trees (BDT) can describe any hierarchy of crisp (non-fuzzy) two-way decisions [15]. Given an input data set of vectors $\mathbf{X} = \{\boldsymbol{x}_1, \boldsymbol{x}_2, \ldots, \boldsymbol{x}_n\}$, where $\boldsymbol{x}_i = [x_{i,1}, x_{i,2}, \ldots, x_{i,m}]^T$, a BDT can be employed to learn the classification that corresponds to any set of labels $\Lambda = \{\lambda_1, \lambda_2, \ldots, \lambda_n\}$. The classification learned by the BDT can be perfect if $\boldsymbol{x}_i = \boldsymbol{x}_j$ implies $\lambda_i = \lambda_j$, $\forall i, j \in \{1, 2, \ldots, n\}$. The BDT is built during the learning process. Initially the tree consists of a single node, the root, which has to make a decision regarding all n input vectors. If not all n vectors have the same label, which is likely to be so, then the set of data is not homogeneous, there is a need for a separation. The decision will compare a single feature, the one with index k $(1 \leq k \leq m)$, of the input vectors with a certain threshold α, and the comparison will separate the vectors into two subgroups: those with $x_{i,k} < \alpha$ $(i = 1 \ldots n)$, and those with $x_{i,k} \geq \alpha$ $(i = 1 \ldots n)$. The root will then have two child nodes, each corresponding to one of the possible outcomes of the above decision. The left child will further classify those n_1 input vectors, which satisfied the former condition, while the right child those n_2 ones that satisfied the latter condition. Obviously, we have $n_1 + n_2 = n$. For both child nodes, the procedure is the same as it was for the root. When at a certain point of the learning algorithm, all vectors being classified by a node have the same label λ_p, then the node is declared a leaf node, which is attributed to the class with index p. Another case when a node is declared leaf node is when all vectors to be separated by the node are identical, so there is no possible condition to separate the vectors. In this case, the label of the node is decided by the majority of labels, or if there is no majority, a label should be chosen from the present ones. In our application, this kind of rare cases use the priority list of labels defined as: (1) tumor, (2) edema, (3) negative.

The separation of a finite set of data vectors always terminates in a finite number of steps. The maximum depth of the tree highly depends on the way of establishing the separation condition in each node. The most popular way, also employed in our application, uses entropy based criteria to choose the separation condition. Whenever a node has to establish its separation criterion for a subset of vectors $\overline{\mathbf{X}} \subseteq \mathbf{X}$ containing $\overline{n}$ items with $1 < \overline{n} \leq n$, the following algorithm is performed:

1. Find all those features which have at least 2 different values in $\overline{\mathbf{X}}$.
2. Find all different values for each feature and sort them in increasing order.
3. Set a threshold candidate at the middle of the distance between each consecutive pair of values for each feature.
4. Choose that feature and that threshold, for which the entropy-based criterion

$$E = \overline{n}_1 \log \frac{\overline{n}_1}{\overline{n}} + \overline{n}_2 \log \frac{\overline{n}_2}{\overline{n}} \qquad (2)$$

gives the minimum value, where $\overline{n}_1$ $(\overline{n}_2)$ will be the cardinality of the subset of vectors $\overline{\mathbf{X}}_1$ $(\overline{\mathbf{X}}_2)$, for which the value of the tested feature is less than (greater or equal than) the tested threshold value.

After having the BDT trained, it can be applied for the classification of test data vectors. Any test vector is first fed to the root node, which according to the stored condition and the feature values of the vector, decides towards which child node to forward the vector. This strategy is followed then by the chosen child node, and the vector will be forwarded to a further child. The classification of a vector terminates at the moment when it is forwarded to a leaf node of the tree. The test vector will be attributed to the class indicated by the labeling of the reached leaf node.

2.4   Random Forest

A binary decision tree is an excellent tool, when the task is to accurately learn a certain complicated pattern. For example, it can reproduce every little detail of any MRI volume applied as training data, while keeping the maximum depth below one hundred. However, this marvellous property drags along a serious danger of overfitting. Learning all the small details of the train data builds a serious obstacle for the decision tree in making correct decisions concerning major properties of the test data. This is why, we followed the recipe of Breiman [16], and built forests of binary decision trees, using randomly chosen subsets of the learning data, and randomly chosen subset of features for each tree separately. Each tree in a random forest is a weak classifier. A large set of trees trained with randomly chosen data will make a single decision on a majority basis. In the current stage of this research, we tested how accurate decisions can be made by random forests trained by the data coming from a single MRI volume. There were several important questions to answer:

1. What is the right number of trees in a random forest? Too few trees are not likely to be accurate, while too many redundant ones will not be runtime efficient.
2. What is the right number of feature vectors to train each tree in the random forest? Again, to few vectors are not expected to lead to accurate decision, while too many vectors bring the risk to overfitting.
3. How to make a random forest accurate and effective, when being trained with data coming from several MRI volumes?

2.5   Post-Processing

Random forests are expected to identify the most part of vectors describing tumor pixels. Since negative pixels belong to a great variety of normal tissues (e.g. white matter, gray matter, cerebro-spinal fluid), some of them might be classified as tumor or edema. To be able to discard such cases, we proposed and implemented a posterior validation scheme for all pixels that are labeled as tumor or edema by the random forest. For each such pixel, we defined a 250-pixel neighborhood (all pixels situated at Euclidean distance below $\sqrt{15}$ units, and counted how many of the neighbors are classified as tumor or edema. Those having a number of such neighbors below the predefined neighborhood threshold, are relabeled as negative pixels during post-processing. The appropriate value of the threshold is to be established as well.

2.6   Evaluation of Accuracy

The Jaccard index (JI) is a normalized score of accuracy, computed as

$$JI = \frac{TP}{TP + FP + FN},$$
(3)

where TP stands for the number of true positives, FP for the false positives, and FN for false negatives. Further on, the Dice score (DS) can be computed as

$$DS = \frac{2TP}{2TP + FP + FN} = \frac{2JI}{1 + JI}. \tag{4}$$

Both indices score 1 in case of an ideal clustering, while a fully random result is indicated by a score close to zero.

3   Results and Discussion

Twelve volumes from the BRATS 2012/13 data set were selected for the evaluation of the proposed methodology:

$$\mathcal{V} = \{HG01, HG02, \ldots HG07, HG09, HG11, HG13, HG14, HG15\}. \tag{5}$$

Let us denote by $DS(i \rightarrow j)$ $(i, j \in \mathcal{V})$ the Dice score given by the random forest trained with data chosen from volume i while tested on the whole volume j. Considering the size of the set $\mathcal{V}$, there are $12 \times 11 = 132$ possible $i \neq j$ scenarios, and 12 ones with $i = j$. We performed all possible such tests with various settings of main parameters like number of trees in the forest and number of samples used for the training of each tree. At the training of individual decision trees, an equal number of random samples were chosen from each of the three tissue types. For example, the so-called 100-sample training refers to the use of a total number of $3 \times 100 = 300$ feature vectors.

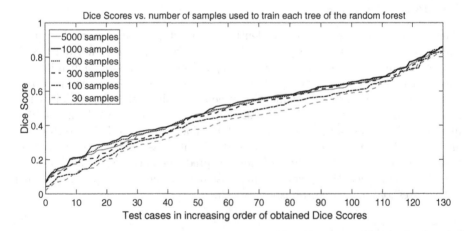

Fig. 1. Classification accuracy benchmarks in case of random forests containing 100 trees, each tree trained with $3 \times (30 \text{ to } 5000)$ feature vectors. Exceptionally, training with HG13 used 4000 samples instead of 5000, because it has less than 5000 tumor pixels.

Figure 1 exhibits the obtained Dice scores in case of trees trained by sample sizes varying from 30 to 5000 items. Each forest in this experiment consisted of 100 trees. For each type of trees, the obtained DS($i \rightarrow j$) ($i \neq j$) Dice scores were sorted in increasing order. The obtained curves indicate that the sample size can strongly influence the classification accuracy. Generally the larger the training sample, the more accurate the decisions, but at a certain level above 1000 samples per tissue type, traces of overfitting are observed. Table 1 shows numerical values of the obtained average Dice scores. The last column also reveals how accurate the classification can be when tested on the same volume that was used for training. Obviously, overfitting does not disturb classification accuracy on the train data set.

Table 1. Averaged accuracy benchmark scores obtained by forests of 100 trees each

Samples from each tissue type	Mean Dice score DS($i \rightarrow i$)	Percentage of Dice score obtained for 1000 samples	Mean Dice score DS($i \rightarrow i$)
5000 samples	0.4916	98.33 %	0.8124
1000 samples	0.5030	100.00 %	0.7810
600 samples	0.4976	98.93 %	0.7702
300 samples	0.4809	95.62 %	0.7626
100 samples	0.4510	89.66 %	0.7354
30 samples	0.4286	85.21 %	0.6912

Table 2. Dice scores obtained when training with one volume and testing on another, trees of 1000 samples per tissue type, and 100 trees in the forest

Volume i	Train data selected from volume i Testing on each volume $j \neq i$				Testing on volume j When trained on each volume $i \neq j$			
	Average	SD	Maximum	Minimum	Average	SD	Maximum	Minimum
HG01	0.5216	0.1722	0.7822	0.2133	0.5857	0.1627	0.8304	0.2508
HG02	0.6191	0.1280	0.8262	0.3500	0.4637	0.2117	0.7416	0.1013
HG03	0.5721	0.1767	0.8615	0.2945	0.4532	0.2309	0.8082	0.1579
HG04	0.5107	0.1983	0.8304	0.2043	0.3902	0.1502	0.6377	0.1474
HG05	0.3872	0.2188	0.6652	0.0552	0.3848	0.1235	0.5279	0.1519
HG06	0.5648	0.1537	0.8386	0.3113	0.5918	0.0955	0.7273	0.6980
HG07	0.5815	0.1537	0.8270	0.3368	0.4796	0.1666	0.7052	0.2110
HG09	0.2662	0.1626	0.5126	0.0684	0.4659	0.1230	0.5789	0.2043
HG11	0.4939	0.2064	0.8564	0.2111	0.5718	0.1320	0.7090	0.2863
HG13	0.5630	0.1992	0.8082	0.2758	0.3368	0.2351	0.6995	0.0552
HG14	0.4721	0.1804	0.7299	0.1248	0.5628	0.1621	0.8270	0.3340
HG15	0.4837	0.1359	0.6426	0.2255	0.7493	0.1144	0.8615	0.5000

Table 2 gives a detailed statistical report on obtained DS($i \rightarrow j$) values, for each individual image volume. The left panel summarizes benchmark values for cases when the given volume served as training data set, and the forest was tested on all other volumes. The right panel reports testing on the given volume, having forests trained with all other volumes separately. The two panels are far from being symmetric, as the best performing train data sets were HG02, HG07, and HG03, while highest accuracy benchmarks were obtained when testing on volumes HG15, HG06, and HG01. The minimum values indicate that data from a single volume cannot train a forest for high quality classification. On the other hand, the maximum values show that data from each volume can contribute to the classification accuracy, and for each test data set there exist possible training sets that yield acceptable classification.

Another aspect that deserves to be remarked and analysed is the poor symmetry of the obtained Dice scores. Having obtained a high value for a certain $DS(i \rightarrow j)$ does not necessarily mean that $DS(j \rightarrow i)$ will also have a high value. In order to numerically characterize the symmetry of the obtained results, we propose to compute the Averaged Symmetry Criterion (ASC), defined as:

$$ASC = \exp \left(\frac{1}{|\mathcal{V}|(|\mathcal{V}| - 1)} \sum_{i,j \in \mathcal{V}; \, i \neq j} \left| \log \frac{DS(i \rightarrow j)}{DS(j \rightarrow i)} \right| \right) , \tag{6}$$

where $|\mathcal{V}|$ stands for the cardinality of $\mathcal{V}$, namely 12 in our case. ASC values obtained for various train samples sizes are reported in Table 3. Dices scores seem to be closest, but still very far from symmetry at sample sizes that assure highest accuracy.

For certain couples of different volumes (i, j), we performed 30 repeated training and testing processes. The goal was to monitor the variability of Dice scores $DS(i \rightarrow j)$ obtained due to the random samples used for training. Figure 2 presents the outcome of repetitive evaluation. Seemingly using less samples means higher variance in benchmark results.

The applied post-processing scheme led to relevant improvement of classification accuracy. Figure 3 shows the histogram of all $DS(i \rightarrow j)$ values before and after post-processing. Here the train data consisted of 600 randomly chosen samples per tissue type for each of the 100 trees in the forest. Dice scores after post-processing reported here are the maximum values obtained by choosing the optimal neighborhood threshold for each individual case. However, this cannot be done automatically. We need to establish either an acceptable constant value of the neighborhood threshold, or to define a strategy that sets the threshold while testing.

Figure 3 also reports the effect of the post-processing. On the left side each individual test is represented, showing the DS before and the maximum DS

Table 3. The relation between $DS(i \rightarrow j)$ and $DS(j \rightarrow i)$, for $i, j \in \mathcal{V}$ and $i \neq j$

Train sample size per tissue type	30	100	300	600	1000	5000
Symmetry benchmark (ASC)	2.065	1.960	1.792	1.705	1.670	1.710

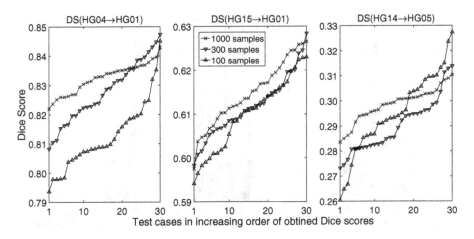

Fig. 2. Reproducibility benchmark: outcome of repeated training on random data samples from a given volume and testing on another volume.

after post-processing. The single curve in the middle plot presents the average effect of post-processing for each possible value of the neighborhood threshold, indicating that it is possible to choose such a threshold value between 190 and 200, for which the average DS rises from 0.502 to 0.583. The bottom right side of Fig. 3 shows those 25 test cases, which were most favorably affected by the post-processing.

Figure 4 shows the outcome of tumor segmentation without and with post-processing, by presenting detected and missed tumor pixels in several consecutive slices of volume HG11. The forest used here consisted of 100 trees, and each tree was trained using 600 samples of each tissue type, randomly selected from volume HG15. In this image, black pixels are the true positive ones, while gray shades represent false positives and false negatives. Post-processing in this certain case rose the Dice score from 0.5339 to 0.8036, which was achieved by discarding lots of false positives, mostly in slices where the real tumor was not present. Even this result could be further improved by implementing another post-processing step that would detect non-tumor (gray) pixels inside the tumor (among black pixels).

The size of tumors that are present in the volumes included in $\mathcal{V}$ varied from $4.5\,\mathrm{cm}^3$ in volume HG13 to $110\,\mathrm{cm}^3$ in volume HG14. The segmentation accuracy of tumors depends on the size of the tumor, as indicated in Fig. 5. The post-processing seems to help more in case of small tumors, so it has a vital role in detecting early stage tumors.

The experiments carried out during this study showed us that a random forest trained with samples from a single volume cannot perform acceptably in all cases. On the other hand, each tested volume had one or more corresponding train volumes that assured fine detection and accurate segmentation of the tumor. The latter allows us to envision a complex random forest that will be suitable for a great majority of cases, which will be reliable enough for clinical deployment.

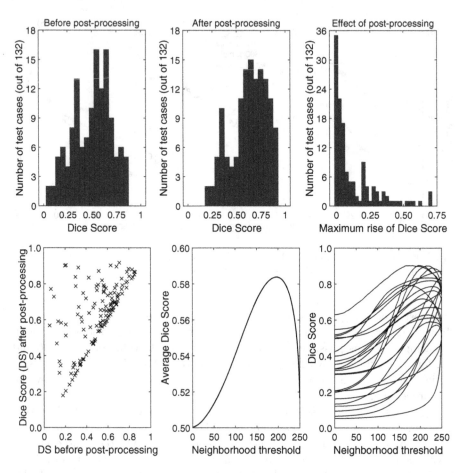

Fig. 3. Effects of the proposed post-processing. Histogram of Dice scores DS($i \rightarrow j$): (top left) before post-processing; (top middle) after post-processing; (top right) histogram of the differences caused by neighborhood-based post-processing; (bottom left) Individual DS($i \rightarrow j$) values before and after post-processing; (bottom middle) The evolution of average Dice score plotted against the value of the neighborhood threshold; (bottom right) Variation of individual Dice scores DS($i \rightarrow j$) plotted against the value of the neighborhood threshold, those 25 which are most affected by post-processing.

The future random forest solution will probably contain clusters of trees, where different clusters will be trained each using its dedicated reduced number of volumes. Clusters of trees will give their own opinion concerning test cases, and the forest will have the role to aggregate these individual opinions and produce the final positive or negative diagnosis. This preliminary study has shown that a random forest based learning algorithm, even if trained with a much more reduced number of features than other random forest based solutions (e.g. [10,11]), can be suitable to detect the presence of the tumor.

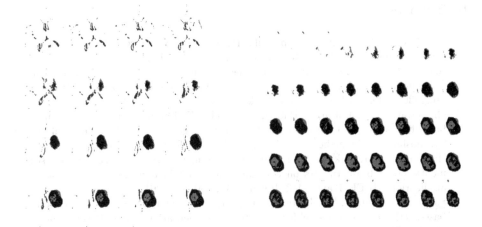

Fig. 4. (left) Detected tumor without post-processing; (right) Detected tumor with neighborhood-based post-processing. Without validating each pixel classified as tumor, several scattered false positives are present in the volume.

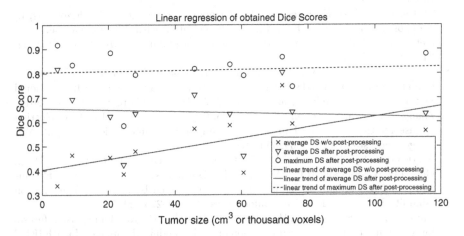

Fig. 5. Obtained average and maximum Dice scores with or without post-processing, plotted against the size of the tumor in the test volume. Linear trends are also indicated.

4   Conclusion

In this paper we presented an automatic tumor detection and segmentation algorithm employing random forests and binary decision trees, in its preliminary stage of implementation. The proposed methodology already reliably detects tumors of 2 cm diameter. It is likely to obtain fine segmentation accuracy in the future using a complex random forest trained with data from dozens of volumes.

References

1. Gordillo, N., Montseny, E., Sobrevilla, P.: State of the art survey on MRI brain tumor segmentation. Magn. Reson. Imaging **31**, 1426–1438 (2013)
2. Asman, A.J., Landman, B.A.: Out-of-atlas labeling: a multi-atlas approach to cancer segmentation. In: 9th IEEE International Symposium on Biomedical Imaging, pp. 1236–1239. IEEE Press, New York (2012)
3. Ghanavati, S., Li, J., Liu, T., Babyn, P.S., Doda, W., Lampropoulos, G.: Automatic brain tumor detection in magnetic resonance images. In: 9th IEEE International Symposium on Biomedical Imaging, pp. 574–577. IEEE Press, New York (2012)
4. Hamamci, A., Kucuk, N., Karamam, K., Engin, K., Unal, G.: Tumor-Cut: segmentation of brain tumors on contranst enhanced MR images for radiosurgery applicarions. IEEE Trans. Med. Imaging **31**, 790–804 (2012)
5. Sahdeva, J., Kumar, V., Gupta, I., Khandelwal, N., Ahuja, C.K.: A novel content-based active countour model for brain tumor segmentation. Magn. Reson. Imaging **30**, 694–715 (2012)
6. Njeh, I., Sallemi, L., Ben Ayed, I., Chtourou, K., Lehericy, S., Galanaud, D., Ben Hamida, A.: 3D multimodal MRI brain glioma tumor and edema segmentation: a graph cut distribution matching approach. Comput. Med. Imaging Graph. **40**, 108–119 (2015)
7. Svolos, P., Tsolaki, E., Kapsalaki, E., Theodorou, K., Fountas, K., Fezoulidis, I., Tsougos, I.: Investigating brain tumor differentiation with diffusion and perfusion metrics at 3T MRI using pattern recognition techniques. Magn. Reson. Imaging **31**, 1567–1577 (2013)
8. Yu, C.P., Ruppert, G., Collins, R., Nguyen, D., Falcao, A., Liu, Y.: 3D blob based brain tumor detection and segmentation in MR images. In: 11th IEEE International Symposium on Biomedical Imaging, pp. 1192–1197. IEEE Press, New York (2012)
9. Zhang, N., Ruan, S., Lebonvallet, S., Liao, Q., Zhou, Y.: Kernel feature selection to fuse multi-spectral MRI images for brain tumor segmentation. Comput. Vis. Image Understand **115**, 256–269 (2011)
10. Tustison, N.J., Shrinidhi, K.L., Wintermark, M., Durst, C.R., Kandel, B.M., Gee, J.C., Grossman, M.C., Avants, B.B.: Optimal symmetric multimodal templates and concatenated random forests for supervised brain tumor segmentation (simplified) with ANTsR. Neuroinformatics **13**, 209–225 (2015)
11. Pinto, A., Pereira, S., Dinis, H., Silva, C.A., Rasteiro, D.: Random decision forests for automatic brain tumor segmentation on multi-modal MRI images. In: 4th IEEE Portuguese BioEngineering Meeting, pp. 1–5. IEEE Press, New York (2015)
12. Szilágyi, L., Lefkovits, L., Iantovics, B., Iclanzan, D., Benyó, B.: Automatic brain tumor segmentation in multispectral MRI volumetric records. In: Arik, S., Huang, T., Lai, W.K., Liu, Q. (eds.) ICONIP 2015. LNCS, vol. 9492, pp. 174–181. Springer, Heidelberg (2015). doi:10.1007/978-3-319-26561-2_21
13. Menze, B.H., Jakab, A., Bauer, S., Kalpathy-Cramer, J., Farahani, K., Kirby, J., et al.: The multimodal brain tumor image segmentation benchmark (BRATS). IEEE Trans. Med. Imaging **34**, 1993–2024 (2015)
14. Vovk, U., Pernuš, F., Likar, B.: A review of methods for correction of intensity inhomogeneity in MRI. IEEE Trans. Med. Imaging **26**, 405–421 (2007)
15. Akers, S.B.: Binary decision diagrams. IEEE Trans. Comput. **C-27**, 509–516 (1978)
16. Breiman, L.: Random forests. Mach. Learn. **45**, 5–32 (2001)

Author Index

Printed in the United States
By Bookmasters